Semiconductors for Micro and Nanotechnology— An Introduction for Engineers

Semiconductors for Micro and Nanotechnology— An Introduction for Engineers

Jan G. Korvink and Andreas Greiner

Authors:
Prof. Dr. Jan G. Korvink
IMTEK-Institute for Microsystem Technology
Faculty for Applied Sciences
Albert Ludwig University Freiburg
D-79110 Freiburg
Germany

Dr. Andreas Greiner
IMTEK-Institute for Microsystem Technology
Faculty for Applied Sciences
Albert Ludwig University Freiburg
D-79110 Freiburg
Germany

Library of Congress Card No.: applied for.
British Library Cataloguing-in-Publication Data:
A catalogue record for this book is available from the British Library.

Die Deutsche Bibliothek — CIP-Cataloguing-in-Publication Data
A catalogue record for this book is available from Die Deutsche Bibliothek.

ISBN 3-527-30257-3

Printed on acid-free paper

Printing: Strauss Offsetdruck GmbH, Mörlenbach
Bookbinding: Litges & Dopf Buchbinderei GmbH, Heppenheim
Printed in the Federal Republik of Germany

Dedicated to

Micheline Pfister,
Sean and Nicolas Vogel,
and in fond memory of
Gerrit Jörgen Korvink

Maria Cristina Vecchi,
Sarah Maria Greiner
und im Gedenken an
Gertrud Maria Greiner

Contents

Chapter 4 The Electromagnetic System 143

Chapter 5 Statistics 169

Chapter 6 Transport Theory 191

Preface

This book addresses the engineering student and practising engineer. It takes an engineering-oriented look at semiconductors. Semiconductors are at the focal point of vast number of technologists, resulting in great engineering, amazing products and unheard-of capital growth. The work horse here is of course silicon. Explaining how semiconductors like silicon behave, and how they can be manipulated to make microchips that work—this is the goal of our book.

We believe that semiconductors can be explained consistently without resorting 100% to the complex language of the solid state physicist. Our approach is more like that of the systems engineer. We see the semiconductor as a set of well-defined subsystems. In an approximately top-down manner, we add the necessary detail (but no more) to get to grips with each subsystem: The physical crystal lattice, and charge carriers in lattice like potentials. This elemental world is dominated by statistics, making strange observations understandable: This is the glue we need to put the systems together and the topic of a further chapter. Next we show the theory needed to predict the behavior of devices made in silicon or other semiconducting materials, the building blocks of modern electronics. Our book wraps up the tour with a practical engineering note: We look at how the various sub-systems interact to produce the observable behavior of the semiconductor. To enrich the subject matter, we tie up the theory with concise boxed topics interspersed in the text.

There are many people to thank for their contributions, and for their help or support. To the Albert Ludwig University for creating a healthy research environment, and for granting one of us (Korvink) sabbatical leave. To Ritsumeikan University in Kusatsu, Japan, and especially to Prof. Dr. Osamu Tabata, who hosted one of us (Korvink) while on sabbatical and where one chapter of the book was written. To the ETH Zurich and especially to Prof. Dr. Henry Baltes, who hosted one of us (Korvink) while on sabbatical and where the book project was wrapped up. To Prof. Dr. Evgenii Rudnyi, Mr. Takamitsu Kakinaga, Ms. Nicole Kerness and Mr. Sadik Hafizovič for carefully reading through the text and finding many errors. To the anonymous reviewers for their invaluable input. To Ms. Anne Rottler for inimitable administrative support. To VCH-Wiley for their deadline tolerance, and especially to Dr. Jörn Ritterbusch and his team for support. To Micheline and Cristina for enduring our distracted glares at home as we fought the clock (the calendar) to finish, and for believing in us.

Jan G. Korvink and Andreas Greiner,
Freiburg im Breisgau,
February 2002

Chapter 1 Introduction

Semiconductors have complex properties, and in the early years of the twentieth century these were mainly discovered by physicists. Many of these properties have been harnessed, and have been exploited in ingenious microelectronic devices. Over the years the devices have been rendered manufacturable by engineers and technologists, and have spawned off both a multi-billion € (or $, or ¥) international industry and a variety of other industrial mini-revolutions including software, embedded systems, the internet and mobile communications. Semiconductors still lie at the heart of this revolution, and silicon has remained its champion, warding off the competitors by its sheer abundance, suitability for manufacturing, and of course its tremendous head-start in the field. Silicon is the working material of an exciting, competitive world, presenting a seemingly endless potential for opportunities.

Chapter Goal The goal of this chapter is to introduce the reader to the field of semiconductors, and to the purpose and organization of the book.

Chapter Roadmap

In this chapter we first explain the conceptual framework of the book. Next, we provide some popular definitions that are in use in the field. Lastly, we indicate some of the sources of information on new inventions.

1.1 The System Concept

This book is about semiconductors. More precisely, it is about semiconductor properties and how to understand them in order to be exploited for the design and fabrication of a large variety of microsystems. Therefore, this book is a great deal about silicon as a paradigm for semiconductors. This of course implies that it is also about other semiconductor systems, namely for those cases where silicon fails to show the desired effects due to a lack of the necessary properties or structure. Nevertheless, we will not venture far away from the paradigmatic material silicon, with its overwhelming advantage for a wide field of applications with low costs for fabrication. To quote the Baltes theorem [1.1]:

> To prove your idea, put in on silicon.
> Since you will need circuitry, make it with CMOS.
> If you want to make it useful, get it packaged.

The more expensive fabrication becomes, the less attractive the material is for the design engineer. Designers must always keep in mind the cost and resources in energy and personnel that it takes to handle materials that need additional nonstandard technological treatment. This is not to say that semiconductors other than silicon are unimportant, and there are many beautiful applications. But most of today's engineers encounter silicon CMOS as a process with which to realize their ideas for microscopic systems. Therefore, most of the emphasis of this book lies in the explanation of the properties and behavior of silicon, or better said, "the semiconductor system silicon".

A semiconductor can be viewed as consisting of many subsystems. For one, there are the individual atoms, combining to form a chunk of crystalline material, and thereby changing their behavior from individual systems to a composite system. The atomic length scale is still smaller than typical length scales that a designer will encounter, despite the fact that sub-nanometer features may be accessible through modern experimental techniques. The subsystems that this book discusses emerge when silicon atoms are assembled into a crystal with unique character. We mainly discuss three systems:

- the particles of quantized atom vibration, or phonons;
- the particles of quantized electromagnetic radiation, or photons;
- the particles of quantized charge, or electrons.

There are many more, and the curious reader is encouraged to move on to other books, where they are treated more formally. The important feature of these subsystems is that they interact. Each interaction yields effects useful to the design engineer.

Why do we emphasize the system concept this much? This has a lot to do with scale considerations. In studying nature, we always encounter scales of different order and in different domains. There are length scales that play a significant role. Below the nanometer range we observe single crystal layers and might even resolve single atoms. Thus we become aware that the crystal is made of discrete constituents. Nevertheless, on a micrometer scale—which corresponds to several thousands of monatomic layers—the crystal appears to be a continuous medium. This means that *at certain length scales* a homogenous isotropic continuum description is sufficient. Modern down-sizing trends might force us to take at least anisotropy into account, which is due to crystal symmetry, if not the detailed structure of the crystal lattice including single defects. Nanotechnology changes all of this. Here we are finally designing at the atomic length scale, a dream that inspired the early twentieth century.

Almost everything becomes quantized, from thermal and electrical resistance to interactions such as the Hall effect.

Time scale considerations are at least as important as length scales. They are governed by the major processes that take place within the materials. The shortest technological time scales are found in electron-electron scattering processes and are on the order of a few femtoseconds, followed by the interaction process of lattice vibrations and electronic systems with a duration of between a few hundreds of femtoseconds to a picosecond. Direct optical transitions from the conduction band to the valence band lie in the range of a few nanoseconds to a few microseconds. For applications in the MHz (10^6 Hz) and GHz (10^9 Hz) regime the details of the electron-electron scattering process are of minor interest and most scattering events may be considered to be instantaneous. For some quantum mechanical effects the temporal resolution of scattering is crucial, for example the intra-collisional field effect.

The same considerations hold for the energy scale. Acoustic electron scattering may be considered elastic, that is to say, it doesn't consume energy. This is true only if the system's resolution lies well above the few meV of any scattering process. At room temperature (300 K) this is a good approximation, because the thermal energy is of the order of 25.4 meV. The level of the thermal energy implies a natural energy scale, at which the band gap energy of silicon of about 1.1 eV is rather large. For high energy radiation of several keV the band gap energy again is negligible.

The above discussion points out the typical *master* property of a composite system: A system reveals a variety of behavior at different length (time, energy, ...) scales. This book therefore demands caution to be able to account for the semiconductor as a system, and to explain its building blocks and their interactions in the light of scale considerations.

1.2 Popular Definitions and Acronyms

The microelectronic and microsystem world is replete with terminology and acronyms. The number of terms grows at a tremendous pace, without regard to aesthetics and grammar. Their use is ruled by expedience. Nevertheless, a small number have survived remarkably long. We list only a few of the most important, for those completely new to the field.

1.2.1 Semiconductors versus Conductors and Insulators

A semiconductor such as silicon provides the technologist with a very special opportunity. In its pure state, it is almost electrically insulating. Being in column IV of the periodic table, it is exceptionally balanced, and comfortably allows one to replace the one or other atom of its crystal with atoms from column III or V (which we will term P and N type doping). Doing so has a remarkable effect, for silicon then becomes conductive, and hence the name “semiconductor”. Three important features are easily controlled. The density of “impurity” atoms can vary to give a tremendously wide control over the conductivity (or resistance) of the bulk material. Secondly, we can decide whether electrons, with negative charge, or holes, with positive charge, are the dominant mechanism of current flow, just by changing to an acceptor or donor atom, i.e., by choosing P or N type doping. Finally, we can “place” the conductive pockets in the upper surface of a silicon wafer, and with a suitable geometry, create entire electronic circuits.

If silicon is exposed to a hot oxygen atmosphere, it forms amorphous silicon dioxide, which is a very good insulator. This is useful to make capacitor devices, with the SiO_2 as the dielectric material, to form the gate insulation for a transistor, and of course to protect the top surface of a chip.

Silicon can also be grown as a doped amorphous film. In this state we lose some of the special properties of semiconductors that we will explore in this book. In the amorphous form we almost have metal-like behavior, and indeed, semiconductor foundries offer both real metals (aluminium, among others) and polysilicon as "metallic" layers.

1.2.2 The Diode Family

The simplest device that one can make using both P and N doping is the diode. The diode is explained in Section 7.6.4. The diode is a one-way valve with two electrical terminals, and allows current to flow through it in only one direction. The diode provides opportunities for many applications. It is used to contact metal wires to silicon substrates as a Shottkey diode. The diode can be made to emit light (LEDs). Diodes can detect electromagnetic radiation as photo-detectors, and they form the basis of semiconductor lasers. Not all of these effects are possible using silicon, and why this is so is also explained later on.

1.2.3 The Transistor Family

This is the true fame of silicon, for it is possible to make this versatile device in quantities unheard of elsewhere in the engineering world. Imagine selling a product with more than 10^9 working parts! Through CMOS (complimentary metal oxide semiconductor) it is possible to create reliable transistors that require extraordinary little power (but remember that very little times 10^9 can easily amount to a lot). The trend in miniaturization, a reduction in lateral dimensions, increase in operation speed, and reduction in power consumption, is unparalleled in engineering history. Top that up with a parallel manufacturing step that does not essentially depend on the number of working parts, and the stage is set for the revolution that we have witnessed.

The transistor is useful as a switch inside the logic gates of digital chips such as the memories, processors and communications chips of modern computers. It is also an excellent amplifier, and hence found everywhere where high quality analog circuitry is required. Other uses include magnetic sensing and chemical sensing.

1.2.4 Passive Devices

In combination with other materials, engineers have managed to miniaturize every possible discrete circuit component known, so that it is possible to create entire electronics using just one process: resistors, capacitors, inductors and interconnect wires, to name the most obvious.

For electromagnetic radiation, waveguides, filters, interferometers and more have been constructed, and for light and other forms of energy or matter, an entirely new industry under the name of microsystems has emerged, which we now briefly consider.

1.2.5 Microsystems: MEMS, MOEMS, NEMS, POEMS, etc.

In North America, the acronym MEMS is used to refer to micro-electromechanical systems, and what is being implied are the devices at the length scale of microelectronics that include some non-electrical signal, and very often the devices feature mechanical moving parts and electrostatic actuation and detection mechanisms, and these mostly couple with some underlying electrical circuitry. A highly successful CMOS MEMS, produced by Infineon Technologies, is shown in Figure 1.1. The device,

Figure 1.1. MEMS device. a) Infineon's surface micromachined capacitive pressure sensor with interdigitated signal conditioning Type KP120 for automotive BAP and MAP applications. b) SEM photograph of the pressure sensor cells compared with a human hair. Image © Infineon Technologies, Munich [1.2].

placed in a low-cost SMD package, is used in MAP and BAP tire pressure applications. With an annual production running to several millions, it is currently sold to leading automotive customers [1.2].

MEMS has to date spawned off two further terms that are of relevance to us, namely MOEMS, for micro-opto-electro-mechanical systems, and NEMS, for the inevitable nano-electro-mechanical systems.

MOEMS can include entire miniaturized optical benches, but perhaps the most familiar example is the digital light modulator chip sold by Texas Instruments, and used in projection display devices, see Figure 1.2.

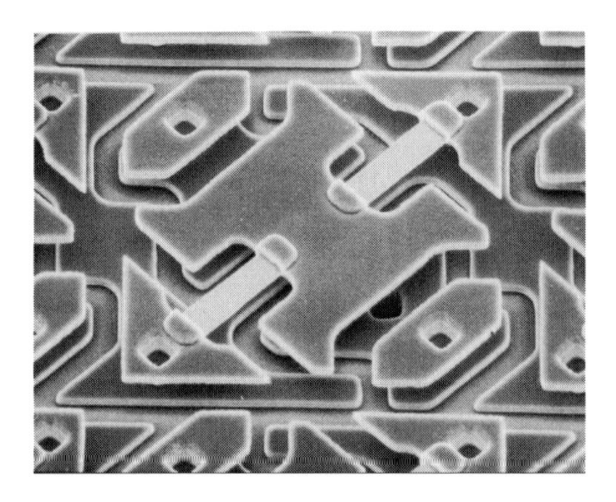

Figure 1.2. MOEMS device. This 30 μm by 30 μm device is a single pixel on a chip that has as many pixels as a modern computer screen display. Each mirror is individually addressable, and deflects light from a source to a systems of lenses that project the pixel onto a screen. Illustration © Texas Instruments Corp., Dallas [1.3].

As for NEMS, the acronym of course refers to the fact that a critical dimension is now no longer the large micrometer, but has become a factor 1000 smaller. The atomic force microscope cantilever [1.4] may appear to be a MEMS-like device, but since it resolves at the atomic diameter scale, it is a clear case of NEMS, see Figure 1.3. Another example is the distributed mirror cavity of solid-state lasers made by careful epitaxial growth of many different semiconductor layers, each layer a few nanometers thick. Of major commercial importance is the submicron microchip electronic device technology. Here the lateral size of a transistor gate is the key size, which we know has dropped to below 100

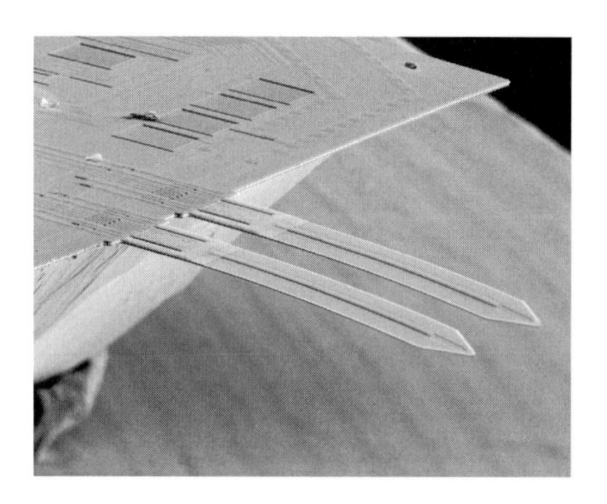

Figure 1.3. NEMS devices. Depicted are two tips of an atomic force microscope, made in CMOS, and used to visualize the force field surrounding individual atoms. Illustration © Physical Electronics Laboratory, ETH Zurich, Switzerland [1.4].

nm in university and industrial research laboratories. Among NEMS we count the quantum wire and the quantum dot, which have not yet made it to the technological-commercial arena, and of course any purposefully-designed and functional molecular monolayer film.

POEMS, or polymer MEMS, are microstructures made of polymer materials, i.e., they completely depart from the traditional semiconductor-based devices. POEMS are usually made by stereo micro-lithography through a photo-polymerization process, by embossing a polymer substrate, by milling and turning, and by injection moulding. This class of devices will become increasingly important because of their potentially low manufacturing cost, and the large base of materials available.

In Japan, it is typical to refer to the whole field of microsystems as Micromachines, and manufacturing technology as Micromachining. In Europe, the terms Microsystems, Microtechnology or Microsystem Technology have taken root, with the addition of Nanosystems and the inevitable Nanosystem Technology following closely. The European naming convention is popular since it is easily translated into any of a large number of languages (German: *Mikrosystemtechnik*, French: *Microtechnique*, Italian: *tecnologia dei microsistemi*, etc.).

1.3 Sources of Information

We encourage every student to regularly consult the published literature, and in particular the following journals:

IOP

- Journal of *Micromechanics and Microengineering*
- Journal of *Nanotechnology*

IEEE

- Journal of *Microelectromechanical Systems*
- Journal of *Electron Devices*
- Journal of *Sensors*
- Journal of *Nanosystems*

Other

- Wiley-VCH *Sensors Update*, a journal of review articles
- MYU Journal of *Sensors and Materials*
- Elsevier Journal of *Sensors and Actuators*
- Springer Verlag Journal of *Microsystem Technology*
- Physical Review
- Journal of Applied Physics

Of course there are more sources than the list above, but it is truly impossible to list everything relevant. Additional sources on the world-wide-web are blossoming (see e.g. [1.5]), as well as the emergence of standard texts on technology, applications and theory. A starting point is best taken from the lists of chapter references. Two useful textbook references are Sze's book on the physics of semiconductor devices [1.6] and Middelhoek's book on silicon sensors [1.7].

1.4 Summary for Chapter 1

Silicon is a very important technological material, and understanding its behavior is a key to participating in the largest industry ever created. To understand the workings of the semiconductor silicon, it helps to approach it as a system of interacting subsystems.

The subsystems comprise the crystal lattice and its quantized vibrations—the phonons, electromagnetic radiation and its quantized form—the photons, and the loosely bound quantized charges—the electrons. The interactions between these systems is a good model with which to understand most of the technologically useful behavior of silicon. To understand the ensuing topics, we require a background in particle statistics. To render the ideas useful for exploitation in devices such as diodes, transistors, sensors and actuators, we require an understanding of particle transport modelling.

These topics are now considered in more detail in the six remaining chapters of the book.

1.5 References for Chapter 1

1.1 Prof. Dr. Henry Baltes, Private communication

1.2 Prof. Dr. Christofer Hierold, Private communication.

1.3 See e.g. http://www.dlp.com

1.4 D. Lange, T. Akiyama, C. Hagleitner, A. Tonin, H. R. Hidber, P. Niedermann, U. Staufer, N. F. de Rooij, O. Brand, and H. Baltes, *Parallel Scanning AFM with On-Chip Circuitry in CMOS Technology*, Proc. IEEE MEMS, Orlando, Florida (1999) 447-452

1.5 See e.g. http://www.memsnet.org/

1.6 S. M. Sze, *Physics of Semiconductor Devices*, 2nd Ed., John Wiley and Sons, New York (1981)

1.7 Simon Middelhoek and S. A. Audet, *Silicon Sensors*, Academic Press, London (1989)

Chapter 2 The Crystal Lattice System

In this chapter we start our study of the semiconductor system with the crystals of silicon (Si), adding some detail on crystalline silicon dioxide (SiO_2) and to a lesser extent on gallium arsenide (GaAs). All three are regular lattice-arrangements of atoms or atoms. For the semiconductors silicon and gallium arsenide, we will consider a model that completely de-couple the behavior of the atoms from the valence electrons, assuming that electronic dynamics can be considered as a perturbation to the lattice dynamics, a topic dealt with in Chapter 3. For all the electrons of the ionic crystal silicon dioxide, as well as the bound electrons of the semiconductors, we here assume that they obediently follow the motion of the atoms.

We will see that by applying the methods of classical, statistical and quantum mechanics to the lattice, we are able to predict a number of observable constitutive phenomena of interest—i.e., we are able to explain macroscopic measurements in terms of microscopic crystal lattice mechanics. The effects include an approximation for the elastic coefficients of continuum theory, acoustic dispersion, specific heat, thermal expansion and heat conduction. In fact, going beyond our current goals, it is possible to similarly treat dielectric, piezoelectric and elastooptic effects. However, the predictions are of a qualitative nature in the majority of cases, mainly because the interatomic potential of covalently bonded atoms is so hard to come by. In fact, in a sense the potential is reverse engineered, that is, using measurements of the crystal, we fit parameters that improve the quality of the models to make them in a sense "predictive".

Chapter Goal

Our goal for this chapter is to explain the observed crystal data with preferably a single comprehensive model that accounts for all effects.

Chapter Roadmap

Our road map is thus as follows. We start by stating some of the relevant observable data for the three materials Si, SiO_2 and $GaAs$, without more than a cursory explanation of the phenomena.

Our next step is to get to grips with the concept of a crystal lattice and crystal structure. Beyond this point, we are able to consider the forces that hold together the static crystal. This gives us a method to describe the way the crystal responds, with stress, to a strain caused by stretching the lattice. Then we progress to vibrating crystal atoms, progressively refining our method to add detail and show that phonons, or quantized acoustic pseudo "particles", are the natural result of a dynamic crystal lattice.

Considering the phonons in the lattice then leads us to a description of heat capacity. Moving away from basic assumptions, we consider the anharmonic crystal and find a way to describe the thermal expansion. The section following presents a cursory look at what happens when the regular crystal lattice is locally deformed through the introduction of foreign atoms. Finally, we leave the infinitely-extended crystal model and briefly consider the crystal surface. This is important, because most microsystem devices are build on top of semiconductor wafers, and so are repeatedly subject to the special features and limitations that the surface introduces.

2.1 Observed Lattice Property Data

Geometric Structure

The geometric structure of a regular crystal lattice is determined using x-ray crystallography techniques, by recording the diffraction patterns of x-ray photons that have passed through the crystal. From such a recorded pattern (see Figure 2.1 (a)), we are able to determine the reflection planes formed by the constituent atoms and so reconstruct the relative positions of the atoms. This data is needed to proceed with a geometric (or, strictly speaking, group-theoretic) characterization of the crystal lattice's symmetry properties. We may also use an atomic force microscope (AFM) to map out the force field that is exerted by the constituent atoms on the surface of a crystal. From such contour plots we can reconstruct the crystal structure and determine the lattice constants. We must be careful, though, because we may observe special surface configurations in stead of the actual bulk crystal structure, see Figure 2.1 (b).

Elastic Properties

The relationship between stress and strain in the linear region is via the elastic property tensor, as we shall shortly derive in Section 2.3.1. To measure the elastic parameters that form the entries of the elastic property tensor, it is necessary to form special test samples of exact geometric shape that, upon mechanical loading, expose the relation between stress and strain in such a way that the elastic coefficients can be deduced from the measurement. The correct choice of geometry relies on the knowledge of the crystal's structure, and hence its symmetries, as we shall see in Section 2.2.1. The most common way to extract the mechanical properties of crystalline materials is to measure the direction-dependent velocity of sound inside the crystal, and by diffracting x-rays through the crystal (for example by using a synchrotron radiation source).

Figure 2.1. (a) The structure of a silicon crystal as mapped out by x-ray crystallography in a Laue diagram. (b) The famous $\langle 111 \rangle$ 7×7 *-reconstructed surface of silicon as mapped out by an atomic force microscope. The bright dots are the Adatoms that screen the underlying lattice. Also see Figure 2.30.*

Dispersion Curves

Dispersion curves are usually measured by scattering a neutron beam (or x-rays) in the crystal and measuring the direction-dependent energy lost or gained by the neutrons. The absorption or loss of energy to the crystal is in the form of phonons.

Thermal Expansion

Thermal expansion measurements proceed as for the stress-strain measurement described above. A thermal strain is produced by heating the sample to a uniform temperature. Armed with the knowledge of the elastic parameters, the influence of the thermal strain on the velocity of sound may then be determined.

The material data presented in the following sections and in Table 2.1 was collected from references [2.5, 2.6]. Both have very complete tables of measured material data, together with reference to the publications where the data was found.

Table 2.1. Lattice properties of the most important microsystem base materials[a].

Property	**Crystalline Silicon** Si	**LPCVD Poly-Si**	**Thermal Oxide** SiO_2	α **-Quartz**	**Process** Si_3N_4	**Gallium Arsenide** GaAs
Atomic Weight	Si: 28.0855	Si: 28.0855 .	Si: 28.0855 O: 15.994	Si: 28.0855 O: 15.994	Si: 28.0855 N: 14.0067	Ga: 69.72 As: 74.9216
Crystal class/ Symmetry	Carbon-like, Cubic symmetry	Poly-crystalline	Amorphous	Trigonal symmetry	Amorphous	Zinc-blende Cubic symmetry
Density (kg/m^3)	2330	2330	< 2200	2650	3100	5320
Elastic moduli (Gpa) Y: Young modulus B: Bi-axial modulus S: Shear modulus C_i : Tensor coefficients.	Pure: C_{11} : 166 C_{12} : 63.9 C_{44} : 79.6 p-Type: C_{11} : 80.5 C_{12} : 115 C_{44} : 52.8 n-Type: C_{11} : 97.1 C_{12} : 54.8 C_{44} : 172	Y: 130 – 174 S: 69	Y: 72 – 75	Y: 87	Y: 97 – 320 B: 249 - 311	Pure: C_{11} : 118.1 C_{12} : 53.2 C_{44} : 59.4

Table 2.1. Lattice properties of the most important microsystem base materials[a].

Property	**Crystalline Silicon** Si	**LPCVD Poly-Si**	**Thermal Oxide** SiO_2	α **-Quartz**	**Process** Si_3N_4	**Gallium Arsenide** GaAs
Hardness (Gpa)	⟨100⟩ : 5.1 – 13 ; ⟨111⟩ : 11.7	10.5 – 12.5	14.4 – 18	8.2	-	-
Lattice parameter ( ) X, C: axes	5.43	5.43 , polycrystal	Amorphous	X_i 4.9127 C 5.4046	Amorphous	5.65
Melting point ($°C$)	1412	1412 .	1705	-	1902	-
Poisson ratio $\tilde{v}$	⟨100⟩ : 0.28 ⟨111⟩ : 0.36	0.2 – 0.3	0.17 - 0.22	0.169	0.26	0.31
Specific heat C_p ($J/\text{kg}K$	702.24	702.24	740	740	750	350
Thermal conductivity (W/mK)	150	150	1.1 - 1.5	1.4	18	46 (12)
Thermal expansion coefficient α (T^{-1})	2.33×10^{-6}	2.33×10^{-6}	0.4×10^{-6} - 0.55×10^{-6}	XY cut 14.3 Z cut 7.8	2.7×10^{-6}	6.86×10^{-6}

a. The tabulated values for amorphous process materials are foundry-dependent and are provided only as an indication of typical values. Also, many of the measurements on crystalline materials are for doped samples and hence should be used with care. Properties depend on the state of the material, and a common choice is to describe them based on the temperature and the pressure during the measurements. For technological work, we require the properties under operating conditions, i.e., at room temperature and at 1 atmosphere of pressure.

2.1.1 Silicon

A semiconductor quality Silicon ingot is a gray, glassy, face-centered crystal. The element is found in column IV of the periodic table. It has the same crystal structure as diamond, as illustrated in Figure 2.2. Silicon

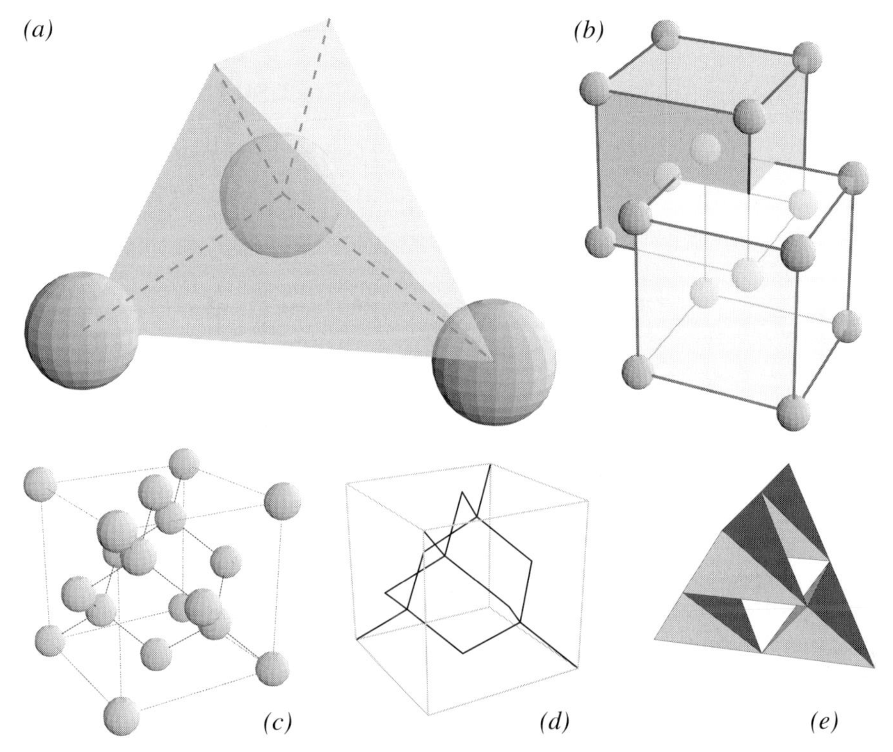

Figure 2.2. The diamond-like structure of the Silicon crystal is caused by the tetrahedral arrangement of the four sp^3 bond-forming orbitals of the silicon atom, symbolically shown (a) as stippled lines connecting the ball-like atomic nuclei in this tetrahedral repeating unit. (b) When the atoms combine to form a crystal, we observe a structure that may be viewed as a set of two nested cubic lattices, or (c) a single face-centered cubic lattice with a basis. (d) The structure of the sp^3 hybrid bonds. (e) The fcc unit cell can also be viewed as four tetrahedra.

has a temperature dependent coefficient of thermal expansion in K^{-1} described by

$$\alpha(T) = (3.725(1 - \exp(-5.88\times10^{-3}(T - 124))) + 5.548\times10^{-4}T)\times10^{-6} \tag{2.1}$$

and a lattice parameter (the interatomic distance in) that varies with temperature as

$$a(T) = 5.4304 + 1.8138\times10^{-5}(T - 298.15) + 1.542\times10^{-9}(T - 298.15)^2 \tag{2.2}$$

We will later take a more detailed look at the thermal strain $\varepsilon(T - T_0) = \alpha(T)(T - T_0)$. Both $\alpha(T)$ and $a(T)$ are plotted in Figure 2.3. Both of

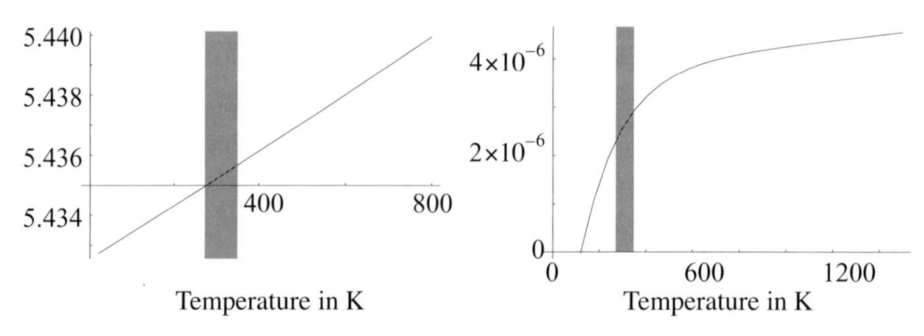

Figure 2.3. The thermal expansion properties of Silicon. Shown on the left is the temperature dependence of the lattice parameter (the size of a unit cell), and to its right is the temperature dependence of the thermal expansion coefficient. The typical engineering temperature range for silicon electronic devices is indicated by the background gray boxes.

these properties are also dependent on the pressure experienced by the material, hence we should write $\alpha(T, p)$ and $a(T, p)$. It is important to note that "technological" silicon is doped with foreign atoms, and will in general have material properties that differ from the values quoted in Table 2.1, but see [2.6] and the references therein. Silicon's phonon dispersion diagram is shown in Figure 2.4.

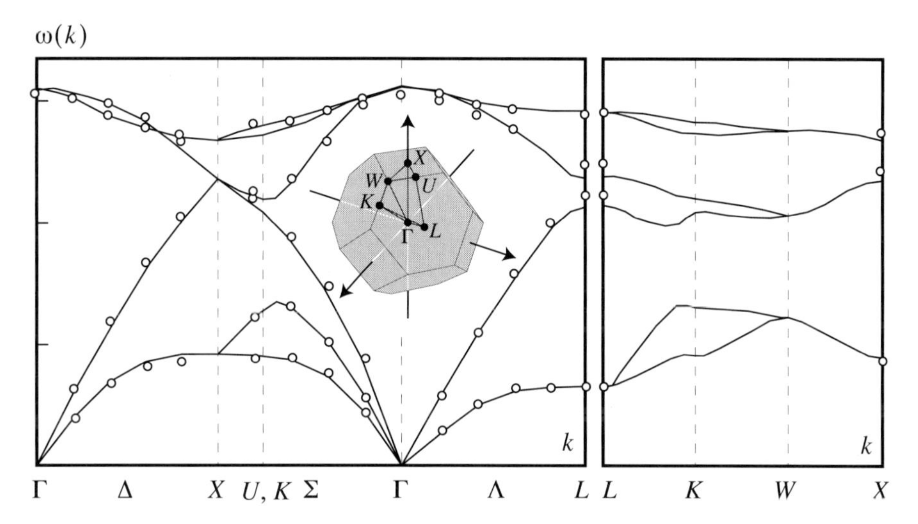

Figure 2.4. The measured and computed dispersion diagrams of crystalline silicon. The vertical axis represents the phonon frequency, the horizontal axis represents straight-line segments in **k***-space between the main symmetry points of the Brillouin zone, which is shown as an insert. Figure adapted from [2.5].*

Silicon is the carrier material for most of today's electronic chips and microsystem (or microelectromechanical system (MEMS), or micromachine) devices. The ingot is sliced into wafers, typically 0.5 mm thick and 50 to 300 mm in diameter. Most electronic devices are manufactured in the first 10 μm of the wafer surface. MEMS devices can extend all the way through the wafer. Cleanroom processing will introduce foreign dopant atoms into the silicon so as to render it more conducting. Other processes include subtractive etching steps, additive deposition steps and modifications such as oxidation of the upper layer of silicon. Apart from certain carefully chosen metal conductors, the most common materials used in conjunction with silicon are oven-grown or deposited thermal silicon oxides and nitrides (see the following sections), as well as poly-crystalline silicon.

IC process quality LPCVD poly-crystalline silicon (Poly-Si) has properties that depend strongly on the foundry of origin. It is assumed to be isotropic in the plane of the wafer, and is mainly used as a thin film thermal and electrical conductor for electronic applications, and as a structural and electrode material for MEMS devices.

2.1.2 Silicon Dioxide

Crystalline silicon dioxide is better known as fused quartz. It is unusual to obtain quartz from a silicon-based process, say CMOS, because the production of crystalline quartz usually requires very high temperatures that would otherwise destroy the carefully produced doping profiles in the silicon. Semiconductor-related silicon dioxide is therefore typically amorphous.

Since quartz has a non-cubic crystal structure, and therefore displays useful properties that are not found in high-symmetry cubic systems such as silicon, yet are of importance to microsystems, we also include it in our discussion. We consider α-quartz, one of the variants of quartz that is

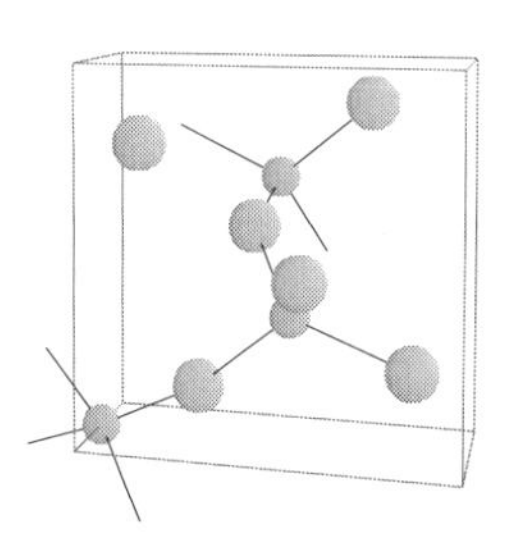

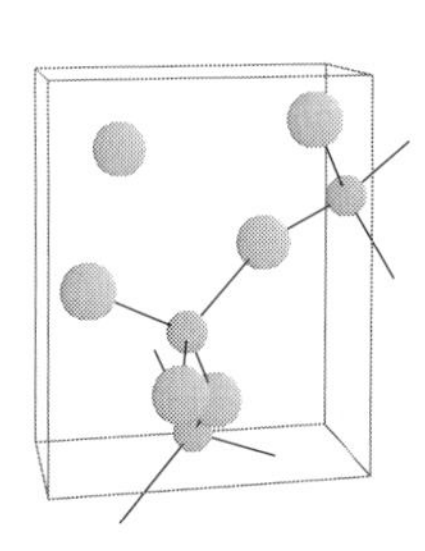

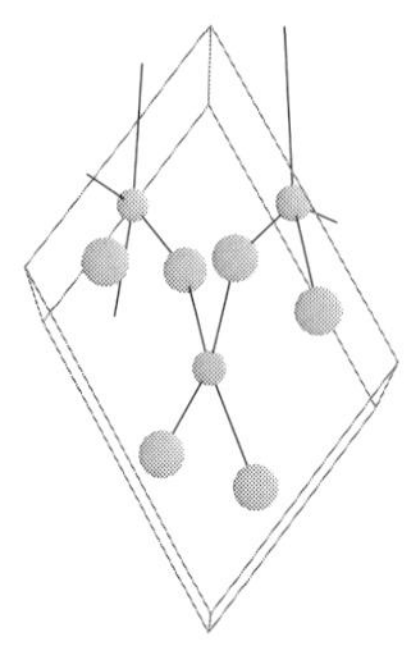

Figure 2.5. Three perspective views of the trigonal unit cell of α-Quartz. From left to right the views are towards the origin along the X_1, X_2 and the C axes. The six large spheres each with two bonds represent oxygen atoms, the three smaller spheres each with four tetrahedral bonds represent silicon atoms.

stable below 573^oC, with trigonal crystal symmetry. The unit cell of the quartz crystal is formed by two axes, called X_1 and X_2, at 60° to each other, see Figure 2.5. Quartz is non-centro-symmetric and hence piezo-electric. It also has a handedness as shown in Figure 2.6.

Figure 2.6. α-Quartz is found as either a right or a left-handed structure, as indicated by the thick lines in the structure diagram that form a screw through the crystal. In the figure, 8 unit cells are arranged in a $(2 \times 2 \times 2)$ block.

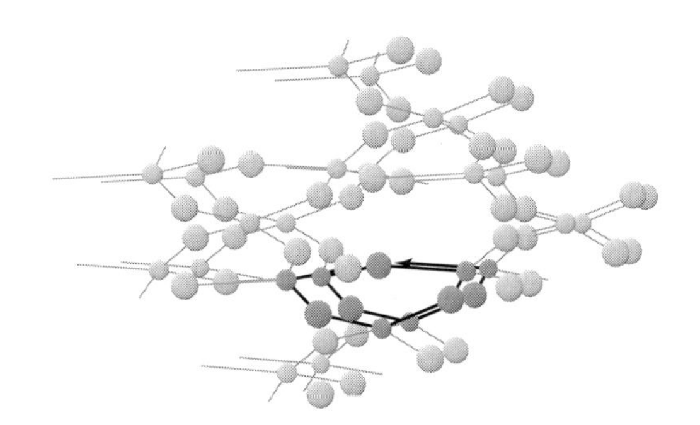

2.1.3 Silicon Nitride

Crystalline silicon nitride (correctly known as tri-silicon tetra-nitride) is not found on silicon IC wafers because, as for silicon dioxide, very high temperatures are required to form the pure crystalline state, see Figure 2.7. These temperature are not compatible with silicon foundry processing. In fact, on silicon wafers, silicon nitride is usually found as an amorphous mixture that only approaches the stochiometric relation of Si_3N_4, the specific relation being a strong function of process parameters and hence is IC-foundry specific. In the industry, it is variously referred to as “nitride”, “glass” or “passivation”, and may also contain amounts of oxygen.

2.1.4 Gallium Arsenide

Crystalline gallium arsenide (GaAs) is a “gold-gray” glassy material with the zinc-blende structure. When bound to each other, both gallium and arsenic atoms form tetrahedral bonds. In the industry, GaAs is referred to as a III-V (three-five), to indicate that it is a compound semiconductor

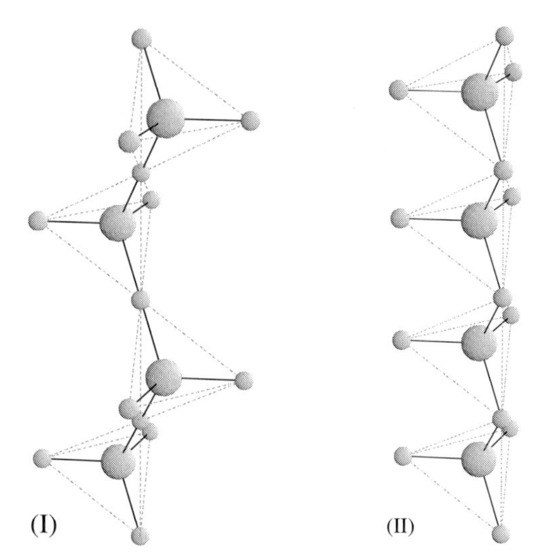

Figure 2.7. Silicon nitride appears in many crystalline configurations. The structure of α and of β Si_3N_4 are based on vertical stacks of SiN_4 tetrahedra, as shown in (I) and (II) respectively.

whose constituents are taken from the columns III and V of the periodic table. The atoms form into a zinc-blende crystal, structurally similar to the diamond-like structure of silicon, but with gallium and arsenic atoms alternating, see Figure 2.8.

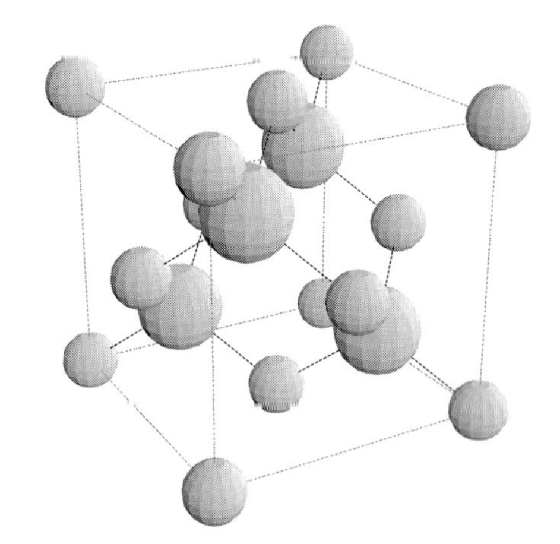

Figure 2.8. The zinc-blende structure of gallium arsenide. The two atom types are represented by spheres of differing diameter. Also see Figure 2.2.

Gallium arsenide is mainly used to make devices and circuits for the all-important opto-electronics industry, where its raw electronic speed or the ability to act as an opto-electronic lasing device is exploited. It is not nearly as popular as silicon, though, mainly because of the prohibitive processing costs. Gallium arsenide has a number of material features that differ significantly from Silicon, and hence a reason why we have included it in our discussion here. Gallium arsenide's phonon dispersion diagram is shown in Figure 2.9.

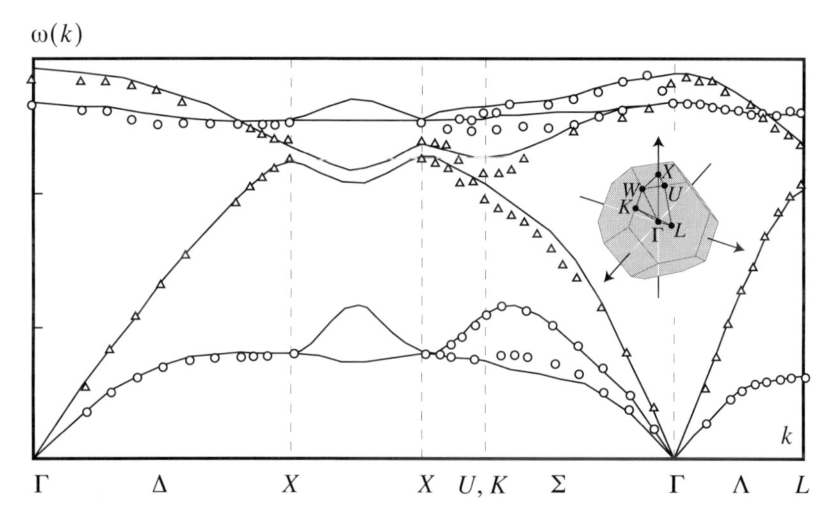

*Figure 2.9. The measured and computed dispersion diagrams of crystalline gallium arsenide. The vertical axis represents the phonon frequency, the horizontal axis represents straight-line segments in **k**-space between the main symmetry points of the Brillouin zone, which is shown as an insert. Figure adapted from [2.5].*

2.2 Crystal Structure

As we have seen, crystals are highly organized regular arrangements of atoms or ions. They differ from amorphous materials, which show no regular lattice, and poly-crystalline materials, which are made up of adjacent irregularly-shaped crystal grains, each with random crystal orientation. From observations and measurements we find that it is the regular crystalline structure that leads to certain special properties and behavior of the associated materials. In this section we develop the basic ideas that enable us to describe crystal structure analytically, so as to exploit the symmetry properties of the crystal in a systematic way.

2.2.1 Symmetries of Crystals

Consider a regular rectangular arrangement of points on the plane. The points could represent the positions of the atoms that make up a hypothetical two-dimensional crystal lattice. At first we assume that the atoms are equally spaced in each of the two perpendicular directions, say by a pitch of a and b. More general arrangements of lattice points are the rule.

Translational Invariance

Consider a vector $\boldsymbol{a}$ that lies parallel to the horizontal lattice direction and with magnitude equal to the pitch a. Similarly, consider vector $\boldsymbol{b}$ in the other lattice direction with magnitude equal to the pitch b. Then, starting at point $\boldsymbol{p}_i$, we can reach any other lattice point $\boldsymbol{q}_j$ with $\boldsymbol{q}_j = \alpha_j\boldsymbol{a} + \beta_j\boldsymbol{b}$, where α_j and β_j are integers. Having reached another interior point $\boldsymbol{q}_j$, the vicinity is the same as for point $\boldsymbol{p}_i$, and hence we say that the lattice is *invariant to translations* of the form $\alpha_j\boldsymbol{a} + \beta_j\boldsymbol{b}$, see the example in Figure 2.10.

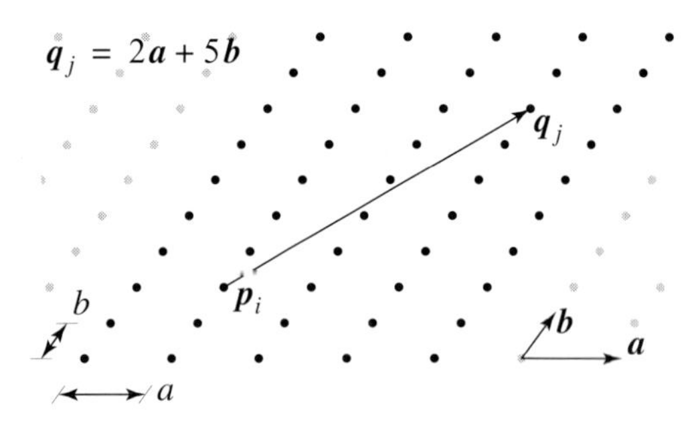

Figure 2.10. In this 2-dimensional infinitely-extending regular lattice the 2 lattice vectors are neither perpendicular nor of equal length. Given a starting point, all lattice points can be reached through $\boldsymbol{q}_j = \alpha_j\boldsymbol{a} + \beta_j\boldsymbol{b}$. *The vicinities of* $\boldsymbol{p}_i$ *and* $\boldsymbol{q}_i$ *are similar.*

Rotational Symmetry

If we consider the vicinity of an interior point $\boldsymbol{p}_i$ in our lattice, and let us assume that we have very many points in the lattice, we see that by rotating the lattice in the plane about point $\boldsymbol{p}_i$ by an angle of $180°$, the vicinity of the point $\boldsymbol{p}_i$ remains unchanged. We say that the lattice is *invariant to rotations* of $180°$. Clearly, setting $a = b$ and $\boldsymbol{a} \perp \boldsymbol{b}$ makes the lattice invariant to rotations of $90°$ as well. Rotational symmetry in lattices are due to rotations that are multiples of either $60°$, $90°$ or $180°$, see Figure 2.11. If the underlying lattice has a rotational symmetry, we will expect the crystal's material properties to have the same symmetries.

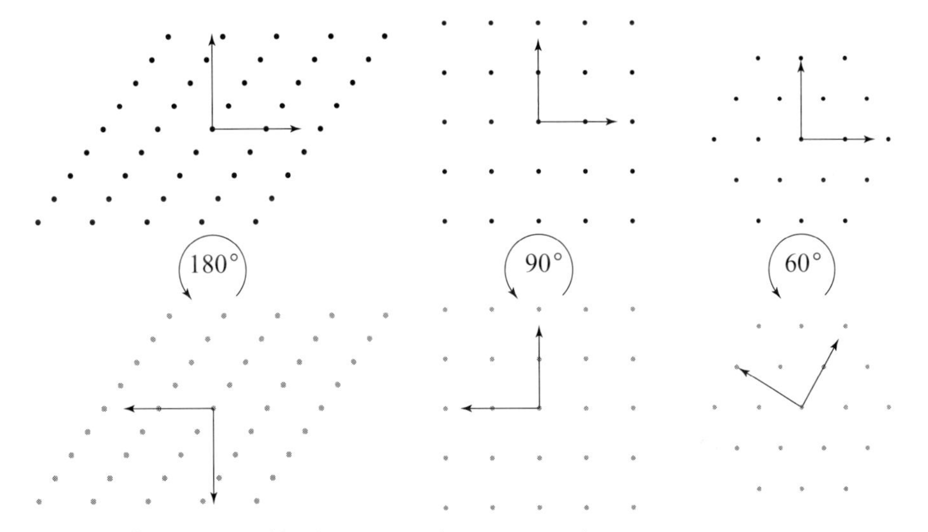

Figure 2.11. Illustrations of lattice symmetries w.r.t. rotations.

Bravais Lattice

With the basic idea of a lattice established, we now use the concept of a Bravais lattice to model the symmetry properties of a crystal's structure. A Bravais lattice is an infinitely extending regular three-dimensional array of points that can be constructed with the parametrized vector

$$\boldsymbol{q}_j = \alpha_j \boldsymbol{a} + \beta_j \boldsymbol{b} + \gamma_j \boldsymbol{c} \tag{2.3}$$

where $\boldsymbol{a}$, $\boldsymbol{b}$ and $\boldsymbol{c}$ are the non-coplanar lattice vectors and α_j, β_j and γ_j are arbitrary (positive and negative) integers. Note that we do not assume that the vectors $\boldsymbol{a}$, $\boldsymbol{b}$ and $\boldsymbol{c}$ are perpendicular. The Bravais lattice is inherently symmetric with respect to translations $\boldsymbol{q}_j$: this is the way we construct it. The remainder of the symmetries are related to rotations and reflections. There are 7 crystal systems: hexagonal, trigonal, triclinic, monoclinic, orthorhombic, tetragonal and cubic. In addition, there are 14 Bravais lattice types, see Figure 2.12. These are grouped into the following six lattice systems in decreasing order of geometric generality (or increasing order of symmetry): triclinic, monoclinic, orthorhombic, tetragonal, Hexagonal and cubic. The face-centered cubic (fcc) diamond-like lattice structure of silicon is described by the symmetric arrangement of vectors shown in Figure 2.13 (I). The fcc lattice is symmetric w.r.t. 90° rotations about all three coordinate axes. Gallium arsenide´s zinc-blende bcc-structure is similarly described as for silicon. Because of the presence of two constituent atoms, GaAs does not allow the same translational symmetries as Si.

Primitive Unit Cell

We associate with a lattice one or more *primitive unit cells*. A primitive unit cell is a geometric shape that, for single-atom crystals, effectively contains one lattice point. If the lattice point is not in the interior of the primitive cell, then more than one lattice point will lie on the boundary of the primitive cell. If, for the purpose of illustration, we associate a sphere with the lattice point, then those parts of the spheres that overlap with the inside of the primitive cell will all add up to the volume of a single sphere, and hence we say that a single lattice point is enclosed. Primitive cells seamlessly tile the space that the lattice occupies, see Figure 2.14.

Wigner-Seitz Unit Cell

The most important of the possible primitive cells is the *Wigner-Seitz* cell. It has the merit that it contains all the symmetries of the underlying Bravais lattice. Its definition is straightforward: The Wigner-Seitz cell of lattice point $\boldsymbol{p}_i$ contains all spatial points that are closer to $\boldsymbol{p}_i$ than to any other lattice point $\boldsymbol{q}_j$. Its construction is also straight-forward: Considering lattice point $\boldsymbol{p}_i$, connect $\boldsymbol{p}_i$ with its neighbor lattice points $\boldsymbol{q}_j$. On each connection line, construct a plane perpendicular to the connecting line at a position halfway along the line. The planes intersect each other

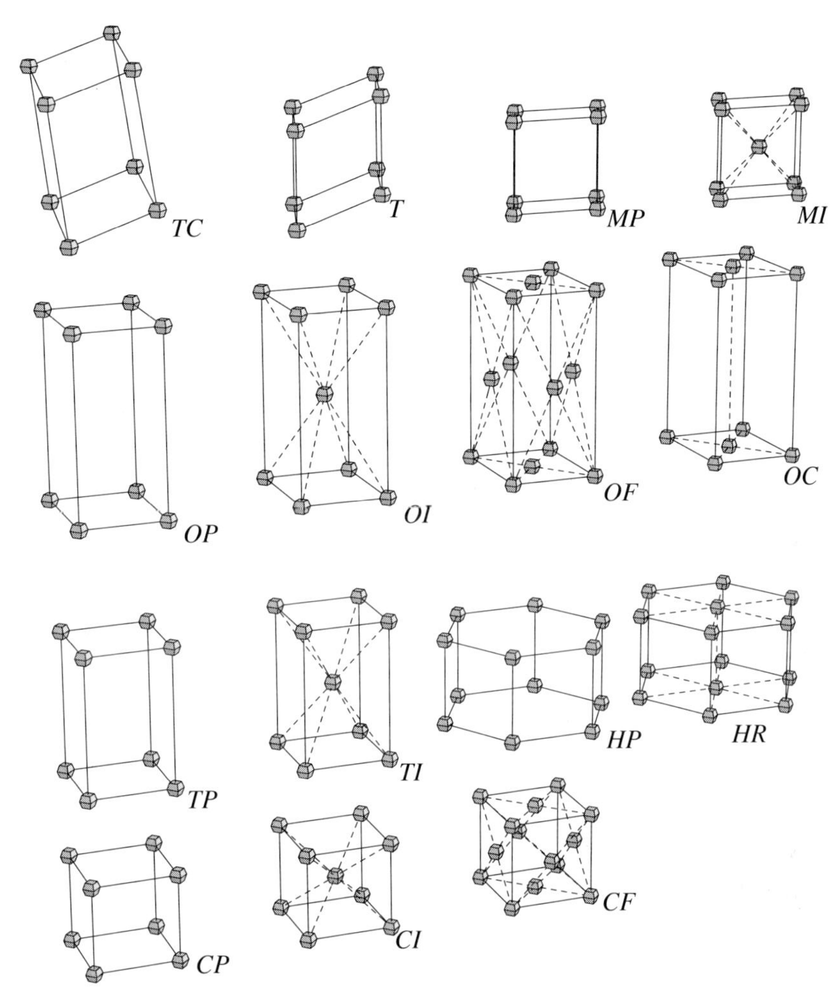

Figure 2.12. The fourteen Bravais lattice types. CP: Cubic P; CI: Body-centered cubic; CF: Face-centered cubic; $a = b = c$, $\alpha = \beta = \gamma = 90°$. *TP: Tetragonal; TI: Body-centered tetragonal;* $a = b \neq c$, $\alpha = \beta = \gamma = 90°$. *HP: Hexagonal; HR: Hexagonal R;* $a = b \neq c$, $\alpha = \beta = 90°$, $\gamma = 120°$ *(this not a separate Bravais lattice type). OP: Orthorhombic; OI: Body-centered orthorhombic; OF: Face-centered orthorhombic; OC: C-orthorhombic;* $a \neq b \neq c$, $\alpha = \beta = \gamma = 90°$. *MP: Monoclinic; MI: Face-centered monoclinic;* $a \neq b \neq c$, $\alpha = \beta = 90° \neq \gamma$. *T: Trigonal;* $a = b = c$, $\alpha \neq \beta \neq \gamma$. *TC: Triclinic;* $a \neq b \neq c$, $\alpha \neq \beta \neq \gamma$. *a, b, c refer to lattice pitches;* α, β, γ *to lattice vector angles.*

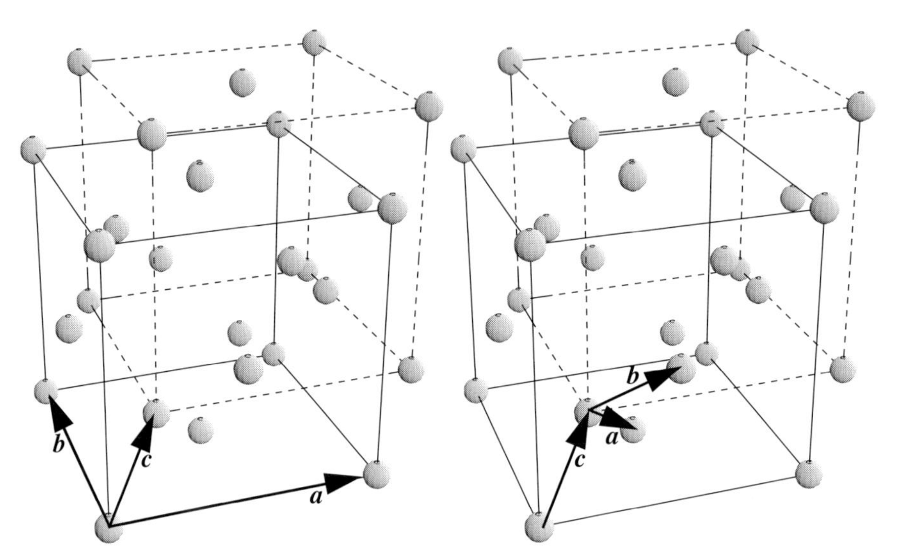

Figure 2.13. The equivalent lattice vectors sets that define the structures of the diamond-like fcc structure of Si and the zinc-blende-like structure of GaAs . The vectors are either the set: $\boldsymbol{a} = (2, 0, 0)$, $\boldsymbol{b} = (0, 2, 0)$, $\boldsymbol{c} = (0.5, 0.5, 0.5)$ *or the set:* $\boldsymbol{a} = (0.5, 0.5, -0.5)$, $\boldsymbol{b} = (0.5, -0.5, 0.5)$, $\boldsymbol{c} = (0.5, 0.5, 0.5)$.

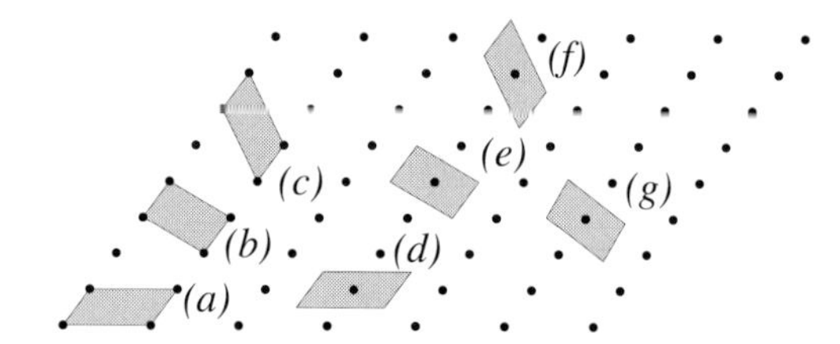

Figure 2.14. Illustrations of some of the many primitive unit cells for a 2-dimensional lattice. Of these, only (g) is a Wigner-Seitz cell. Figure adapted from [2.10].

and, taken together, define a closed volume around the lattice point $\boldsymbol{p}_i$. The smallest of these volumes is the Wigner-Seitz cell, illustrated in Figure 2.15.

Reciprocal Lattice

The spatial Bravais crystal lattice is often called the direct lattice, to refer to the fact that we can associate a reciprocal lattice with it. In fact, in studying the properties of the crystal lattice, most data will be referred to

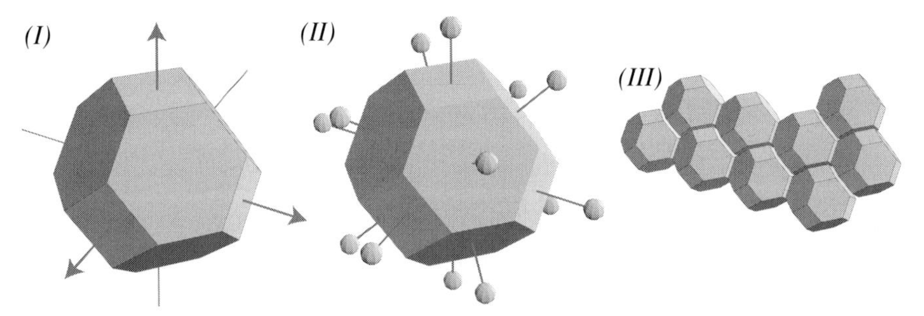

Figure 2.15. The Wigner-Seitz cell for the bcc lattice. (I) shows the cell without the atomic positions. (II) includes the atomic positions so as to facilitate understanding the method of construction. (III) The Wigner-Seitz cell tiles the space completely.

the reciprocal lattice. Let us consider a plane wave (see Box 2.1), traversing the lattice, of the form

$$e^{i\boldsymbol{k}\bullet\boldsymbol{r}}. \tag{2.4}$$

We notice two features of this expression: the vectors $\boldsymbol{k}$ and $\boldsymbol{r}$ are symmetrical in the expression (we can swap their positions without altering the expression); and the crystal is periodic in $\boldsymbol{r}$ space. The vector $\boldsymbol{r}$ is a position in real space; the vector $\boldsymbol{k}$ is called the position vector in reciprocal space, and often also called the wave vector, since it is "defined" using a plane wave. Next, we consider waves that have the same periodicity as the Bravais lattice. The Bravais lattice points lie on the regular grid described by the vectors $\boldsymbol{q}$, so that a periodic match in wave amplitude is expected if we move from one grid position to another, and therefore

$$e^{i\boldsymbol{k}\bullet(\boldsymbol{r}+\boldsymbol{q})} = e^{i\boldsymbol{k}\bullet\boldsymbol{r}}. \tag{2.5}$$

This provides us with a condition for the wave vector $\boldsymbol{k}$ for waves that have the same spatial period as the atomic lattice, because we can cancel out the common factor $e^{i\boldsymbol{k}\bullet\boldsymbol{r}}$ to obtain

$$e^{i\boldsymbol{k}\bullet\boldsymbol{q}} = 1 = e^{0}. \tag{2.6}$$

Equation (2.6) generates the $\boldsymbol{k}$ (wave) vectors from the $\boldsymbol{q}$ (lattice position) vectors, and we see that these are mutually orthogonal. We summarize the classical results of the reciprocal lattice:

Reciprocal Lattice Properties

- The reciprocal lattice is also a Bravais lattice.
- Just as the direct lattice positions are generated with the primitive unit vectors $\boldsymbol{q}_j = \alpha_j \boldsymbol{a} + \beta_j \boldsymbol{b} + \gamma_j \boldsymbol{c}$, the reciprocal lattice can also be so generated using $\boldsymbol{k}_j = \delta_j \boldsymbol{d} + \varepsilon_j \boldsymbol{e} + \zeta_j \boldsymbol{f}$. The relation between the two primitive vector sets is

$$\boldsymbol{d} = 2\pi \frac{\boldsymbol{b} \times \boldsymbol{c}}{\boldsymbol{a} \cdot (\boldsymbol{b} \times \boldsymbol{c})} \tag{2.7a}$$

$$\boldsymbol{e} = 2\pi \frac{\boldsymbol{c} \times \boldsymbol{a}}{\boldsymbol{a} \cdot (\boldsymbol{b} \times \boldsymbol{c})} \tag{2.7b}$$

$$\boldsymbol{f} = 2\pi \frac{\boldsymbol{a} \times \boldsymbol{b}}{\boldsymbol{a} \cdot (\boldsymbol{b} \times \boldsymbol{c})} \tag{2.7c}$$

- The reciprocal of the reciprocal lattice is the direct lattice.
- The reciprocal lattice also has a primitive cell. This cell is called the *Brillouin zone* after its inventor.
- The volume of the direct lattice primitive cell is $v = |\boldsymbol{a} \cdot (\boldsymbol{b} \times \boldsymbol{c})|$.
- The volume of the reciprocal lattice primitive cell is $V = |\boldsymbol{d} \cdot (\boldsymbol{e} \times \boldsymbol{f})|$. From (2.7c) we see that

$$\boldsymbol{d} = \frac{2\pi}{v}(\boldsymbol{b} \times \boldsymbol{c}) \qquad \boldsymbol{e} = \frac{2\pi}{v}(\boldsymbol{c} \times \boldsymbol{a}) \qquad \boldsymbol{f} = \frac{2\pi}{v}(\boldsymbol{a} \times \boldsymbol{b}) \tag{2.8}$$

and hence that

$$V = \frac{(2\pi)^3}{v} \tag{2.9}$$

Miller Indices

The planes formed by the lattice are identified using Miller indices. These are defined on the reciprocal lattice, and are defined as the coordinates of the shortest reciprocal lattice vector that is normal to the plane. Thus, if we consider a plane passing through the crystal, the Miller indi-

Box 2.1. Plane waves and wave-vectors.

We describe a general plane wave with

$$\psi(\boldsymbol{r}, t) = Ae^{i(\boldsymbol{k} \cdot \boldsymbol{r} - \omega t)} . \qquad \text{(B 2.1.1)}$$

Recalling the relation between the trigonometric functions and the exponential function,

$$e^{i\theta} = \cos(\theta) + i\sin(\theta) . \qquad \text{(B 2.1.2)}$$

we see that ψ is indeed a wave-like function, for θ parametrizes an endless circular cycle on the complex plane.

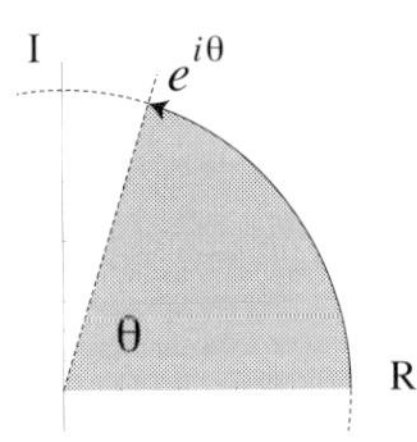

Figure B2.1.1: The exponential function describes a cycle in the complex plane.

Equation (B 2.1.1) is a powerful way of describing a plane wave. Two vectors, the spatial position vector $\boldsymbol{r}$ and the wave vector, or reciprocal position vector $\boldsymbol{k}$, are position arguments to the exponential function. The cyclic angle that it the wave has rotated through is the complete factor $(\boldsymbol{k} \cdot \boldsymbol{r} - \omega t)$. Since $\boldsymbol{r}$ measures the distance along an arbitrary spatial direction and $\boldsymbol{k}$ points along the propagation direction of the wave, $\boldsymbol{k} \cdot \boldsymbol{r}$ gives the component of this parametric angle. $\boldsymbol{k}$ measures rotation angle per distance travelled (a full cycle of 2π is one wavelength), so that $\boldsymbol{k} \cdot \boldsymbol{r}$ counts the number cycles completed along a spatial segment.

If we fix the time t ("freezing" the wave in space and time), then moving along a spatial direction we will experience a wave-like variation of the amplitude of ψ as $\boldsymbol{k} \cdot \boldsymbol{r}$. If we now move in a direction perpendicular to the propagation of the wave $\boldsymbol{k}$, then we will experience no amplitude modulation. Next, staying in the perpendicular direction to the wave propagation at a fixed position, if the time is again allowed to vary, we will now experience an amplitude modulation as the wave moves past us.

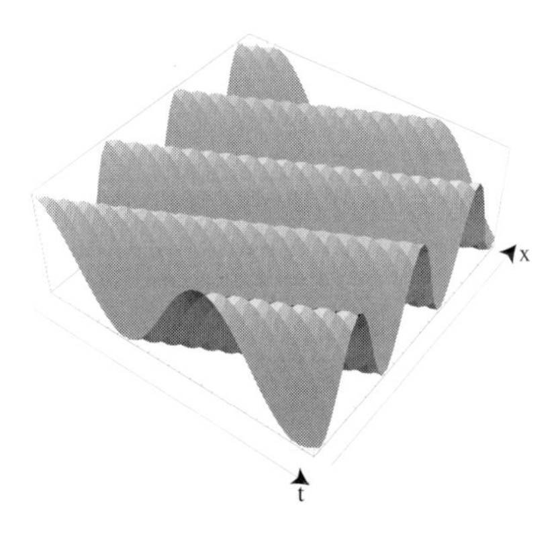

Figure B2.1.2: The appearance of a 1-dimensional wave plotted for t and x as parameters. If t is kept stationary, moving in the x-direction is accompanied by a wave-like variation.

ces are the components of a $\boldsymbol{k}$-space vector (the vector is normal to the plane) that fulfil $\boldsymbol{k} \cdot \boldsymbol{r} = \text{Constant}$, where $\boldsymbol{r}$ lies on the plane of interest.

As an example, consider a cubic lattice with the Miller indices (m, n, p) of a plane that lies parallel to a face of the cube. The indices determine the plane-normal $\boldsymbol{k}$-space vector $\boldsymbol{k} = m\boldsymbol{b}_1 + n\boldsymbol{b}_2 + p\boldsymbol{b}_3$. The plane is fixed by $\boldsymbol{k} \cdot \boldsymbol{r}$ = Constant $= 2\pi(r_1 m + r_2 n + r_3 p)$. Since $\boldsymbol{k} \cdot \boldsymbol{a}_1 = 2\pi m$, $\boldsymbol{k} \cdot \boldsymbol{a}_2 = 2\pi n$ and $\boldsymbol{k} \cdot \boldsymbol{a}_3 = 2\pi p$, the space axis intercepts for the plane are therefore $r_1 = \text{Constant}/2\pi m$, $r_2 = \text{Constant}/2\pi n$ and $r_3 = \text{Constant}/2\pi p$.

Silicon Reciprocal Lattice Shape

We have seen that the silicon crystal can be represented by a face-centered cubic lattice with a basis. This means that its reciprocal lattice is a body-centered lattice with a basis. This has the following implications. The Wigner-Seitz cell for the silicon direct lattice is a rhombic dodecahedron, whereas the first Brillouin zone (the Wigner-Seitz cell of the reciprocal lattice) is a truncated dodecahedron.

2.3 Elastic Properties: The Stressed Uniform Lattice

In a broad sense the geometry of a crystal's interatomic bonds represent the "structural girders" of the crystal lattice along which the forces act that keep the crystal intact. The strength of these directional interatomic forces, and the way in which they respond to small geometrical perturbations; these are the keys that give a crystal lattice its tensorial elastic properties and that enable us to numerically relate applied stress to a strain response. In this section we derive the Hooke law for crystalline, amorphous and poly-crystalline materials based on lattice considerations.

2.3.1 Statics

Atomic Bond Model

It is well known that atoms form different types of bonds with each other. The classification is conveniently viewed as the interaction between a pair of atoms:

- *Ionic*—a "saturated" bond type that is characterized by the fact that one atom ties up the electrons participating in the bond in its outermost shell. This leaves the two atoms oppositely charged. The coulombic (electrostatic) force between the net charges of the constituent "ions" form the bond. This bond is sometimes termed localized, because the electrons are tightly bound to the participating atoms.

- *van der Waals*—a very weak bonding force that is often termed the fluctuating dipole force because it is proportional to an induced dipole between the constituent atoms, and this effective dipole moment has a non-vanishing time average.
- *Valency*—another localized electron-pair bond. Here, the electron-pair of the participating atoms form a hybrid orbital that is equally shared by the two atoms, hence the term covalent. Clearly, ionic and covalent bonds are the two limiting cases of a similar phenomena, so that a bond in-between these limits can also be expected. Valency bonds are quite strong, and account for the hardness and brittle nature of the materials.
- *Metallic*—the electrons participating in the bonding are non-localized. Typically, the number of valence electrons at a point is exceeded by the number of nearest-neighbor atoms. The electrons are therefore shared by many atoms, making them much more mobile, and also accounts for the ductility of the material.

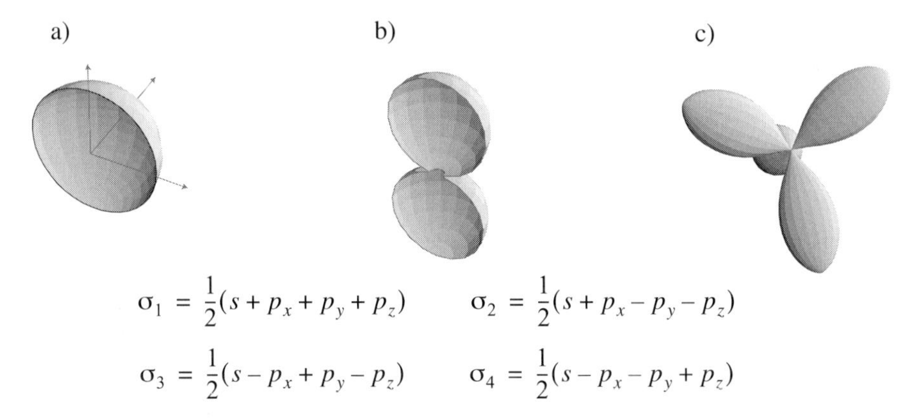

Figure 2.16. The geometry of a tetrahedral bond for Carbon-like atoms can be represented by a hybrid sp^3-function as a superposition of 2s and 2p-orbitals. a) s-orbital. b) p-orbital. c) sp^3-orbital.

The semiconductors that we consider are covalently bonded. We now give a qualitative description of the tetrahedral covalent bond of the atoms of an fcc crystal. Solving the Schrödinger equation (see Box 2.2

Box 2.2. The stationary Schrödinger equation and the Hamiltonian of a Solid [2.4].

In principle, all basic calculations of solid state properties are computed with the Schrödinger equation

$$H\varphi = E\varphi \qquad \text{(B 2.2.1)}$$

The Hamiltonian operator H defines the dynamics and statics of the model, φ represents a state of the model, and E is the scalar-valued energy. Thus the Schrödinger equation is an Eigensystem equations with the energy plying the role of an eigenvalue and the state the role of an eigen-function.

The Hamiltonian is usually built up of contributions from identifiable subcomponents of the system. Thus, for a solid-state material we write that

$$H = H_e + H_i + H_{ei} + H_x \qquad \text{(B 2.2.2)}$$

i.e., the sum of the contributions from the electrons, the atoms, their interaction and interactions with external influences (e.g., a magnetic field). Recall that the Hamiltonian is the sum of kinetic and potential energy terms. Thus, for the electrons we have

$$H_e = T_e + U_{ee}$$
$$= \sum_{\alpha} \frac{p_{\alpha}^2}{2m} + \frac{1}{8\pi\varepsilon_0} \sum_{\alpha\beta}{}' \frac{e^2}{|\boldsymbol{r}_{\alpha} - \boldsymbol{r}_{\beta}|} \qquad \text{(B 2.2.3)}$$

The second term is the Coulombic potential energy due to the electron charges. Note that the sum is primed: the sum excludes terms where $\alpha = \beta$.

For the atoms the Hamiltonian looks similar to that of the electron

$$H_i = T_i + U_{ii}$$
$$= \sum_{\alpha} \frac{P_{\alpha}^2}{2M} + \frac{1}{2} \sum_{\alpha\beta}{}' V_i(\boldsymbol{R}_{\alpha} - \boldsymbol{R}_{\beta}) \qquad \text{(B 2.2.4)}$$

For the electron-atom interaction we associate only a potential energy

$$H_{ei} = U_{ei} = \sum_{\alpha\beta} V_{ei}(\boldsymbol{r}_{\alpha} - \boldsymbol{R}_{\beta}) \qquad \text{(B 2.2.5)}$$

Equation (B 2.2.1) is hardly ever solved in all its generality. The judicious use of approximations and simplifications have yielded not only tremendous insight into the inner workings of solid state materials, but have also been tremendously successful in predicting complex phenomena.

We will return to this topic in the next chapter, where we will calculate the valence band structure of silicon to remarkable accuracy.

and Chapter 3) for a single atom yields the orthogonal eigenfunctions that correspond to the energy levels of the atom, also known as the orbitals. The spherical harmonic functions shown in Table 2.2 are such eigenfunctions. The tetrahedral bond structure of the Si atom can be made, through a superposition of basis orbitals, to form the hybrid sp^3 orbital

Table 2.2. The spherical harmonic functions $\left|Y_l^m(\theta, \phi)^2\right|$, *Also see Table 3.1 and Figure 3.9 in Chapter 3.*

l	m	Plot	l	m	Plot	l	m	Plot	l	m	Plot
0	0	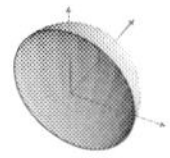									
1	0		1	1							
2	0		2	1		2	2				
3	0		3	1		3	2		3	3	
4	0		4	1		4	2		4	3	

that we associate with its valence electrons. The construction is illustrated in Figure 2.16.

The valence electron orbitals may overlap to form bonds between atoms. The interpretation is straightforward: a valence electron's orbital represents the probability distribution of finding that electron in a specific region of space. In a first approximation, we let the orbitals simply overlap and allow them to interfere to form a new shared orbital. Bonding takes place if the new configuration has a lower energy than the two separate atom orbitals. The new, shared, hybrid orbital gives the electrons of the two binding atoms a relatively high probability of occupying the space between the atoms.

Figure 2.17 shows what happens when atoms bond to form a diamond-

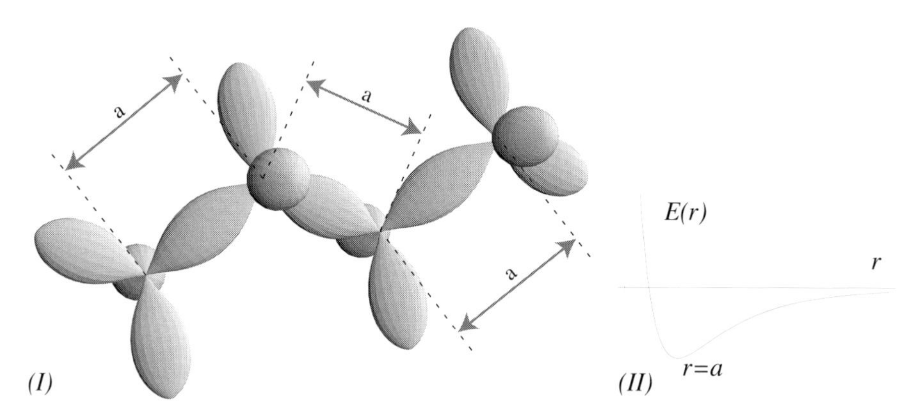

Figure 2.17. When atoms with a tetrahedral bond structure form a covalently-bonded crystal lattice, the valence electrons are localized about the nearest-neighbor atoms. We visualize these (I) as the overlap positions of the sp3-hybrid orbitals, shown here for four atoms in a diamond-like lattice. We assume that the distance a between the atomic cores, the lattice constant, is the position of minimum energy for the bond (II).

like crystal. The energy of the atomic arrangement is lowered to a minimum when all atoms lie approximately at a separation equal to the lattice constant a (the lattice constant is the equilibrium distance that separates atoms of a lattice). The sp^3 orbitals of neighboring atoms overlap to form new shared orbitals. The electrons associated with these new states are effectively shared by the neighboring atoms, but localized in the bond. The bonding process does not alter the orbitals of the other electrons significantly.

The potential energy plot of Figure 2.17 (II) is necessarily only approximate, yet contains the necessary features for a single ionic bond. In fact, it is a plot of the Morse potential, which contains a weaker attractive and a very strong, repulsive constituent, localized at the core

$$E(\boldsymbol{r}) = D_e(1 - \exp[-\beta(\boldsymbol{r} - \boldsymbol{r}_{\text{eq}})])^2 \tag{2.10}$$

The potential energy for bond formation D_e and the equilibrium bond length $\boldsymbol{r}_{\text{eq}}$ depend on the constituent atoms, and may be obtained by experiment. The parameter β controls the width of the potential well, i.e., the range of the interparticle forces. For covalent crystals, such as silicon, this picture is too simplistic. The tetrahedral structure of the orbitals forms bonds that can also take up twisting moments, so that, in addition to atom pair interactions, also triplet and perhaps even larger sets of interacting atoms should be considered. A three-atom interaction model that accounts well for most thermodynamic quantities of Si appears to be the Stillinger-Weber potential [2.8], which, for atom i and the interaction with its nearest neighbors j, k, l and m is

$$\begin{aligned} E_{\text{Si-Si}}(\boldsymbol{r}_i, \boldsymbol{r}_j, \boldsymbol{r}_k, \boldsymbol{r}_l, \boldsymbol{r}_m) &= \frac{1}{2}[E_2(\boldsymbol{r}_{ij}) + E_2(\boldsymbol{r}_{ik}) + E_2(\boldsymbol{r}_{il}) + E_2(\boldsymbol{r}_{im})] \\ &+ E_3(\boldsymbol{r}_{ij}, \boldsymbol{r}_{ik}, \theta_{jik}) + E_3(\boldsymbol{r}_{im}, \boldsymbol{r}_{ik}, \theta_{mik}) + E_3(\boldsymbol{r}_{il}, \boldsymbol{r}_{im}, \theta_{lim}) \\ &+ E_3(\boldsymbol{r}_{ij}, \boldsymbol{r}_{im}, \theta_{jim}) + E_3(\boldsymbol{r}_{ij}, \boldsymbol{r}_{il}, \theta_{jil}) + E_3(\boldsymbol{r}_{il}, \boldsymbol{r}_{ik}, \theta_{lik}) \end{aligned} \tag{2.11}$$

In this model, the two-atom interaction is modelled as

$$E_2(\boldsymbol{r}_{ij}) = \begin{cases} \mathcal{E}G(He^{-p} - |\boldsymbol{r}_{ij}/a|^{-q})\exp((|\boldsymbol{r}_{ij}/a| - c)^{-1}) & |\boldsymbol{r}_{ij}/a| < c \\ 0 & |\boldsymbol{r}_{ij}/a| > c \end{cases} \tag{2.12}$$

and the three-atom interaction by

$$E_3(\boldsymbol{r}_{ij}, \boldsymbol{r}_{ik}, \theta_{jik}) = \begin{cases} \mathcal{E}\lambda\exp\left[\gamma\left(\left(\left|\frac{\boldsymbol{r}_{ij}}{a}\right| - c\right)^{-1} + \left(\left|\frac{\boldsymbol{r}_{ik}}{a}\right| - c\right)^{-1}\right)\right] \times \left(\cos(\theta_{jik}) + \frac{1}{3}\right)^2 & ,|\boldsymbol{r}_{ij}/a| < c \\ 0 & \text{otherwise} \end{cases} \tag{2.13}$$

For silicon, this model works well with the following parameters: $G = 7.0495563$, $H = 0.60222456$, $p = 4$, $q = 0$, $\lambda = 21$, $\gamma = 1.2$, the cut-off radius $c = 1.8$, the "lattice" constant $a = 0.20951 nm$ and the bond energy $\mathcal{E} = 6.9447\times10^{-19} J/\mathrm{ion}$. Note that in the above $\boldsymbol{r}_{ij} = \boldsymbol{r}_i - \boldsymbol{r}_j$.

Linearization We assume that the atoms of the crystal always remain in the vicinity of their lattice positions, and that the distance they displace from these positions is "small" when measured against the lattice constant a, see Figure 2.18. This is a reasonable assumption for a solid crystal at typical

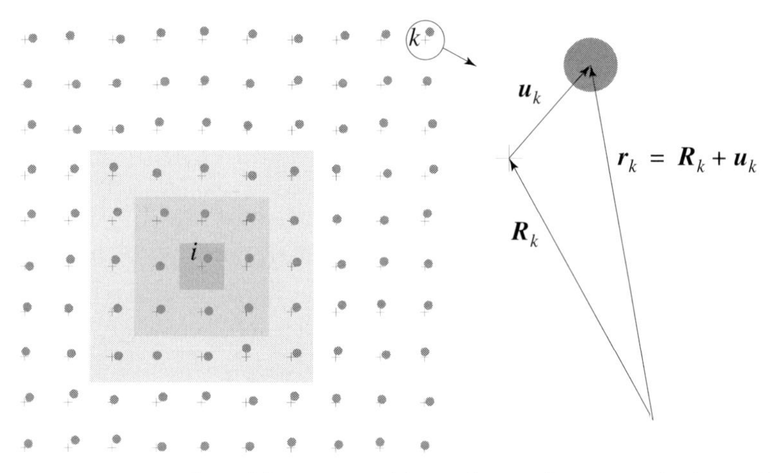

Figure 2.18. Instantaneous snapshot of the atom positions of a regular square lattice with respect to their average lattice site positions. On the right is shown the relation between the lattice position vector $\boldsymbol{R}_k$, the atom position vector $\boldsymbol{r}_k$ and the atom displacement vector $\boldsymbol{u}_k$ for atom k. The shading around atom i indicates how the inter-atom interaction strength falls off as a function of distance.

operating temperatures; the covalent and ionic bonds are found to be strong enough to make it valid.

Linearized Potential Energy

To illustrate the derivation of the elastic energy in terms of the strain and the elastic constants of the crystal, we will use an interatomic potential that only involves two-atom interactions. For a more general derivation, see e.g. [2.1]. Denote the relative vector position of atom i by $\boldsymbol{u}_i$, and its absolute position by $\boldsymbol{r}_i = \boldsymbol{R}_i + \boldsymbol{u}_i$, where $\boldsymbol{R}_i$ localizes the regular lattice site. For a potential binding energy E between a pair of atoms as depicted in Figure 2.17 (II), we form the potential energy of the whole crystal with

$$U = \frac{1}{2}\sum_{i=1}^{N}\sum_{j=1}^{N} E(\boldsymbol{r}_i - \boldsymbol{r}_j) = \frac{1}{2}\sum_{i=1}^{N}\sum_{j=1}^{N} E(\boldsymbol{R}_i - \boldsymbol{R}_j + \boldsymbol{u}_i - \boldsymbol{u}_j) = \frac{1}{2}\sum_{i=1}^{N}\sum_{j=1}^{N} E(\boldsymbol{R}_{ij} + \boldsymbol{u}_{ij}) \quad (2.14)$$

where $\boldsymbol{R}_{ij} = \boldsymbol{R}_i - \boldsymbol{R}_j$ and $\boldsymbol{u}_{ij} = \boldsymbol{u}_i - \boldsymbol{u}_j$. The factor $1/2$ arises because the double sum counts each atom pair twice. We expand the energy about the lattice site $\boldsymbol{R}_{ij}$ using the Taylor expansion for vectors

$$E(\boldsymbol{R}_{ij} + \boldsymbol{u}_{ij}) = E(\boldsymbol{R}_{ij}) + (\boldsymbol{u}_{ij} \cdot \nabla)E(\boldsymbol{R}_{ij}) + \frac{1}{2}(\boldsymbol{u}_{ij} \cdot \nabla)^2 E(\boldsymbol{R}_{ij}) + \dots \quad (2.15)$$

because of the assumption that the atom displacements $\boldsymbol{u}_i$ and hence their differences $\boldsymbol{u}_{ij}$ are small. The terms $(\boldsymbol{u}_{ij} \cdot \nabla)^n E(\boldsymbol{R}_{ij})$ must be read as $(\boldsymbol{u}_{ij} \cdot \nabla)^n$ operating n times on the position-dependent energy E evaluated at the atom position $\boldsymbol{R}_{ij}$. Applying (2.15) to (2.14) we obtain

$$U = \frac{1}{2}\sum_{i=1}^{N}\sum_{j=1}^{N} E(\boldsymbol{R}_{ij}) + \frac{1}{2}\sum_{i=1}^{N}\sum_{j=1}^{N} \boldsymbol{u}_{ij} \bullet \nabla E(\boldsymbol{R}_{ij}) + \frac{1}{4}\sum_{i=1}^{N}\sum_{j=1}^{N} (\boldsymbol{u}_{ij} \bullet \nabla)^2 (E(\boldsymbol{R}_{ij})) + \dots \quad (2.16)$$

The first term in (2.16) is a constant for the lattice, and is denoted by U_o. The second term is identically zero, because the energy gradient is evaluated at the rest position of each atom where by definition is must be zero

$$\frac{1}{2}\sum_{i=1}^{N}\sum_{j=1}^{N}\boldsymbol{u}_{ij}\bullet\nabla E(\boldsymbol{R}_{ij}) = 0 \tag{2.17}$$

This leaves us with the zeroth and second and higher order terms

$$U = U_o + \frac{1}{4}\sum_{i=1}^{N}\sum_{j=1}^{N}(\boldsymbol{u}_{ij}\bullet\nabla)^2 E(\boldsymbol{R}_{ij}) + \dots \tag{2.18}$$

The first term in (2.18) is the energy associated with the atoms at the lattice positions, i.e., at rest, and represents the datum of energy for the crystal. The second term is the harmonic potential energy, or the small-displacement potential energy. When it is expanded, we obtain

$$U^h = \frac{1}{4}\sum_{i=1}^{N}\sum_{j=1}^{N}\boldsymbol{u}_{ij}\bullet\nabla(\nabla E(\boldsymbol{R}_{ij}))\bullet\boldsymbol{u}_{ij} \tag{2.19}$$

Elasticity Tensor

Equation (2.19) illustrates two terms, the deformation felt between two sites $\boldsymbol{u}_{ij}$, and the second derivative of the interatomic potential at zero deformation $\nabla(\nabla E(\boldsymbol{R}_{ij}))$. The second derivative is the "spring constant" of the lattice, and the deformation is related to the "spring extension". We can now go a step further and write the harmonic potential energy in terms of the elastic constants and the strain.

Moving towards a continuum view, we will write the quantities in (2.19) in terms of the position $\boldsymbol{R}_i$ alone. Consider the term $\nabla[\nabla E(\boldsymbol{R}_{ij})] = \nabla[\nabla E(\boldsymbol{R}_i - \boldsymbol{R}_j)]$. We expect $E(\boldsymbol{R}_{ij})$ to fall off rapidly away from $\boldsymbol{R}_i$ when $\boldsymbol{R}_j$ is far removed, so that for the site $\boldsymbol{R}_i$, most of the terms in (2.19) in the sum over j are effectively zero. To tidy up the notation, we denote $\boldsymbol{D}(\boldsymbol{R}_i)_{\alpha\beta}$ as the components of the dynamical tensor $\boldsymbol{D}(\boldsymbol{R}_i) = \nabla[\nabla E(\boldsymbol{R}_{ij})]$. The remainder of the sites $\boldsymbol{R}_j$ are close to $\boldsymbol{R}_i$, making $\boldsymbol{R}_{ij}$ small, so that we are justified in making an expansion of the components of $\boldsymbol{D}(\boldsymbol{R}_{ij})$ about $\boldsymbol{R}_i$

$$\boldsymbol{D}(\boldsymbol{R}_{ij})_{\alpha\beta} = \boldsymbol{D}(\boldsymbol{R}_i)_{\alpha\beta} + \nabla\{\boldsymbol{D}(\boldsymbol{R}_i)_{\alpha\beta}\}\bullet\boldsymbol{R}_{ij} + \dots \tag{2.20}$$

for which we only keep the leading term. We now apply a similar series expansion to the term $\boldsymbol{u}_j$, because we expect it to vary little in the vicinity of $\boldsymbol{R}_i$, and from now on we do not consider atom sites that are far away, to obtain

$$\boldsymbol{u}_j = \boldsymbol{u}_i + \boldsymbol{R}_{ij} \bullet \nabla \boldsymbol{u}_i(\boldsymbol{R}_i) + \dots \tag{2.21}$$

Taking the first right-hand-side term of (2.20) and the first two terms of (2.21), and inserting these into (2.19), we obtain

$$U^h = -\frac{1}{4}\sum_{i=1}^{N}\sum_{j=1}^{N}(\boldsymbol{R}_{ij} \bullet \nabla \boldsymbol{u}_i)^T \bullet \boldsymbol{D}(\boldsymbol{R}_i) \bullet (\boldsymbol{R}_{ij} \bullet \nabla \boldsymbol{u}_i) \tag{2.22}$$

The result of (2.22) can now be rewritten again in terms of the original quantities, to give

$$\begin{aligned} U^h &= -\frac{1}{4}\sum_{i=1}^{N}(\nabla \boldsymbol{u}_i)^T : \sum_{j=1}^{N}\{\boldsymbol{R}_{ij}\boldsymbol{D}(\boldsymbol{R}_i)\boldsymbol{R}_{ij}\} : \nabla \boldsymbol{u}_i \\ &= -\frac{1}{2}\sum_{i=1}^{N}(\nabla \boldsymbol{u}_i)^T : \boldsymbol{F}(\boldsymbol{R}_i) : \nabla \boldsymbol{u}_i \end{aligned} \tag{2.23}$$

The rank four elastic material property tensor $\boldsymbol{F}(\boldsymbol{R}_i)$ in the vicinity of the lattice site $\boldsymbol{R}_i$ is defined as

$$\boldsymbol{F}(\boldsymbol{R}_i) = \frac{1}{2}\sum_{j=1}^{N}(\boldsymbol{R}_{ij}\boldsymbol{D}(\boldsymbol{R}_i)\boldsymbol{R}_{ij}) \tag{2.24}$$

in terms of the crystal constituent positions and the resulting net inter-atom binding energy. We expect this tensor to be translationally invariant with respect to the lattice. Therefore, we can now move from a discrete crystal description to the continuum, by considering the crystal as a collection of primitive cells of volume V with an average "density" of elastic material property in any particular unit cell to be $\boldsymbol{E} = \boldsymbol{F}/V$, so that we can replace the sum in (2.23) by an integral to obtain

$$U^h = -\frac{1}{2}\int_{\Omega}(\nabla \boldsymbol{u}_i)^T \bullet \boldsymbol{E} \bullet \nabla \boldsymbol{u}_i dV \tag{2.25}$$

Strain

We now wish to bring the gradient of deformation $\nabla \boldsymbol{u}$, a vector-valued field over the crystal, in relation to the strain, which is a rank two tensor-valued field. We consider Figure 2.19, where we follow the behavior of

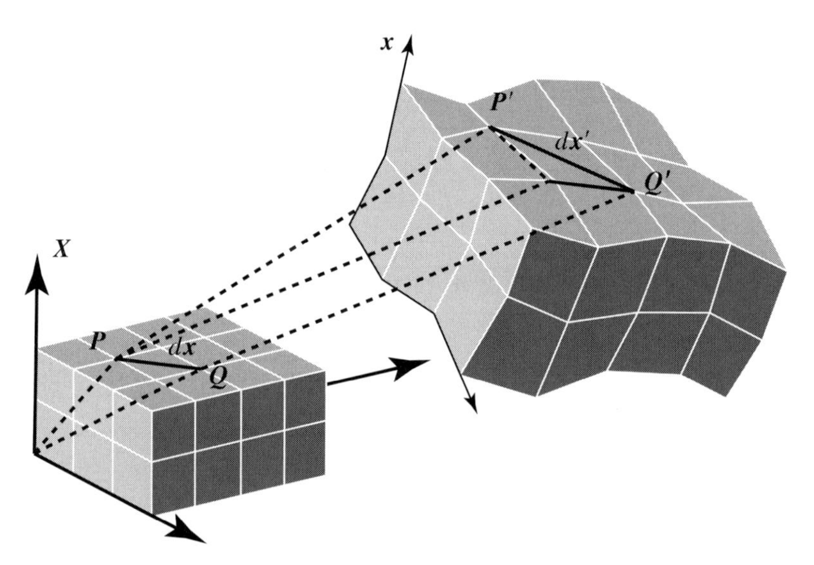

Figure 2.19. An arbitrary body before and after deformation. The points $\boldsymbol{P}$ *and* $\boldsymbol{Q}$ *are "close" to each other.*

two points $\boldsymbol{P}$ and $\boldsymbol{Q}$ before and after deformation. The points are chosen to be close to each other, and we assume that the body has deformed elastically, i.e., without cracks forming, and without plastically yielding. To a very good approximation, the movement of each point in the body can then be written as a linear transformation

$$\boldsymbol{P}' = \alpha_o + (\delta + \alpha) \bullet \boldsymbol{P} \tag{2.26}$$

where α_o represents a pure translation and α a rotation plus stretch. Next, we consider what happens to the line element $\boldsymbol{dx}$ under the deformation field $\boldsymbol{u}$

$$\begin{aligned} \boldsymbol{dx}' &= \boldsymbol{Q}' - \boldsymbol{P}' = \alpha_o + (\delta + \alpha)^T \bullet \boldsymbol{Q} - (\alpha_o + (\delta + \alpha)^T \bullet \boldsymbol{P}) \\ &= \boldsymbol{dx} + \alpha^T \bullet \boldsymbol{dx} = \boldsymbol{dx} + \boldsymbol{du} \end{aligned} \tag{2.27}$$

We can rewrite $\boldsymbol{du}$ via the chain rule as $\boldsymbol{du} = (\nabla \boldsymbol{u})^T \bullet \boldsymbol{dx}$, thereby making the association $\alpha = \nabla \boldsymbol{u}$, to find that we can now rewrite the expression for $\boldsymbol{dx}'$ only in terms of the original quantities that we measure

$$\boldsymbol{dx}' = \boldsymbol{dx} + (\nabla \boldsymbol{u})^T \bullet \boldsymbol{dx} \tag{2.28}$$

We are interested in how the line element deforms (stretches), and for this we form an expression for the square of its length

$$\begin{aligned} (\boldsymbol{dx}')^2 &= (\boldsymbol{dx})^T \bullet \boldsymbol{dx} + [(\nabla \boldsymbol{u})^T \bullet \boldsymbol{dx}]^T \bullet \boldsymbol{dx} + (\boldsymbol{dx})^T \bullet [(\nabla \boldsymbol{u})^T \bullet \boldsymbol{dx}] \\ &\quad + [(\nabla \boldsymbol{u})^T \bullet \boldsymbol{dx}]^T \bullet [(\nabla \boldsymbol{u})^T \bullet \boldsymbol{dx}] \\ &= (\boldsymbol{dx})^T \bullet [\delta + \nabla \boldsymbol{u} + (\nabla \boldsymbol{u})^T + (\nabla \boldsymbol{u})^T \bullet \nabla \boldsymbol{u}] \bullet \boldsymbol{dx} \end{aligned} \tag{2.29}$$

Consider the argument $[\delta + \nabla \boldsymbol{u} + (\nabla \boldsymbol{u})^T + (\nabla \boldsymbol{u})^T \bullet \nabla \boldsymbol{u}]$. If the stretch is small, which we have assumed, then we can use $\sqrt{1+s} \cong 1 + s/2$, which defines the strain ε by

$$\varepsilon = \frac{1}{2}[\nabla \boldsymbol{u} + (\nabla \boldsymbol{u})^T + (\nabla \boldsymbol{u})^T \bullet \nabla \boldsymbol{u}] \approx \frac{1}{2}[\nabla \boldsymbol{u} + (\nabla \boldsymbol{u})^T] \tag{2.30}$$

The components of the small strain approximation may be written as

$$\varepsilon_{\alpha\beta} = \frac{1}{2}\left[\frac{\partial u_\alpha}{\partial x_\beta} + \frac{\partial u_\beta}{\partial x_\alpha}\right] \tag{2.31}$$

The small strain is therefore a symmetric rank two tensor.

Crystal Energy in Terms of Strain

We return our attention to equation (2.25), where our goal is to rewrite the harmonic crystal energy U^h in terms of the crystal's strain. Clearly, U^h should not be dependent on our choice of coordinate axes, and is therefore invariant with respect to rigid-body rotations. We now choose a deformation field $\boldsymbol{u} = \theta \times \boldsymbol{R}$ which simply rotates the crystal atoms from their lattice positions by a constant defined angle θ. The gradient of this field is $\nabla \boldsymbol{u} = \nabla\theta \times \boldsymbol{R} + \theta \times \nabla \boldsymbol{R} = 0$, so that the energy associated with pure rotations is clearly zero. The gradient of $\boldsymbol{u}$ can always be written as the sum of a symmetric and an anti-symmetric part

$$\begin{aligned}\nabla \boldsymbol{u} = \nabla \boldsymbol{u}_s + \nabla \boldsymbol{u}_a &= [\nabla \boldsymbol{u} + (\nabla \boldsymbol{u})^T]/2 + [\nabla \boldsymbol{u} - (\nabla \boldsymbol{u})^T]/2 \\ &= (\varepsilon + \kappa)\end{aligned} \tag{2.32}$$

Note that $\varepsilon = \varepsilon^T$ and $\kappa = -\kappa^T$. We substitute ε and κ into (2.25) to obtain

$$U^h = -\frac{1}{2}\int_{\Omega}(\varepsilon^T{:}\boldsymbol{E}{:}\varepsilon + \varepsilon^T{:}\boldsymbol{E}{:}\kappa + \kappa^T{:}\boldsymbol{E}{:}\varepsilon + \kappa^T{:}\boldsymbol{E}{:}\kappa)dV \tag{2.33}$$

for the harmonic energy of the crystal.

Independent Elastic Constants

The elastic tensor **E** has inherent symmetries that can be exploited to simplify (2.33). We first write an expression for the components of the tensor **E**, centered at site i, in cartesian coordinates

$$\boldsymbol{E}(\boldsymbol{R}_i)_{\alpha\beta\gamma\mu} = \frac{1}{2V}\sum_{j=1}^{N}\left((\boldsymbol{R}_{ij})_\alpha\left(\frac{\partial}{\partial x_\beta}\left[\frac{\partial E(\boldsymbol{R}_i)}{\partial x_\mu}\right]\right)(\boldsymbol{R}_{ij})_\gamma\right) \tag{2.34}$$

Consider the argument of the sum. Clearly, **E** is symmetric with respect to the indices β and μ, since the order of differentiation of the energy with respect to a spatial coordinate is arbitrary. Furthermore, swapping α and γ also has no effect. Now look at the terms $\varepsilon^T{:}\boldsymbol{E}{:}\kappa$ and $\kappa^T{:}\boldsymbol{E}{:}\varepsilon$ in (2.33). They vanish because, due the above symmetries of $\boldsymbol{E}$, $\varepsilon^T{:}\boldsymbol{E}{:}\kappa = -\kappa^T{:}\boldsymbol{E}{:}\varepsilon$. Finally, because the anti-symmetric strain κ represents a pure rotation, the last term $\kappa^T{:}\boldsymbol{E}{:}\kappa$ in (2.33) must also vanish. As a consequence, the tensor **E** must be symmetric with respect to its first and second pairs of indices as well, because it now inherits the symmetry of the symmetric strain. Thus we obtain the familiar expression for the elastic energy as

$$U^h = -\frac{1}{2}\int_V [\varepsilon^T : \boldsymbol{E}(\boldsymbol{R}) : \varepsilon] dV$$
$$= -\frac{1}{2}\int_V \left(\varepsilon^T : \left[\frac{1}{2V}\sum_{j=1}^{N} (\boldsymbol{R}-\boldsymbol{R}_j)\boldsymbol{D}(\boldsymbol{R})(\boldsymbol{R}-\boldsymbol{R}_j)\right] : \varepsilon\right) dR^3 \tag{2.35}$$

Of the possible 81 independent components for the fourth rank tensor **E**, only 21 remain. We can organize these into a 6×6 matrix **C** if the following six index pair associations are made:

$$E_{(ij)(kl)} \equiv C_{(m)(n)} \qquad ij \to m \qquad kl \to n$$
$$\left.\begin{matrix} 11 \to 1 & 22 \to 2 & 33 \to 3 \\ 23 \to 4 & 31 \to 5 & 12 \to 6 \end{matrix}\right\} \tag{2.36}$$

In order to obtain analog relations using the reduced index formalism, we have to specify the transformation of the stress and strain as well. For the stress we can use the same index mapping as in (2.36)

$$\sigma_{ij} \equiv s_m \qquad ij \to m$$
$$\left.\begin{matrix} 11 \to 1 & 22 \to 2 & 33 \to 3 \\ 23 \to 4 & 31 \to 5 & 12 \to 6 \end{matrix}\right\} \tag{2.37}$$

For the strain, however, we need to combine the index mapping with a scaling

$$\begin{matrix} e_1 = \varepsilon_{11} & e_2 = \varepsilon_{22} & e_3 = \varepsilon_{33} \\ e_4 = 2\varepsilon_{23} & e_5 = 2\varepsilon_{31} & e_6 = 2\varepsilon_{12} \end{matrix} \tag{2.38}$$

In this way, we retain the algebraic form for the energy terms

$$U^h = -\frac{1}{2}\int_V [e^T \bullet \boldsymbol{C}(\boldsymbol{R}) \bullet e] dR^3 = -\frac{1}{2}\int_V [e^T \bullet s] dR^3 \qquad (2.39)$$

The further reduction of the number of independent elastic coefficients now depends on the inherent symmetries of the underlying Bravais lattice. Silicon has a very high level of symmetry because of its cubic structure. The unit cell is invariant to rotations of 90° about any of its coordinate axes. Consider a rotation of the x-axis of 90° so that $x \to x$, $y \to z$ and $z \to -y$. Since the energy will remain the same, we must have that $C_{22} = C_{33}$, $C_{55} = C_{66}$ and $C_{21} = C_{31}$. By a similar argument for rotations about the other two axes, we obtain that $C_{11} = C_{22} = C_{33}$, $C_{21} = C_{31} = C_{23}$ and $C_{44} = C_{55} = C_{66}$. Since the other matrix entries experience an odd sign change in the transformed coordinates, yet symmetry of $\boldsymbol{C}$ is required, they must all be equal to zero. To summarize, Si and other cubic-symmetry crystals have an elasticity matrix with the following structure (for coefficient values, consult Table 2.1) but with only three independent values:

$$\boldsymbol{C} = \begin{bmatrix} C_{11} & C_{12} & C_{12} & & & \\ C_{12} & C_{11} & C_{12} & & & \\ C_{12} & C_{12} & C_{11} & & & \\ & & & C_{44} & & \\ & & & & C_{44} & \\ & & & & & C_{44} \end{bmatrix} \qquad (2.40)$$

$\boldsymbol{C}$ can be inverted to produce $\boldsymbol{S}$ with exactly the same structure and the following relation between the constants:

$$\left.\begin{array}{cc} S_{44} = C_{44}^{-1} & (S_{11} - S_{12}) = (C_{11} - C_{12})^{-1} \\ (S_{11} + 2S_{12}) = (C_{11} + 2C_{12})^{-1} & \end{array}\right\} \qquad (2.41)$$

The bulk modulus B and compressibility K of the cubic material is given by

$$B = \frac{1}{3}(C_{11} + 2C_{12}) = \frac{1}{K} \tag{2.42}$$

Isotropic materials have the same structure as (2.40) but with only two independent constants. In the literature, however, no less that three notations are common. Often engineers use the Young modulus E and Poisson ratio ν, whereas we find the Lamé constants λ and μ often used in the mechanics literature. The relations between the systems are summarized by:

$$\left.\begin{array}{ll}
C_{12} = \lambda & C_{44} = \dfrac{C_{11} - C_{12}}{2} = \mu \\
C_{11} = \lambda + 2\mu & C_{12} = \dfrac{E\nu}{(1+\nu)(1-2\nu)} \\
C_{44} = \dfrac{E}{2(1+\nu)} & C_{11} = \dfrac{E(1-\nu)}{(1+\nu)(1-2\nu)} \\
E = \dfrac{\mu(3\lambda + 2\mu)}{\lambda + \mu} & \nu = \dfrac{\lambda}{2(\lambda + \mu)} \\
E = \dfrac{C_{44}(3C_{11} - 4C_{44})}{C_{11} - C_{44}} & \nu = \dfrac{C_{11} - 2C_{44}}{2(C_{11} - C_{44})}
\end{array}\right\} \tag{2.43}$$

The α-Quartz crystal has the following elastic coefficient matrix structure (for coefficient values, consult Table 2.1):

$$C_{\text{Quartz}} = \begin{bmatrix}
C_{11} & C_{12} & C_{12} & C_{14} & & \\
C_{12} & C_{11} & C_{12} & -C_{14} & & \\
C_{12} & C_{12} & C_{11} & & & \\
C_{14} & -C_{14} & & C_{44} & & \\
 & & & & C_{44} & \\
 & & & & & C_{44}
\end{bmatrix} \tag{2.44}$$

2.4 The Vibrating Uniform Lattice

The elastic, spring-like nature of the interatomic bonds, together with the massive atoms placed at regular intervals; these are the items we isolate for a model of the classical mechanical dynamics of the crystal lattice (see Box 2.3 for brief details on Lagrangian and Hamiltonian mechanics). Here we see that the regular lattice displays unique new features unseen elsewhere: acoustic dispersion is complex and anisotropic, acoustic energy is quantized, and the quanta, called phonons, act like particles carrying information and energy about the lattice.

2.4.1 Normal Modes

As we saw in Section 2.3, an exact description of the forces between the atoms that make up a crystal are, in general, geometrically and mathematically very complex. Nevertheless, certain simplifications are possible here and lead both to an understanding of what we otherwise observe in experiments, and often to a fairly close approximation of reality. We assume that:

- The atoms that make up the lattice are close to their equilibrium positions, so that we may use a harmonic representation of the potential binding energy about the equilibrium atom positions. The spatial gradient of this energy is then the position-dependent force acting on the atom, which is zero when each atom resides at its equilibrium position.
- The lattice atoms interact with their nearest neighbors only.
- The lattice is infinite and perfect. This assumption allows us to limit our attention to a single Wigner-Seitz cell by assuming translational symmetry.
- The bound inner shell electrons move so much faster than the crystal waves that they follow the movement of the more massive nucleus that they are bound to adiabatically.

Box 2.3. A Brief Note on Hamiltonian and Lagrangian Mechanics.

Hamilton's Principle. Hamilton's principle states that the variation of the action A, also called the variational indicator, is a minimum over the path chosen by a mechanical system when proceeding from one known configuration to another

$$\delta A\big|_{t_1}^{t_2} = 0 \qquad \text{(B 2.3.1)}$$

The Action. The action is defined in terms of the lagrangian L and the generalized energy sources $\Xi_j \xi_j$ of the system

$$A = \int_{t_1}^{t_2}\left[L + \sum_{j=1}^{n} \Xi_j \xi_j\right] dt \qquad \text{(B 2.3.2)}$$

where Ξ_j is the generalized force and ξ_j the generalized displacement

The Lagrangian. The mechanical Lagrangian of a system is the difference between its kinetic co-energy $T^*(\dot{\xi}_j, \xi_j, t)$ (the kinetic energy expressed in terms of the system's velocities) and its potential energy $U(\xi_j, t)$

$$L_M(\dot{\xi}_j, \xi_j, t) = T^*(\dot{\xi}_j, \xi_j, t) - U(\xi_j, t) \qquad \text{(B 2.3.3)}$$

In general, the lagrangian for a crystal is written in terms of three contributions: the mechanical (or matter), the electromagnetic field and the field-matter interaction

$$L = \int_V (\mathfrak{L}_M + \mathfrak{L}_F + \mathfrak{L}_I) dV \qquad \text{(B 2.3.4)}$$

$$\mathfrak{L}_M = \frac{1}{2}(\rho \dot{u} + \varepsilon \bullet C \bullet \varepsilon)$$

$$\mathfrak{L}_F = \frac{1}{2}(\boldsymbol{E} + \boldsymbol{B}) \qquad \mathfrak{L}_I = \left(\frac{\dot{\boldsymbol{A}}}{c} - q\psi\right) \qquad \text{(B 2.3.5)}$$

Generalized Energy Sources. This term groups all external or non-conservative internal sources (sinks) of energy in terms of generalized forces Ξ_j and the generalized displacements ξ_j.

Generalized Displacements and Velocities. For a system, we establish the m independent scalar degrees of freedom $\boldsymbol{d} \in \Re^m$ required to describe its motion in general. Then we impose the p constraint equations $\boldsymbol{B} \bullet \boldsymbol{d} = 0$ that specify the required kinematics (the admissible path in $\Re^m$). This reduces the number of degrees of freedom by p, and we obtain the $n = m - p$ generalized scalar displacements $\xi \in \Re^n \supset \Re^m$ of the system. The generalized velocities are simply the time rate of change of the generalized displacements, $\dot{\xi} = d\xi/dt$.

Lagrange's Equations. An immediate consequence of (B 2.3.2) is that the following equations hold for the motion

$$\frac{d}{dt}\left(\frac{\partial L}{\partial \dot{\xi}_j}\right) - \frac{\partial L}{\partial \xi_j} = \Xi_j \qquad \text{(B 2.3.6)}$$

These are known as the Lagrange equations of motion. For a continuum, we can rewrite (B 2.3.6) for the Lagrange density as

$$\frac{d}{dt}\frac{\partial \mathfrak{L}}{\partial \dot{x}_j} = \frac{\partial \mathfrak{L}}{\partial x_j} - \frac{d}{dX_k}\frac{\partial \mathfrak{L}}{\partial(\partial x_j/\partial X_k)} \qquad \text{(B 2.3.7)}$$

In the continuum crystal lagrangian the variables X represent the material coordinates of the undeformed crystal; x the spatial coordinates of the deformed crystal. The Lagrange equations are most convenient, because they allow us to add detail to the energy expressions, so as to derive the equations of motion thereafter in a standard way.

- The valence electrons form a uniform cloud of negative space charge that interacts with the atoms.

Normal Modes

Normal modes are the natural eigen-shapes of the mechanical system. We already know these from musical instruments: for example from the shapes of a vibrating string (one-dimensional), the shapes seen on the stretched surface of a vibrating drum (two-dimensional), or a vibrating bowl of jelly (three-dimensional). Normal modes are important, because they are orthogonal and span the space of the atomic movements and thereby describe all possible motions of the crystal. To obtain the normal modes of the crystal, we will assume time-dependent solutions that have the same geometric periodicity of the crystal.

We will now derive an expression for the equations of motion of the crystal from the mechanical Lagrangian $L_M = T^* - U^h$ (see Box 2.3.) From the previous section we have that $U^h = 1/2\int_V(\varepsilon^T:\boldsymbol{E}:\varepsilon)dV$. The kinetic co-energy is added up from the contributions of the individual atoms

$$T^* = \frac{1}{2}\sum_{i=1}^{n} m\dot{\boldsymbol{r}}_i^2 \qquad (2.45)$$

As for the potential energy, this sum can also be turned into a volume integral, thereby making the transition to a continuum theory. We consider a primitive cell of volume V

$$T^* = \frac{1}{2}\sum_{i=1}^{n}\left(\frac{m}{V}\right)\dot{\boldsymbol{r}}_i^2 V = \frac{1}{2}\sum_{i=1}^{n}\rho\dot{\boldsymbol{r}}_i^2 V \cong \frac{1}{2}\int_V \rho\dot{\boldsymbol{r}}_i^2 dV \qquad (2.46)$$

Thus we have for the mechanical Lagrangian that

$$L_M = \int_V \mathcal{L}_M dV = \frac{1}{2}\int_V(\rho\dot{\boldsymbol{r}}_i^2 - \varepsilon^T:\boldsymbol{E}:\varepsilon)dV, \text{ with}$$

$$\mathcal{L}_M = \frac{1}{2}(\rho\dot{\boldsymbol{r}}_i^2 - \varepsilon^T:\boldsymbol{E}:\varepsilon) \qquad (2.47)$$

The continuum Lagrange equations read [2.7]

$$\frac{d}{dt}\frac{\partial\mathcal{L}}{\partial\dot{x}_j} = \frac{\partial\mathcal{L}}{\partial x_j} - \frac{d}{dX_k}\frac{\partial\mathcal{L}}{\partial(\partial x_j/\partial X_k)} \qquad (2.48)$$

where we have chosen the coordinates X to describe the undeformed configuration of the crystal, and x the coordinates in the deformed crystal, see Figure 2.19.

1D Monatomic Dispersion Relation

The complexity in applying the lagrangian formulation to a general 3D crystal can be avoided by considering a 1D model system that demonstrates the salient features of the more involved 3D system. A large 1D lattice of N identical bound atoms are arranged in the form of a ring by employing the Born-von Karmann boundary condition, i.e., $u(N+1) = u(N)$. From equation (2.31), the strain in the 1D lattice is simply

$$\varepsilon_{11} = \frac{1}{2}(\nabla u + \nabla u^T) = \frac{du}{dX} \tag{2.49}$$

which gives a potential energy density of

$$\mathcal{U} = \frac{1}{2}E\left(\frac{du}{dX}\right)^2 \tag{2.50}$$

where u is the displacement of the atom from its lattice equilibrium site and E is the linear Young modulus of the interatomic bond. Note that we only consider nearest-neighbour interactions. The kinetic co-energy density is

$$\mathcal{T}^* = \frac{1}{2}\rho\dot{u}^2 \tag{2.51}$$

where the mass density is $\rho = m/a$ for an atomic mass m and interatomic spacing a. The lagrangian density for the chain is the difference between the kinetic co-energy density and the potential energy density, $\mathcal{L} = \mathcal{T}^* - \mathcal{U}$. Note that $u = X - x$ and hence that $\nabla u = \mathrm{Id} - \nabla x$ and $\dot{u} = -\dot{x}$. We insert the lagrangian density in equation (2.48) to obtain

$$\rho\ddot{u} = -E\frac{d}{dX}\left(\frac{du}{dX}\right) \tag{2.52}$$

Since the potential energy is located in the bonds and not the lattice site, and therefore depends on the positions of the neighboring lattice sites, the term d^2u/dX^2 is replaced by its lattice equivalent for the lattice site i $(2u_i - u_{i-1} - u_{i+1})/a^2$. This finite difference formula expresses the fact that the curvature of u at the lattice site i depends on the next-neighbor lattice positions. This step is necessary for a treatment of waves with a wavelength of the order of the interatomic spacing. If we use the simpler site relation, we only obtain the long wavelength limit of the dispersion relation, indicated by the slope lines in Figure 2.20. We now look for solutions, periodic in space and time, of the form

$$u(X, t)\big|_{X = ia} = \exp[j(kX - \omega t)]\big|_{X = ia} \tag{2.53}$$

which we insert into equation (2.52) and cancel the common exponential

$$-\rho\omega^2 = -\frac{E}{a^2}(2 - \exp[kia] - \exp[-kia]) \tag{2.54}$$

Reorganizing equation (2.54), we obtain

$$\omega = \sqrt{\frac{2E(1 - \cos[ka])}{\rho a^2}} = 2\sqrt{\frac{E}{\rho a^2}}\left|\sin\left[\frac{ka}{2}\right]\right| \tag{2.55}$$

Equation (2.55) is plotted in Figure 2.20 on the left, and is the dispersion relation for a monatomic chain. The curve is typical for an acoustic wave in a crystalline solid, and is interpreted as follows. In the vicinity where ω is small, the dispersion relation is linear (since $\sin[ka/2] \approx ka/2$) and the wave propagates with a speed of $\sqrt{E/\rho}$ as a linear acoustic wave. As the frequency increases, the dispersion relation flattens off, causing the speed of the wave $\partial\omega/\partial k$ to approach zero (a standing wave resonance).

1D Diatomic Dispersion Relation

Crystals with a basis, i.e., crystals with a unit cell that contains different atoms, introduce an important additional feature in the dispersion curve. We again consider a 1D chain of atoms, but now consider a unit cell containing two different atoms of masses m and M.

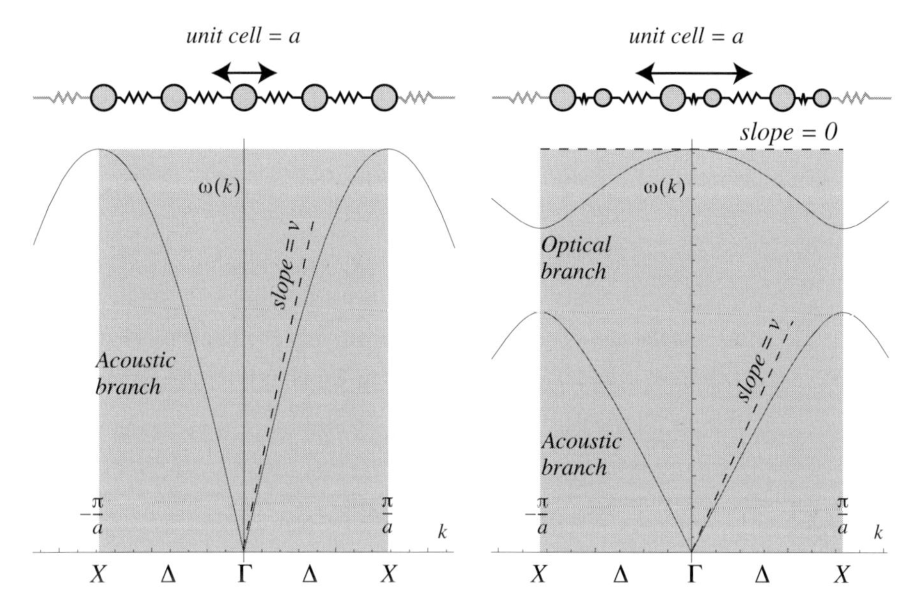

Figure 2.20. The monatomic and diatomic one-dimensional chain lattices lend themselves to analytical treatment. Depicted are the two computed dispersion curves with schematic chains and reciprocal unit cell indicated as a gray background box. Γ is the symmetry point at the origin, Δ the symmetry point at the reciprocal cell boundary. Only nearest-neighbor interactions are accounted for.

The kinetic co-energy is added up from the contributions of the individual atoms

$$T^* = \frac{1}{2}\sum_{i\alpha}^{n,\,2} m_\alpha \dot{u}_{i\alpha}^2 \tag{2.56}$$

The index α counts over the atoms in an elementary basis cell, and the index i counts over the lattice cells. Similarly, the bond potential energy is dependent on the stretching of the inter-atom bonds

$$U = \frac{1}{4}\sum_{i\alpha,\,\beta j}^{n,\,2,\,n,\,2} \frac{E_{\alpha i\beta j}}{a}(u_{i\alpha} - u_{j\beta})^2 \tag{2.57}$$

The restriction to next-neighbor interactions and identical interatomic force constants yields the following equations of motion when the expressions (2.56) and (2.57) are inserted into the Lagrange equations

$$m\ddot{u}_{im} = -\frac{E}{a}(2u_{im} - u_{iM} - u_{(i-1)M}) \tag{2.58a}$$

$$M\ddot{u}_{iM} = -\frac{E}{a}(2u_{iM} - u_{(i+1)m} - u_{im}) \tag{2.58b}$$

At rest the atoms occupy the cell positions (due to identical static forces) $(i - 1/4)a$ and $(i + 1/4)a$. Hence we choose an harmonic atom displacement ansatz for each atom of the form

$$u_{im} = \frac{1}{\sqrt{m}} c_m \exp\left\{ j\left[ka\left(i - \frac{1}{4}\right) - \omega t\right]\right\} \tag{2.59a}$$

$$u_{iM} = \frac{1}{\sqrt{M}} c_M \exp\left\{ j\left[ka\left(i + \frac{1}{4}\right) - \omega t\right]\right\} \tag{2.59b}$$

The second ansatz can be written in terms of u_{im}

$$u_{iM} = \sqrt{\frac{m}{M}}\left(\frac{c_M}{c_m}\right)\exp\left(j\frac{ka}{2}\right)u_{im} \tag{2.60}$$

Equations (2.59b) and (2.60) are now inserted into (2.58a). Eliminating common factors and simplifying, we obtain an equation for the amplitudes c_m and c_M

$$\begin{bmatrix} \left(\frac{2E}{a\sqrt{m}} - \omega^2\sqrt{m}\right) & -\frac{2E}{a\sqrt{M}}\cos\left(\frac{ka}{2}\right) \\ -\frac{2E}{a\sqrt{m}}\cos\left(\frac{ka}{2}\right) & \left(\frac{2E}{a\sqrt{M}} - \omega^2\sqrt{M}\right) \end{bmatrix} \begin{bmatrix} c_m \\ c_M \end{bmatrix} = 0 \tag{2.61}$$

This is an eigensystem equation for ω, and its non-trivial solutions are obtained by requiring the determinant of the matrix to be zero. Performing this, we obtain

$$\omega^2 = \frac{E}{a}\left[\left(\frac{1}{m} + \frac{1}{M}\right) \mp \sqrt{\left(\frac{1}{m} + \frac{1}{M}\right)^2 - \frac{4}{mM}\sin\left(\frac{ka}{2}\right)^2}\right] \tag{2.62}$$

The two solutions are plotted in Figure 2.20 on the right.

Discussion

The case when the two masses of the unit-cell atoms differ only by a small amount, i.e., $M = m + \mu$ with μ small, is instructive. The optical and acoustic branches approach then each other at the edge of the reciprocal cell, i.e., at $k = \mp\pi/a$. As the mass difference μ goes to zero, the lattice becomes a monatomic lattice with lattice constant $a/2$, so that the branches touch each other at the reciprocal cell edge.

The interpretation of the two branches is as follows. For the lower branch, all the atoms move in unison just as for an acoustic wave, hence the name *acoustic branch*. In fact, it appears as a center-of-mass oscillation. For the upper or *optical branch*, the center of mass is stationary, and the atoms of a cell only move relative to each other. Its name refers to the fact that for ionic crystals, this mode is often excited by optical interactions.

2D Square Lattice Dispersion Relation

The next construction shows the richness in structure that appears in the dispersion relation when an additional spatial dimension and one level of interaction is added to the 1D monatomic lattice. It serves as an illustration that the anisotropy of the interatomic binding energy enables more involved crystal vibrational modes and hence additional branches in the dispersion curves. At the same time it shows that most of the essential features of the expected structure is already clear from the simple 1D models.

The model considers a 2D monatomic square lattice, and includes the interaction of the four nearest and four next-nearest neighboring lattice atoms, i.e., 8 interactions in all. Furthermore, the forces are assumed to be linear (the harmonic assumption) and the next-nearest neighbor spring stiffnesses are taken to be $f = 0.6$ times the stiffnesses of the nearest neighbor bonds.

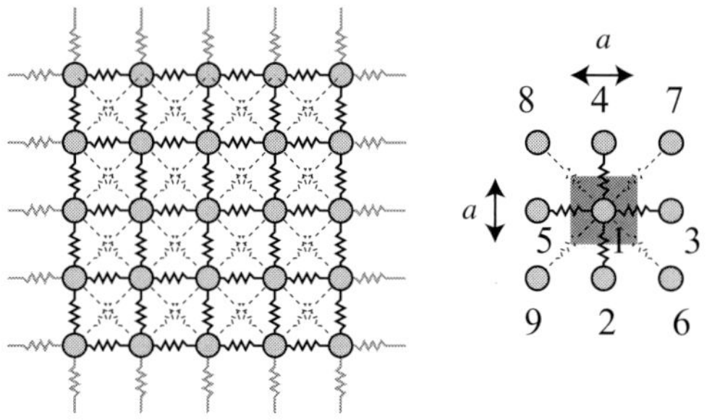

Figure 2.21. Model for a 2D monatomic square lattice. The atom numbers are markers for the derivation of the equations of motion.

Following the discussion for the linear lattices, we formulate the equations of motion as

$$m\ddot{u}_{\alpha i} = -\sum_{\beta=1}^{2} \frac{\partial U}{\partial u_{\beta i}} = -\sum_{j=1}^{n} \sum_{\beta=1}^{2} D_{\alpha\beta ij} u_{\beta j}, \; \alpha = 1, 2, \quad i = 1, 2, \ldots, n \tag{2.63}$$

where the dynamical matrix $D_{\alpha\beta ij}$ represents the stiffness of the bond between atom sites i and j and between the directions α and β. The current model has a nine-atom 2D interaction, for which we can use an 18×18 stiffness matrix. The nonzero unique elements of the matrix are

$$d_1 = D_{1115} = D_{2214} = D_{1113} = D_{2212} = -\frac{E}{a} \tag{2.64a}$$

$$d_2 = D_{1119} = D_{2218} = D_{1117} = D_{2216} = -\frac{fE}{a} \tag{2.64b}$$

Equation (2.63) thus becomes

$$m\ddot{\boldsymbol{u}}_1 = \frac{E}{a}(4\boldsymbol{u}_1 - \boldsymbol{u}_5 - \boldsymbol{u}_3 - \boldsymbol{u}_2 - \boldsymbol{u}_4) + \frac{fE}{a}(4\boldsymbol{u}_1 - \boldsymbol{u}_9 - \boldsymbol{u}_6 - \boldsymbol{u}_7 - \boldsymbol{u}_8) \tag{2.65}$$

We make an harmonic ansatz for the atom displacement of the form

$$u_p = c\exp\{j[(pk_xa + qk_ya) - \omega t]\} \tag{2.66}$$

We can write the ansatz for any neighboring atom in terms of the central atom. For a parallel and diagonal atom we obtain

$$\begin{aligned} u_{i,j} &= c\exp\{j\{[(i-1)k_xa + jk_ya] - \omega t\}\} = u_{i,j}e^{-jk_xa} \\ u_{i-1,j-1} &= c\exp\{j\{[(i-1)k_xa + (j-1)k_ya] - \omega t\}\} \\ &= u_{i,j}e^{-j(k_xa + k_ya)} \end{aligned} \tag{2.67}$$

which, when inserted into (2.65) becomes

$$\begin{aligned} -\omega^2 mc &= c\frac{E}{a}[(4 - e^{-jk_xa} - e^{jk_xa} - e^{-jk_ya} - e^{jk_ya}) \\ &+ f(4 - e^{-j(k_xa + k_ya)} - e^{-j(-k_xa + k_ya)} - e^{j(k_xa + k_ya)} - e^{-j(k_xa - k_ya)})] \end{aligned} \tag{2.68}$$

This single parameter eigenvalue equation has two solutions

$$\begin{aligned} \omega^2 &= \frac{E}{am}[4 - e^{-jk_xa} - e^{jk_xa} - e^{-jk_ya} - e^{jk_ya} \\ &+ f(4 - e^{-j(k_xa + k_ya)} - e^{-j(-k_xa + k_ya)} - e^{j(k_xa + k_ya)} - e^{-j(k_xa - k_ya)})] \end{aligned} \tag{2.69}$$

Using the identity $\cos\phi = (e^{j\phi} + e^{-j\phi})/2$ we transform the equation above to

$$\begin{aligned} \omega &= \mp\left[\frac{E}{am}\{[4 - \cos(2k_xa) - \cos(2k_ya)] \right. \\ &\left. f[4 - \cos(2k_xa)\cos(2k_ya)]\}\right]^{\frac{1}{2}} \end{aligned} \tag{2.70}$$

The dispersion relation, equation (2.70), represents two ω surfaces in (k_x, k_y) space, and are plotted in Figure 2.22. Because not only nearest, but also next nearest neighbor interactions are included, the upper dispersion surface shows a new feature, in that the maximum is not at the boundary of the primitive cell, i.e., its gradient is zero in the interior of the cell.

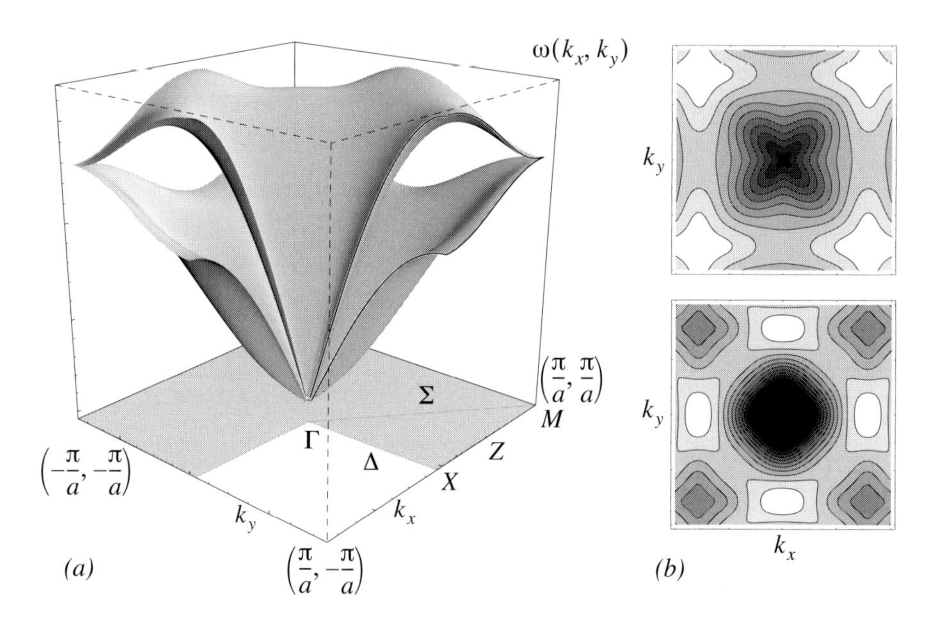

Figure 2.22. Acoustic dispersion branch surfaces $\omega(k_x, k_y)$ for a monatomic lattice with nearest and next-nearest-atom linear elastic interaction. (a) One quarter of the surfaces are omitted to illustrate the inner structure. (b) The contours of the upper and lower surfaces. The inter-atom forces have a ratio of $\beta = 0.6$.

By now the general method should become clear: the equations of motion, (2.63), written in terms of the discrete inter-atomic forces $\boldsymbol{F}_0 = \sum_{-\infty}^{\infty} \Gamma_p \cdot \boldsymbol{u}_p$ acting on a typical lattice atom 0, with a harmonic ansatz for the motion $\boldsymbol{u}_p(\boldsymbol{k}) = \boldsymbol{A} \exp\{j[\boldsymbol{k} \cdot \boldsymbol{r}_p - \omega t]\}$, leads to an eigenvalue equation. For a many-atom unit cell we obtain

$$\omega^2 \boldsymbol{B} = D \cdot \boldsymbol{B} \tag{2.71}$$

for the amplitude eigenvectors $\boldsymbol{B}(j) = m_j^{1/2} \boldsymbol{A}(j)$ and eigenfrequencies ω. How to compute the dynamical matrix D is described, for the carbon lattice, in Box 2.4. Solving (2.71) for different values of $\boldsymbol{k}$ enables us to compute the phonon dispersion diagram shown in Figure 2.26 (but also see Figure 2.4). For Figure 2.4, a 6×6 low order approximation of the dynamical matrix was used that could easily be fitted with the elastic

Box 2.4. Computing the carbon structure's dynamical matrix.

Atom Contributions. The dynamical matrix D describes the effect that perturbing the position of atom i has on the force acting on atom 0, for all the interacting neighbors of atom 0. Clearly this is a direct function of the crystal structure, so that we can write the dynamical matrix in terms of the bonding strengths between the adjacent sites. We now do this for the diamond lattice, depicted in Figure B2.4.1. From the viewpoint of atom 0, we

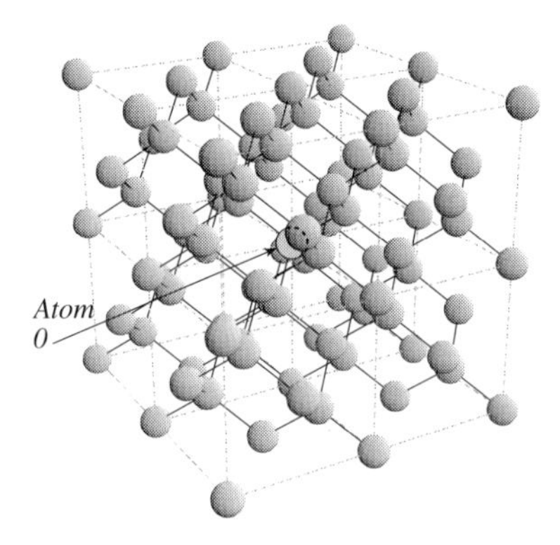

Figure B2.4.1: The structure of the diamond lattice w.r.t. the centered atom 0, marked by an arrow. Bookkeeping becomes critical.

have 4 equally distant neighbors at the positions $\{1/4, 1/4, -1/4\}$, $\{1/4, -1/4, 1/4\}$, $\{1/4, 1/4, -1/4\}$ and $\{-1/4, -1/4, 1/4\}$, at, $r = \sqrt{3}/4$, and so on. The sets of equally distant neighbors are positioned in "onion shells" about atom 0. For each of these sets we can write down a force-constant tensor that is equal to the tensor product of its normalized position vector w.r.t. atom 0. For example for the first atom, the tensor equals

$$\Gamma_1 = \gamma_1 \begin{bmatrix} 1/3 & 1/3 & -1/3 \\ 1/3 & 1/3 & -1/3 \\ -1/3 & -1/3 & 1/3 \end{bmatrix} \tag{B 2.4.1}$$

and can be interpreted as follows: If atom i experiences a unit displacement in direction j, then atom 0 experiences a force in direction k of size $\Gamma_{jk}^{(i0)}$. Each unique separation $|\boldsymbol{r}_l|$ is associated with a force constant γ_l. A complete formulation includes all force constant tensors and hence all neighboring atoms involved in an interaction with atom 0.

The Dynamical Matrix. The net result of all atom interactions is computed from

$$D_{jk} = -\frac{1}{m}\sum_i \Gamma_{jk}^{(i0)} e^{i(\boldsymbol{k}\cdot\boldsymbol{r}_i)} \tag{B 2.4.2}$$

The force constants are evaluated by comparing the long-wavelength dynamical matrix with the values obtained from elasticity theory, (2.80).

constants of silicon. The qualitative agreement with the measured curve is fairly good, as all major features are represented.

Phonon Density of States

A more useful form of the phonon density of states as a function of the phonon frequency $\mathcal{D}(\omega)$ and not as a function of the wave vector coordinates. The density of states in k-space is reasonably straightforward to formulate. For the case of a 1D monatomic lattice of length L with lattice constant a, it is the constant value $w(k) = L/\pi$ for $k \le \pi/a$ and zero otherwise. This amounts to saying that for each of the N atoms in

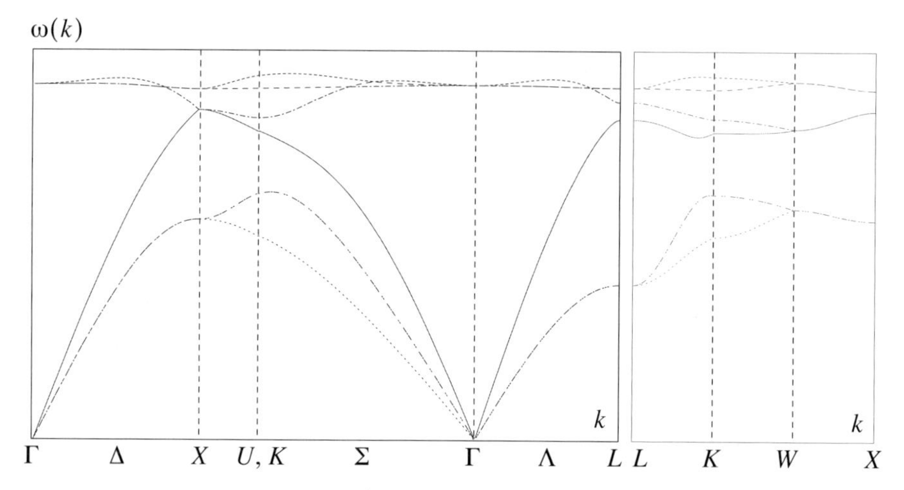

Figure 2.23. A computed phonon dispersion diagram for Silicon by the method described in the text. In this model, a dynamical matrix was used that only incorporates the nearest and next nearest neighbor atoms. In addition, a central force component was included so as to gain more realism. Qualitatively the shape closely resembles the diagram shown in Figure 2.4, which is based on measured values as well as a more realistic computation.

the chain there is one state, and in k-space the density is N states per Brillouin zone volume. We expect the same result for 3D. The number of states $\mathcal{D}(\omega)|_{\bar{\omega}}d\omega$ per frequency interval $d\omega$ and at the frequency $\bar{\omega}$ can be computed using

$$\mathcal{D}(\omega)|_{\bar{\omega}}d\omega = w(k)\frac{\partial k}{\partial \omega}\bigg|_{\bar{\omega}} d\omega = \frac{w(k)}{\partial \omega/\partial k|_{\bar{\omega}}}d\omega \tag{2.72}$$

and so we see that the dispersion relation $\omega(k)$ is a prerequisite to compute the required density of states $\mathcal{D}(\omega) = w(k)/(\partial\omega/\partial k)$. Before we look at a practical method with which to compute $\mathcal{D}(\omega)$ for real crystals, we first discuss an important structural feature, namely the singularities that occur when $\partial\omega/\partial k$ is zero. If we take a look at Figure 2.4, or simpler still, Figure 2.20, we see that at the symmetry points $\partial\omega/\partial k$ is either undefined or zero. Thus we have that $\mathcal{D}(\omega)$ either tends to infinity or shows a discontinuity. These are known as Van Hove singularities after their discoverer [2.14].

Phonon Density of States Computation Method

Since the density of states is constant in k-space, the Walker calculation method [2.15] (also called the root sampling method) requires performing the following steps:

- Form a uniform grid in k-space. Only consider a volume that is non-redundant, i.e., exploit all symmetries in the Brillouin zone of the crystal to reduce the size of the problem. Ensure that no grid points lie directly on a symmetry plane or on a symmetry line, as this would complicate the counting.
- For each point on the grid, solve (2.71) for the eigenfrequencies ω.
- Compute ω_{max}, the highest frequency anywhere on the grid.
- Form a set of histogram bins from 0 to ω_{max} for each separate branch of the computed dispersion curve, and calculate the histograms of branch frequencies found on the grid. That is, each histogram bin counts how many branch states occurred for a frequency lying within the bin's interval.
- Identify the singular points on the histograms. Smoothen the histograms between the singular points and plot the resultant phonon density of states.

The method requires enough k-space points spread evenly over the Brillouin zone and histograms with small-enough bin widths for a realistic result. (The method can also be used to compute the density of electronic states $E(\omega)$, from the electronic band structure $E(\boldsymbol{k})$, as discussed in Section 3.3). Applying the above algorithm to the linear chain of Section 2.4.1, we obtain the curves of Figure 2.24. Doing the same for the 2d lattice results in the density-of-states diagram of Figure 2.25. For the silicon dispersion diagram of Figure 2.23 we have numerically computed the density of states by the root sampling method presented in Figure 2.26.

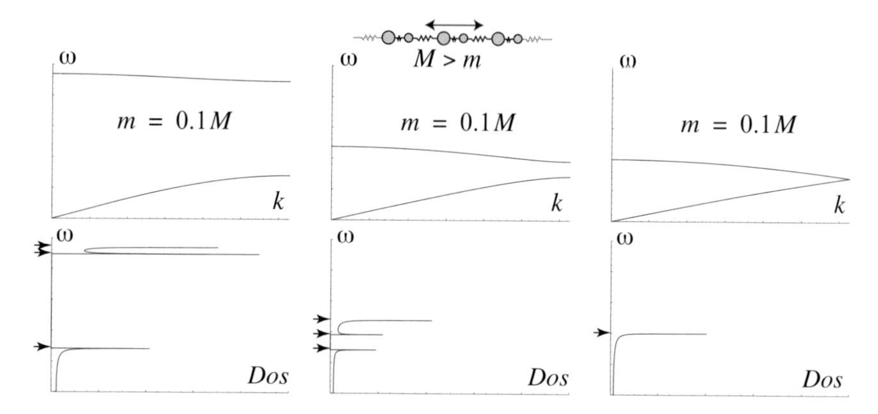

Figure 2.24. The computed phonon dispersion curves and density-of-states for the linear diatomic chain, for three mass ratios. The arrows show the positions of the Van Hove singularities. Also see Figure 2.20. For these plots ω *was divided into* 1000 *bins, and* 100000 *evenly-spaced points in* k *-space were used.*

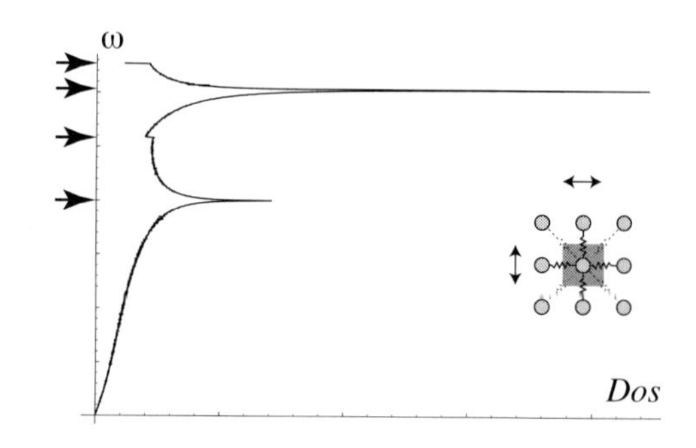

Figure 2.25. The computed phonon density-of-states for the 2D lattice. The arrows show the positions of the Van Hove singularities. Also see Figure 2.22. For this plot ω *was divided into* 1000 *bins, and* 360000 *evenly-spaced points in* k *-space were used.*

As for the dispersion diagram, it shows the correct features of the density of states, but lacks realism due to the inadequacy of the dynamical matrix.

Elastic Waves in Silicon

We now turn our attention to the classical mechanical equation of motion in a solid (2.52) and (2.63) in the long wavelength limit, but ignoring body forces

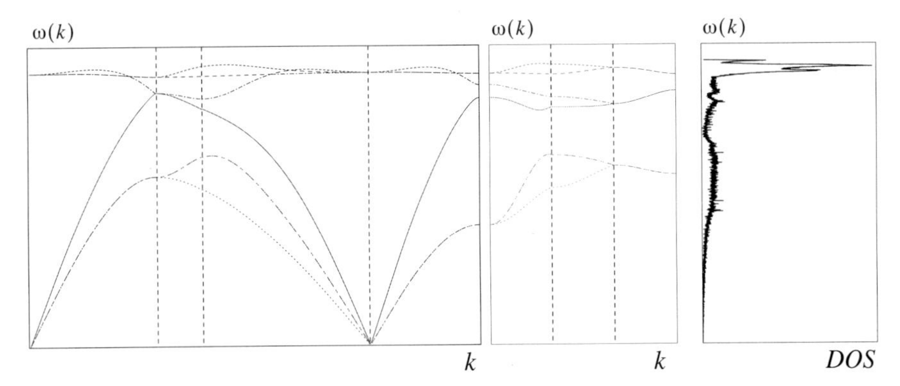

Figure 2.26. The computed phonon density of state diagram for a simplistic silicon model, shown together with the computed phonon dispersion diagram. Notice that peaks in the density of states diagram occur at positions where $\partial\omega/\partial k = 0$ [2.13].

$$\rho\frac{\partial^2 \boldsymbol{u}}{\partial t^2} = \nabla\bullet\sigma = \nabla\bullet(C:\varepsilon) \tag{2.73}$$

Remember that $\boldsymbol{u} = \begin{bmatrix} u_1 & u_2 & u_3 \end{bmatrix}^T$. It is simplest to first evaluate $\sigma = C:\varepsilon$ (here for a cubic-symmetry crystal) in the engineering notation of (2.38) and (2.40)

$$C:\varepsilon$$

$$= \begin{bmatrix} C_{11} & C_{12} & C_{12} & & & \\ C_{12} & C_{11} & C_{12} & & & \\ C_{12} & C_{12} & C_{11} & & & \\ & & & C_{44} & & \\ & & & & C_{44} & \\ & & & & & C_{44} \end{bmatrix} \cdot \begin{bmatrix} \varepsilon_{11} \\ \varepsilon_{22} \\ \varepsilon_{33} \\ 2\varepsilon_{23} \\ 2\varepsilon_{31} \\ 2\varepsilon_{12} \end{bmatrix} = \begin{bmatrix} C_{11}\varepsilon_{11} + C_{12}\varepsilon_{22} + C_{12}\varepsilon_{33} \\ C_{12}\varepsilon_{11} + C_{11}\varepsilon_{22} + C_{12}\varepsilon_{33} \\ C_{12}\varepsilon_{11} + C_{12}\varepsilon_{22} + C_{11}\varepsilon_{33} \\ 2C_{44}\varepsilon_{23} \\ 2C_{44}\varepsilon_{31} \\ 2C_{44}\varepsilon_{12} \end{bmatrix} \tag{2.74}$$

ant then to transform the result back to the usual tensor notation

$$C:\varepsilon$$

$$= \begin{bmatrix} C_{11}\varepsilon_{11} + C_{12}(\varepsilon_{22} + \varepsilon_{33}) & 2C_{44}\varepsilon_{12} & 2C_{44}\varepsilon_{31} \\ 2C_{44}\varepsilon_{12} & C_{11}\varepsilon_{22} + C_{12}(\varepsilon_{11} + \varepsilon_{33}) & 2C_{44}\varepsilon_{23} \\ 2C_{44}\varepsilon_{31} & 2C_{44}\varepsilon_{23} & C_{11}\varepsilon_{33} + C_{12}(\varepsilon_{11} + \varepsilon_{22}) \end{bmatrix} \tag{2.75}$$

We take the divergence of $\nabla\bullet(C:\varepsilon)$, to obtain a first version of the wave equation

$$\rho\frac{\partial^2 u_1}{\partial t^2} = C_{11}\frac{\partial\varepsilon_{11}}{\partial x_1} + C_{12}\frac{\partial(\varepsilon_{22} + \varepsilon_{33})}{\partial x_1} + 2C_{44}\left(\frac{\partial\varepsilon_{12}}{\partial x_2} + \frac{\partial\varepsilon_{31}}{\partial x_3}\right) \tag{2.76a}$$

$$\rho\frac{\partial^2 u_2}{\partial t^2} = C_{11}\frac{\partial\varepsilon_{22}}{\partial x_2} + C_{12}\frac{\partial(\varepsilon_{11} + \varepsilon_{33})}{\partial x_2} + 2C_{44}\left(\frac{\partial\varepsilon_{12}}{\partial x_1} + \frac{\partial\varepsilon_{23}}{\partial x_3}\right) \tag{2.76b}$$

$$\rho\frac{\partial^2 u_3}{\partial t^2} = C_{11}\frac{\partial\varepsilon_{33}}{\partial x_3} + C_{12}\frac{\partial(\varepsilon_{11} + \varepsilon_{22})}{\partial x_3} + 2C_{44}\left(\frac{\partial\varepsilon_{31}}{\partial x_1} + \frac{\partial\varepsilon_{23}}{\partial x_2}\right) \tag{2.76c}$$

With the definition of strain (2.31) we obtain the final form of the bulk lattice wave equation system for a cubic-symmetry crystal

$$\rho\frac{\partial^2 u_1}{\partial t^2} = C_{11}\frac{\partial^2 u_1}{\partial x_1^2} + (C_{12} + C_{44})\left(\frac{\partial^2 u_2}{\partial x_1\partial x_2} + \frac{\partial^2 u_3}{\partial x_1\partial x_3}\right) + C_{44}\left(\frac{\partial^2 u_1}{\partial x_2^2} + \frac{\partial^2 u_1}{\partial x_3^2}\right) \tag{2.77a}$$

$$\rho\frac{\partial^2 u_2}{\partial t^2} = C_{11}\frac{\partial^2 u_2}{\partial x_2^2} + (C_{12} + C_{44})\left(\frac{\partial^2 u_1}{\partial x_2\partial x_1} + \frac{\partial^2 u_3}{\partial x_2\partial x_3}\right) + C_{44}\left(\frac{\partial^2 u_2}{\partial x_1^2} + \frac{\partial^2 u_2}{\partial x_3^2}\right) \tag{2.77b}$$

$$\rho\frac{\partial^2 u_3}{\partial t^2} = C_{11}\frac{\partial^2 u_3}{\partial x_3^2} + (C_{12} + C_{44})\left(\frac{\partial^2 u_1}{\partial x_3\partial x_1} + \frac{\partial^2 u_2}{\partial x_3\partial x_2}\right) + C_{44}\left(\frac{\partial^2 u_3}{\partial x_1^2} + \frac{\partial^2 u_3}{\partial x_2^2}\right) \tag{2.77c}$$

To find the principal bulk modes, we assume a harmonic plane wave solution of the form

$$u = Af(\tau)\text{, with } \tau = t - \frac{k \cdot x}{c} \tag{2.78}$$

and unit vector k pointing in the direction of wave propagation, so that (2.77a) becomes an eigensystem

$$(\Gamma - \rho c^2) \cdot A = 0 \tag{2.79}$$

with the tensor Γ defined for cubic materials such as silicon as

$$\Gamma = \begin{bmatrix} C_{11}k_1^2 + C_{44}(k_2^2 + k_3^2) & (C_{12} + C_{44})k_1k_2 & (C_{12} + C_{44})k_1k_3 \\ (C_{12} + C_{44})k_1k_2 & C_{11}k_2^2 + C_{44}(k_1^2 + k_3^2) & (C_{12} + C_{44})k_2k_3 \\ (C_{12} + C_{44})k_1k_3 & (C_{12} + C_{44})k_2k_3 & C_{11}k_3^2 + C_{44}(k_1^2 + k_2^2) \end{bmatrix} \tag{2.80}$$

Equation (2.79) is also known as the Christoffel equation, and finding its eigenvalues using material parameters for silicon, over a range of angles, results in the slowness plots of Figure 2.27 [2.11]. Along the $[100]$ family of silicon crystal axes, the principal wave velocities are

$$c_1 = \sqrt{C_{11}/\rho} = 8433m/s\,, \tag{2.81a}$$

$$c_2 = c_3 = \sqrt{C_{44}/\rho} = 5845m/s \tag{2.81b}$$

Along the $[110]$ direction, we obtain the following results

$$c_1 = \sqrt{C_{44}/\rho} = 5845m/s \tag{2.82a}$$

$$c_2 = \sqrt{(C_{11} - C_{12})/2\rho} = 4674m/s \tag{2.82b}$$

$$c_3 = \sqrt{(C_{11} + C_{12} + 2C_{44})/2\rho} = 9134m/s \tag{2.82c}$$

2.4.2 Phonons, Specific Heat, Thermal Expansion

The behavior of certain material properties, such as the lattice heat capacity at low and medium temperatures, can only be explained if we assume that the acoustic energy of lattice vibrations is quantized, a result indi-

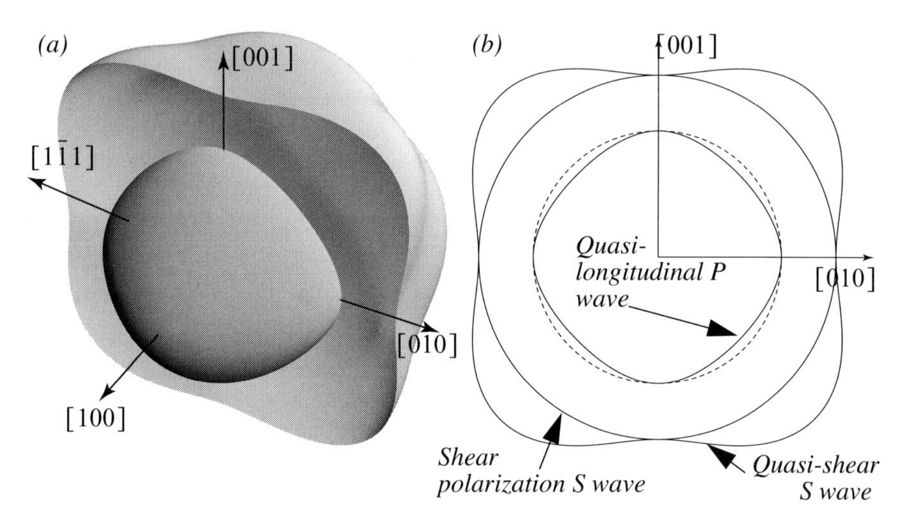

Figure 2.27. The three computed slowness surfaces for silicon. These waves have differing velocities of propagation as shown in the figures. (a) The 3D plot of the quasi waves shows minima along the $[111]$ *axes. (b) Here we see that the shear polarization wave for silicon is isotropic, and hence a circle. The dotted line is a circle, provided for reference. The inverse wave velocities at the crystal axes intersections are* $1.119\times10^{-4} s/m$ *and* $1.711\times10^{-4} s/m$. *Also see Box 2.5.*

cated in the previous section on the normal modes of the crystal. By performing a quantum-mechanical analysis of the lattice, essentially consisting of taking a Hamiltonian of the crystal and finding its stationary state (see 3.2.5), we find that the vibrational energy is quantized for the various states as

$$E_j = \hbar\omega_j . \tag{2.83}$$

The vibrational energy is also a travelling wave (see Box 2.1) which has a distinct momentum, so that

$$\boldsymbol{p} = \hbar\boldsymbol{k} . \tag{2.84}$$

Box 2.5. The characteristic surfaces of waves in anisotropic media [2.11].

For a general anisotropic medium we can write the orientation-dependent strength of a plane wave as

$$\boldsymbol{c}(\boldsymbol{k}) = c(\boldsymbol{k})\boldsymbol{k}\,,\ |\boldsymbol{k}| = 1 \qquad \text{(B 2.5.1)}$$

Thus

$$\boldsymbol{c}(\boldsymbol{k}) \cdot \boldsymbol{k} = c(\boldsymbol{k}) \qquad \text{(B 2.5.2)}$$

It is usual to set

$$\boldsymbol{k} = (\sin\theta\cos\phi,\ \sin\theta\sin\phi,\ \cos\theta) \qquad \text{(B 2.5.3)}$$

Equation (B 2.5.1) describes a closed surface called the velocity surface. We have seen that a general anisotropic medium supports three waves, two of which are quasi-shear transverse waves and one that is a quasi-parallel normal wave. Thus we have a family of three wave surfaces. Inverting equation (B 2.5.1) we obtain

$$\boldsymbol{L}(\boldsymbol{k}) = \boldsymbol{k}/c(\boldsymbol{k}) \qquad \text{(B 2.5.4)}$$

so that $\boldsymbol{c}(\boldsymbol{k}) \cdot \boldsymbol{L}(\boldsymbol{k}) = 1$. Equation (B 2.5.4) describes a closed surface called the slowness surface. Again there are three surfaces corresponding to the different wave mode solutions.

Normality. We now wish to form $\boldsymbol{c}^E(\boldsymbol{k}) \cdot d\boldsymbol{L}(\boldsymbol{k})$, for which we first define the two arguments. The modal wave's Poynting vector is

$$\boldsymbol{P} = -(\sigma \cdot \dot{\boldsymbol{u}}) = \frac{C{:}(\boldsymbol{A} \otimes \boldsymbol{A}) \cdot \boldsymbol{k}}{c(\boldsymbol{k})}\left(\frac{\partial f}{\partial \tau}\right)^2 \qquad \text{(B 2.5.5)}$$

Together with the wave's energy density

$$E_0 = \frac{1}{2}\rho|\boldsymbol{A}|^2\left(\frac{\partial f}{\partial \tau}\right)^2 \qquad \text{(B 2.5.6)}$$

we form the energy velocity vector $\boldsymbol{c}^E$ by

$$\boldsymbol{c}^E(\boldsymbol{k}) = \frac{\boldsymbol{P}}{E_0} = \frac{2C{:}(\boldsymbol{A} \otimes \boldsymbol{A}) \cdot \boldsymbol{k}}{c(\boldsymbol{k})\rho|\boldsymbol{A}|^2} \qquad \text{(B 2.5.7)}$$

Since $d\boldsymbol{L} = \nabla_k \boldsymbol{L} \cdot d\boldsymbol{k}$, we obtain

$$\nabla_k \boldsymbol{L} = \nabla_k\left(\frac{\boldsymbol{k}}{c(\boldsymbol{k})}\right) = \frac{\boldsymbol{\delta}}{c(\boldsymbol{k})} - \frac{\boldsymbol{k} \cdot \nabla_k c(\boldsymbol{k})}{c(\boldsymbol{k})^2} \qquad \text{(B 2.5.8)}$$

But $\nabla_k c(\boldsymbol{k})$ is nothing else than the group velocity of the wave c^G , which in turn equals to $(2(\boldsymbol{A}^T \cdot C){:}\boldsymbol{A} \cdot \boldsymbol{k})/(c(\boldsymbol{k})\rho|\boldsymbol{A}|^2)$, so that we deduce that $\boldsymbol{c}^E(\boldsymbol{k}) \cdot d\boldsymbol{L}(\boldsymbol{k}) = 0$. This means that the energy velocity is normal to the slowness surface.

Complementarity. The scalar product of the energy velocity and the direction vector gives the magnitude of the wave velocity in that direction, i.e., $\boldsymbol{c}^E(\boldsymbol{k}) \cdot \boldsymbol{k} = c(\boldsymbol{k})$. This follows from the Christoffel equation, (2.79). Hence we see that $\boldsymbol{c}^E(\boldsymbol{k}) \cdot \boldsymbol{L}(\boldsymbol{k}) = 1$. Now consider the following construction

$$\begin{aligned} d(\boldsymbol{c}^E(\boldsymbol{k}) \cdot \boldsymbol{L}(\boldsymbol{k})) &= d\boldsymbol{c}^E(\boldsymbol{k}) \cdot \boldsymbol{L}(\boldsymbol{k}) + \\ \boldsymbol{c}^E(\boldsymbol{k}) \cdot d\boldsymbol{L}(\boldsymbol{k}) &= d\boldsymbol{c}^E(\boldsymbol{k}) \cdot \boldsymbol{L}(\boldsymbol{k}) = 0 \end{aligned} \qquad \text{(B 2.5.9)}$$

Inserting the definition for $\boldsymbol{L}(\boldsymbol{k})$, we obtain that $d\boldsymbol{c}^E(\boldsymbol{k}) \cdot \boldsymbol{k} = 0$, or, that the tangent of the velocity surface is normal to the propagation direction. The above findings are now summarized in.

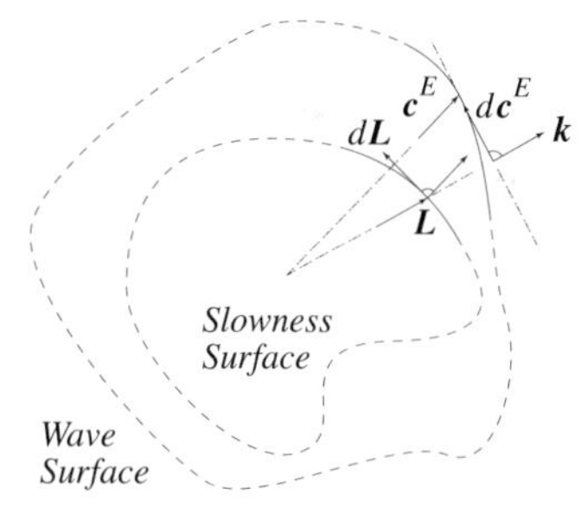

Figure B2.5.1: The basic geometric relations of the characteristic wave surfaces. Also refer to Figure 2.27 for the slowness surface of silicon.

Vibrational energy quanta follow the statistics of bosons, i.e., they are not subject to the Pauli exclusion principle that holds for fermions such as the electrons, so that

$$\langle n_j \rangle = \frac{1}{\exp((E_j - \mu)/kT) - 1} = \frac{1}{\exp((\hbar\omega_j - \mu)/kT) - 1}, \tag{2.85}$$

We will delay any further discussion of particle statistics to Chapter 5. Because of the similarity of this expression with the case of photons (quantized electromagnetic energy, see Section 5.2.4), the quantized crystal vibrations are called phonons.

Phonons

The connection to the normal modes of the previous section is that the excitation of the mth mode (from branch s with wavevector $\boldsymbol{k}$) corresponds to saying that there exist m photons (from branch s with wavevector $\boldsymbol{k}$) in the crystal [2.1].

Heat Capacity

The vibrating crystal lattice is an internal energy store. A measure for this energy storage ability is the heat capacity at constant volume, defined as

$$C_V = T\left(\frac{\partial S}{\partial T}\right)_V = \left(\frac{\partial E}{\partial T}\right)_V \tag{2.86}$$

in terms of the entropy S and the internal energy E. Classical statistical mechanics (SM), which treats the lattice as $3N$ classical linear harmonic oscillators, assigns an average of energy kT to each vibrational degree-of-freedom, from

$$U = \frac{\int e^{\frac{-E}{k_B T}} E d\Gamma}{\int e^{\frac{-E}{k_B T}} d\Gamma} = -\frac{\partial}{\partial\left(\frac{1}{k_B T}\right)} \ln\left[\int e^{\frac{-E}{k_B T}} d\Gamma\right] \tag{2.87}$$
$$= 3Nk_B T = \langle E \rangle_{\text{Classical}}$$

and hence we obtain a simple expression for the heat capacity as

$$C_V = \left(\frac{\partial E}{\partial T}\right)_V = 3Nk_B \approx 24 \ \text{Jmol}^{-1}\text{K}^{-1} \tag{2.88}$$

The expression shows no temperature dependence, which turns out to be wrong for all but the very high temperatures. This classical result is due to Dulong and Petit. From classical mechanics (CM), we obtained the $3N$ normal modes of the crystal via a quadratic expression for the Hamiltonian (the harmonic approximation). In CM, any motion of the crystal is simply a linear combination of the normal modes, each with its characteristic frequency ω, without regard to the energy. A quantum-mechanical (QM) treatment of the linear harmonic oscillator, however, assigns each mode with allowed energy levels $E = (n + 1/2)\hbar\omega$, $n = 1, 2, \ldots$. The internal energy must depend on how energy is absorbed into the quantized phonon levels. We calculate the energy of the ensemble of oscillators as a sum over the QM states

$$U = \frac{\sum_i E_i e^{-\frac{E_i}{k_B T}}}{\sum_i E_i \cdot e^{-\frac{E_i}{k_B T}}} \approx U_{\text{equilibrium}} + \sum_i \frac{1}{2}\hbar\omega(\boldsymbol{k}) + \sum_i \frac{\hbar\omega(\boldsymbol{k})}{\exp\left[-\frac{E_i}{k_B T}\right] - 1} \tag{2.89}$$

from which we can evaluate the specific heat as

$$C_V = \sum_i \frac{\partial}{\partial T} \frac{\hbar\omega(\boldsymbol{k})}{\exp\left[-\frac{E_i}{k_B T}\right] - 1} \tag{2.90}$$

This quantum-corrected model now depends on the temperature, as required. Since we know how to determine $\omega(\boldsymbol{k})$, (2.90) can in principle be evaluated.

A particularly simple QM model was worked out by Debye. It replaces the lattice's dispersion relation with a simplified linear form, $\omega(\boldsymbol{k}) = v_s \boldsymbol{k}$, where v_s is the scalar isotropic speed of sound in the crystal (the equation for the slope(s) in Figure 2.20). In addition, the Debye temperature Θ_D is introduced that characterizes the lowest temperature at which all modes of the crystal are being excited. After some algebraic manipulations (see e.g. [2.1], p. 458) we obtain the expression

$$C_v = 9Nk_B\left(\frac{T}{\Theta_D}\right)^3 \int_0^{\frac{\Theta_D}{T}} \frac{x^4 e^x}{(e^x - 1)^2} dx \quad (2.91)$$

Upon evaluation the Debye relation leads to the following rather awful but exact analytical expression (computed using Mathematica®)

$$\begin{aligned} C_v = -9Nk_B \frac{1}{15\left(\exp\left[\frac{T}{\Theta_D}\right]-1\right)\left(\frac{\Theta_D}{T}\right)^4} \Bigg\{ & 15\exp\left[\frac{T}{\Theta_D}\right] - 4\left(\pi\frac{\Theta_D}{T}\right)^4 + \\ & 4\exp\left[\frac{T}{\Theta_D}\right]\left(\pi\frac{\Theta_D}{T}\right)^4 + 60\frac{\Theta_D}{T}\ln\left[1-\exp\left[\frac{T}{\Theta_D}\right]\right] - \\ & 60\exp\left[\frac{T}{\Theta_D}\right]\frac{\Theta_D}{T}\ln\left[1-\exp\left[\frac{T}{\Theta_D}\right]\right] + \\ & 180\left(\frac{\Theta_D}{T}\right)^2 \mathrm{Li}_2(\exp\left[\frac{T}{\Theta_D}\right]) - 180\exp\left[\frac{T}{\Theta_D}\right]\left(\frac{\Theta_D}{T}\right)^2 \mathrm{Li}_2(\exp\left[\frac{T}{\Theta_D}\right]) - \\ & 360\left(\frac{\Theta_D}{T}\right)^3 \mathrm{Li}_3(\exp\left[\frac{T}{\Theta_D}\right]) - 360\exp\left[\frac{T}{\Theta_D}\right]\left(\frac{\Theta_D}{T}\right)^3 \mathrm{Li}_3(\exp\left[\frac{T}{\Theta_D}\right]) + \\ & 360\left(\frac{\Theta_D}{T}\right)^4 \mathrm{Li}_4(\exp\left[\frac{T}{\Theta_D}\right]) - 360\exp\left[\frac{T}{\Theta_D}\right]\left(\frac{\Theta_D}{T}\right)^4 \mathrm{Li}_4(\exp\left[\frac{T}{\Theta_D}\right]) \Bigg\} \end{aligned} \quad (2.92)$$

where $\mathrm{Li}_m(z)$ is the polylogarithm function. It provides an excellent approximation of the specific heat over all temperature ranges, and is plotted in Figure 2.28.

An-harmonicity

The harmonic potential is in fact a truncated series expansion model of the true crystal inter-atom potential, and as such is not capable of explaining all lattice phenomena satisfactorily. We briefly discuss one important examples here: the expansion of the lattice with temperature.

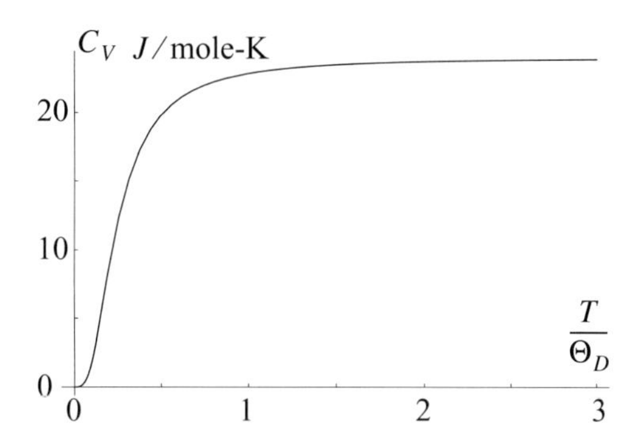

Figure 2.28. The specific heat C_V of a harmonic crystal versus T/Θ_D computed using the Debye interpolation formula. The high temperature limit of DuLong and Petit is approximately 24 $J/$mole-K .

The other important example, the lattice thermal conductivity, will be discussed in Section 7.1.2.

Thermal Expansion

The thermal expansion coefficient is defined by

$$\alpha_T = \frac{1}{V}\left(\frac{\partial V}{\partial T}\right)_{(P)} = \frac{1}{3B}\left(\frac{\partial P}{\partial T}\right)_{(V)} \tag{2.93}$$

where B is the bulk modulus defined as $B = -V(\partial P/\partial V)_{(T)}$, and equal to the inverse of the compressibility, l is some length, T is the temperature, P is the pressure and V is the volume. From thermodynamics, we have that

$$T\left(\frac{\partial S}{\partial T}\right)_{(V)} = \left(\frac{\partial U}{\partial T}\right)_{(V)}, P = -\left(\frac{\partial F}{\partial V}\right)_{(T)} \text{ and } F = U - TS \tag{2.94}$$

for the entropy S, the internal energy U and the Helmholtz free energy F. Combining these, we can write the pressure only in terms of the internal energy, volume and temperature as

$$P = -\frac{\partial}{\partial V}\left\{U - T\int_0^T \frac{\partial}{\partial T'}U(T', V)\frac{dT'}{T'}\right\} \tag{2.95}$$

All that remains is to insert the expression (2.89) for the internal energy into (2.95), and then to insert this result into (2.93). Evaluating this expression is beyond our scope. It is important, however, to note that it we assume a harmonic crystal potential (and hence internal energy) then the thermal expansion coefficient would vanish. If a fourth-order model is used (the third order expansion being unstable) we obtain

$$\alpha = \frac{1}{3B}\sum_k \sum_i \left(-\frac{\partial}{\partial V}\hbar\omega_k\right)\frac{\partial}{\partial T}\left(\frac{1}{\exp\left[-\frac{E_i}{k_B T}\right]-1}\right) \tag{2.96}$$

This expression is often simplified by the introduction of a Grüneisen parameter γ that absorbs the derivatives of the mode frequencies w.r.t the volume, and that exploits the similarities of (2.90) and (2.96), to yield $\alpha = \gamma c_v/3B$.

2.5 *Modifications to the Uniform Bulk Lattice*

Up to now we have considered the perfect crystal to be infinitely extended in space, which enabled us to make a straightforward ansatz for the mechanical behavior of a crystal atom in the potential field of the electrons. In reality semiconductor crystals are finite, subject to defects, and terminated by surfaces. Semiconductor manufacturing is performed for the most part by selectively growing and etching thin film layers at the uppermost surface of a semiconductor wafer, and hence active devices include a variety of material interfaces. Since the real crystal has a modified structure, we will look at the implications for the phonons. Surface phonons are dealt with in Chapter 7.

Point defects

When a single atom in the lattice is substituted by another atom type with different mass and/or a different binding force, the uniform lattice description is no longer valid. Qualitatively, new spatially localized phonon states become possible. These can be one of the following four cases:

- A: at a frequency higher than the optical branch, and,
- B: close enough to the optical branch to cause a splitting of its levels;
- C: between the optical an acoustic branch;
- D: within the acoustic branch.

The cases are illustrated in Figure 2.29. We now reconsider the diatomic

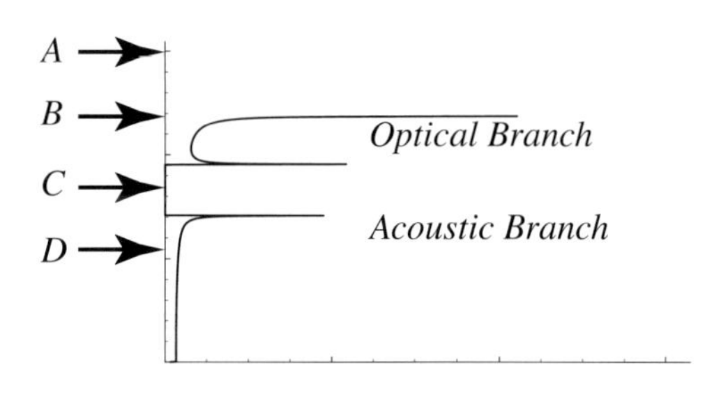

Figure 2.29. Positions of the four localized phonon states, due to point defects [2.12], on a 1D diatomic lattice density-of-states diagram.

lattice, with masses M and m, of Section 2.4.1 but with the mass of one site slightly perturbed, i.e., replaced with a mass μ. The four cases are:

- $\mu < m < M$, mass of type m is replaced;
- $m < \mu < M$, mass of type m is replaced, i.e., in the gap;
- $m < \mu < M$, mass of type M is replaced, i.e., in the gap;
- $m < M < \mu$, mass of type M is replaced.

Introducing an Interface

The assumptions we have made about the bulk crystal lattice do not necessarily hold at the interface, and as a result we can expect certain modifications to the physical effects we have considered so far that should account for its presence. One reason why material surfaces and interfaces are important, is that we usually build devices on crystal surfaces; also, many effects that are useful take place at material interfaces. By an *interface* we denote that transition region between the bulk crystal and some other substance, whether it is a solid, liquid or gas. Usually the interface is only a few atomic diameters thick, and if it separates the crystal from a liquid or gas, we usually denote it as the *surface* of the crystal.

A new crystal surface can be formed by etching into an existing crystal, or by growing a crystal from a seed, or by cleaving an existing crystal along one of its planes. An interface can be created by depositing another material onto the crystal, or by performing a reaction on the crystal surface that results in another compound being formed (e.g., by oxidation).

This section will only introduce a few important concepts. Surface physics is a large and active discipline and the interested reader is encouraged to refer to a specialized text such as [2.2].

Dangling Bond

Immediately after a crystal surface is formed in a vacuum by cleaving, the atoms at the surface have dangling bonds. In time, the atoms rearrange themselves to assume a more energetically favorable configuration. For example, the atoms, still bound to the underlying bulk material, will relax to a lattice constant smaller (or larger) than the bulk value, to reflect the fact that the bonding forces are one-sided in a direction normal to the surface. Furthermore, the surface could buckle in shape so as to enable the dangling bonds to create bonds with each other. In silicon, atoms on the $<111>$ surface can form a so-called 2×1 (or 7×7) reconstruction with π-bonded chains [2.2], see Figure 2.30. In the atmosphere, we can expect various gas atoms and molecules to attach to the dangling bonds. For example, it appears that hydrogen preferentially bonds to silicon surfaces, a fact that influences for example the etch rate of silicon. We will consider a simple model of a crystal surface to describe the modifications to the bulk dispersion relation in Section 7.6.3.

Superlattices

A superlattice is a term used for any lattice-like structure with a lattice constant larger than that of the underlying crystalline material. Thus, e.g., when a crystal surface forms atomic rearrangements due to relaxation effects, a superlattice is formed with a lattice constant that is often twice as large as before. Epitaxially formed heterostructures such as found in

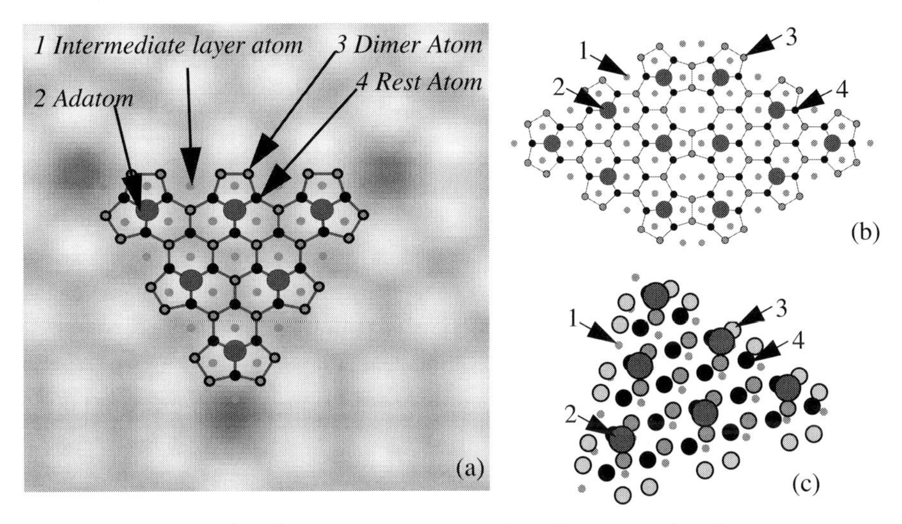

Figure 2.30. (a) The Si $\langle 111 \rangle$ 7×7 -reconstructed *surface imaged with a scanning tunneling microscope. The overlay indicates the various atomic positions in the DAS layer. The pictures (b) and (c) to the right are for comparison purposes, and show the first four silicon atom layers. (c) Shows a 3-dimensional view of the first four atom layers of the lattice.*

quantum-well semiconductor lasers also represent superlattices. We expect the superlattice to have a local influence on the dispersion curve because additional waves are admitted with another wavelength [2.2], [2.16].

2.6 Summary for Chapter 2

The forces between the constituent atoms of a crystal lead to the pseudoparticle of lattice vibrational energy: the phonon. From this view, we obtain the dispersion curves for a solid with its different phonon branches. The acoustic phonons, in the long wavelength limit, are the elastic waves we know from a macroscopic viewpoint, and we completely recover the classical theory of elasticity. The high frequency optical phonons explain why a solid lattice is able to interact with light. With the simpler harmonic model we already obtain useful expressions for the heat capacity and thermal expansion. In Chapter 7 we will see that for a realistic model of the heat conductivity, this model must be extended.

2.7 References for Chapter 2

2.1 Neil W. Ashcroft, N. David Mermin, *Solid State Physics*, Saunders College Publishing, Philadelphia (1988)

2.2 H. Lüth, *Surfaces and Interfaces of Solid Materials,* 3rd Ed., Springer Verlag, Berlin (1995)

2.3 Marie-Catharin Desjonquères, D. Spanjaard, *Concepts in Surface Physics*, 2nd Ed., Springer Verlag, Berlin (1996)

2.4 Otfried Madelung, *Introduction to Solid-State Theory*, Springer-Verlag, Heidelberg (1981)

2.5 Otfried Madelung, *Semiconductors - Basic Data*, Springer-Verlag, Heidelberg (1996)

2.6 Arokia Nathan, Henry Baltes, *Microtransducer CAD*, Springer-Verlag, Vienna (1999)

2.7 D. F. Nelson, *Electric, Optic and Acoustic Interactions in Dielectrics,* John Wiley and Sons, New York (1979)

2.8 Frank H. Stillinger, Thomas A. Weber, *Computer Simulation of Local Order in Condensed Phases of Silicon*, Phys. Rev. B, 31(8) (1985) pp. 5262–5271

2.9 Simon M. Sze, *Physics of Semiconductor Devices*, 2nd Ed., John Wiley and Sons, New York (1981)

2.10 Franklin F. Y. Wang, *Introduction to Solid State Electronics*, 2nd Ed., North-Holland, Amsterdam (1989)

2.11 Abraham I. Beltzer, *Acoustics of Solids*, Springer Verlag, Berlin (1988)

2.12 R. W. H. Stevenson, ed., *Phonons in Perfect Lattices and in Lattices with Point Defects*, Scottish Universities Summer School 1965, Oliver & Boyd, Edinburgh (1966). Chapter 1, *General Introduction*, by C. Kittel, Chapter 2, *Phonons in Perfect Lattices*, by W. Cochran.

2.13 Ian Johnston, Graham Keeler, Roger Rollins and Steven Spick-lemire, *Solid State Physics Simulations*, John Wiley & Sons, New York (1996)

2.14 Léon van Hove, *The Occurrence of Singularities in the Elastic Frequency Distribution of a Crystal*, Phys. Rev. **89**(6) (1953) 1189-1193

2.15 C. B. Walker, *X-Ray Study of Lattice Vibrations in Aluminium*, Phys. Rev. **103**(3) (1956) 547-557

2.16 Peter Y. Yu, Manuel Cordona, *Fundamentals of Semiconductors*, Springer Verlag, Berlin (1996)

Chapter 3 The Electronic System

Since the early days of discoveries on the nature of electrons, (we can note in hindsight), the world hasn't looked back. One notable laboratory at Bell Laboratories, run by William Schockley, has brought us innumerable innovations to control and exploit the behavior of electrons in semiconductors, including the transistor, and establishing many of the key ideas used in semiconductor circuits today that, upon reading in one sitting, lets us amaze at how clear these pioneers already saw the end result. Subsequent industrial innovations have not let us down, engineering ever faster switching transistors and electronic circuits according to the famed "law" of Gordon Moore, whereby miniaturization and speedup doubles every one and a half years. The result: current laptop computers are as powerful as the supercomputer of the author's student days (but much more reliable and comfortable to use!).

Chapter Goal Electrons move through crystals in special ways, and the mechanism is dominated by quantum mechanics. The goal of this chapter is therefore to introduce two topics:

- Basic quantum mechanics;
- The semiconductor electronic system, and the band structure of silicon.

Chapter Roadmap

We start with the free electron, introducing the Schrödinger equation. Next, the electron is bound by a variety of potentials. The hydrogen atom represents the simplest quantum-mechanical model of an atom, with good predictive qualities. This leads up to the periodic potential, a model for the periodically-placed crystal atom potentials. This model naturally leads to the concepts of a forbidden band, and band splitting. The chapter wraps up with and expression for the effective mass of the bound electrons.

In this chapter we take some space to explain the necessary mathematical instruments and physical concepts to understand the quantum nature of phenomena described in the book. You are strongly encouraged to carefully study the mathematical manipulations of wavefunctions, operators and all the objects described in this chapter. Moreover, you should freely work with all the objects and practise to manipulate them. Most of the time you will gain a deeper understanding, and of course there is no risk of doing any harm to them. The worst that can happen is that you might not find any meaningful physical interpretation for what you did. For a detailed understanding of specific topics special literature is given in the references.

3.1 Quantum Mechanics of Single Electrons

Quantum mechanics describes the fundamental properties of electrons [3.1], [3.2]. It tells us that both particle-like and wave-like behavior is possible. In a semiconductor both types of behavior are observable. The particle concept turns out to be an excellent description for a wide range of classical phenomena and the most common applications. Modern devices on the nanometer scale instead demonstrate the quantum nature of electrons in many beautiful ways. Their deeper understanding requires a certain background in quantum mechanics. Therefore, it is necessary to study both classical mechanics and quantum mechanics of electrons. We summarize the quantum mechanical concepts for single electrons, that is, non-interacting electrons. Electrons are subject to Coulomb interactions (between charges) and it is well known that there is no general analytical solution for the motion of a system of more than two interacting particles. A semiconductor usually contains many times more than two electrons in the conduction band. Nevertheless, in most cases the single electron picture turns out to be a good description of electrons in a wide range of applications.

3.1.1 Wavefunctions and their Interpretation

A wave-like nature was first assigned to particles in general by De Broglie in 1924 and was proven by the electron interference experiments of Davisson and Germer in 1927 (see Figure 3.1).

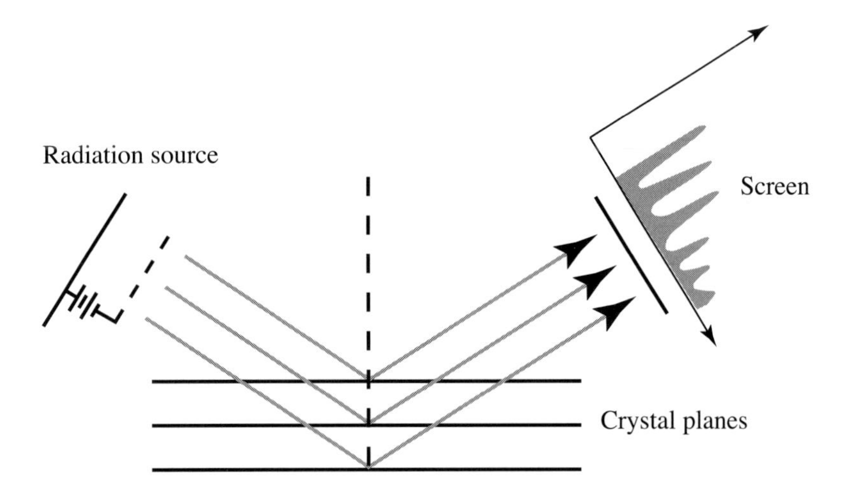

Figure 3.1. Interference experiment of Davisson and Germer. Accelerated electrons hit a crystal surface and are diffracted according to their wavelength. This is to be understood in analogy with the X-ray diffraction for the analysis of crystal lattice structure (see Section 2.2).

Interference Experiment

The interference pattern on the screen shows the dependence of the intensity on the acceleration bias for the electrons, which itself is a measure of the momentum of an electron. The fact that the electron waves follow the Bragg relation gives for the momentum and wavevector the relation $\boldsymbol{p} = \hbar \boldsymbol{k}$, where $\hbar$ is the Planck constant. Thus we describe the electrons as a plane wave in the same way as we did for the lattice waves (Box 2.1). These waves have a frequency ω and a wavevector **k** and are characterized by the functions $\sin(\omega t - \boldsymbol{k}x)$, $\cos(\omega t - \boldsymbol{k}x)$, $\exp(\omega t - \boldsymbol{k}x)$ or linear combinations of them. An arbitrary linear combination $\Psi(\boldsymbol{x}, t)$, in general, is a complex scalar field. On the other hand, the particle's position **x**, its momentum **p**, charge q and many other properties that may be determined by measurement are real quantities. Therefore, we need an interpretation of the wavefunction.

Probability Density

The square modulus of the wavefunction $|\Psi(\boldsymbol{x}, t)|^2$ is interpreted as the probability density of finding an electron at time t at position **x**. Thus for a normalized wavefunction

$$\int_{\Omega} |\Psi(\boldsymbol{x}, t)|^2 dV = \int_{\Omega} \Psi^*(\boldsymbol{x}, t)\Psi(\boldsymbol{x}, t) dV = \langle\Psi|\Psi\rangle = 1 \tag{3.1}$$

holds, because the electron must be found with probability one somewhere in the physical domain Ω. We implicitly assumed that the wave function is square-integrable, i.e., the integral over Ω exists. This is not necessarily the case, as we shall later when dealing with free electrons. $\sqrt{\langle\Psi|\Psi\rangle} = \|\Psi\|$ is called the norm of Ψ and (3.1) is the standard form for the numerical evaluation of the expectation values of operators. The short-hand notation given by the last term in angle brackets is known as the Dirac notation. Once we have defined the probability density

$$\rho(\boldsymbol{x}, \boldsymbol{x}) = \Psi^*(\boldsymbol{x}, t)\Psi(\boldsymbol{x}, t) \tag{3.2}$$

we calculate its moments. Remember that for a real mass density the property corresponding to the first moment with respect to the position vector **x** is the center of mass. For the electron described in terms of a wavefunction it is the average value of its position

$$\langle \hat{x} \rangle = \int_{\Omega} \Psi^*(x, t)\hat{x}\Psi(x, t) = \langle \Psi | \hat{x} \Psi \rangle \tag{3.3}$$

In this case $\hat{x}$ is called the position operator. We shall use ^ to identify operators whenever we are not dealing with a special *representation*. In the position representation $\hat{x}$ is a vector with cartesian components x, y and z.

Expectation Values

The result of (3.3) gives the most probable position where to find a particle. It is called the expectation value of the position operator. It is possible to calculate the *expectation values* of arbitrary functions of $\hat{x}$, such as the variance of the position, which gives us important information about the spreading of the wavefunction

$$\mathrm{Var}(x) = \langle (\Delta \hat{x})^2 \rangle = \langle \Psi | (\hat{x} - \langle \hat{x} \rangle)^2 \Psi \rangle = \langle \Psi | (\hat{x}^2 - \langle \hat{x} \rangle^2) \Psi \rangle \tag{3.4}$$

An illustrative example is shown in Figure 3.2, representing a one-

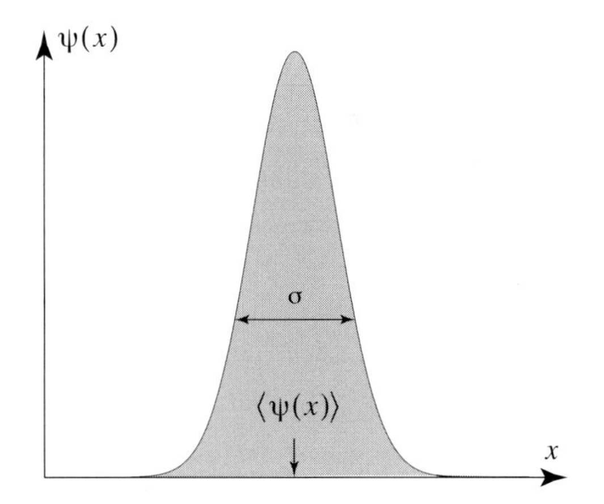

Figure 3.2. A Gaussian-shaped wavefunction with its average value $\langle \psi(x) \rangle$ and variance σ.

dimensional wavefunction at $t = t_0$ given by

$$\Psi(x, t_0) = \psi(x) = \frac{1}{(2\pi)^{1/4}}\sigma^{1/2}\exp\left(-\frac{(x-x_0)^2}{4\sigma^2}\right) \tag{3.5}$$

For the wavefunction of type (3.5) $\sigma^2 = \langle(\Delta\hat{x})^2\rangle$ holds.The spatial wavefunction may be represented as a superposition of plane waves, i.e., as an inverse Fourier transform

$$\psi(x) = \frac{1}{2\pi}\int_{-\infty}^{\infty}\phi(k)\exp(ikx)dk \tag{3.6}$$

The wavevector in (3.6) is the same as found in the momentum relation above.

Position and Momentum Represen-tations

$\psi(x)$ we call the wavefunction in position representation, more precisely its spatial part. $\psi(k)$ we call the wavefunction in momentum representation. For the expectation value of the momentum we write

$$\langle\hat{p}\rangle = \hbar\langle\phi|k\phi\rangle = \int_{-\infty}^{\infty}\phi^*(k)\hat{p}\phi(k)dk = -i\hbar\int_{-\infty}^{\infty}\psi^*(x)\frac{\partial}{\partial x}\psi(x)dx \tag{3.7}$$

The important message in (3.7) is that the result of calculating an expectation value does not depend on the representation of the wavefunction. Note that when representing wavefunctions in position space the momentum is a differential operator $\hat{p} = i\hbar\frac{\partial}{\partial x}$. This implies that the momentum operator and the position operator do not commute, i.e., applying the chain rule for differentiation we see that

$$(\hat{p}_x\hat{x} - \hat{x}\hat{p}_x)\psi(x) = -i\hbar\psi(x) \tag{3.8}$$

holds. In other words, it makes a difference whether $\hat{p}_x\hat{x}$ or $\hat{x}\hat{p}_x$ is applied to $\psi(x)$ and it does not give the same result. (3.8) is called the commutator of the operators $\hat{p}_x$ and $\hat{x}$. If operators do not commute, i.e., the commutator has a non-zero value, this means that the respective physical quantities cannot be measured simultaneously with arbitrary precision.

Uncertainty Principle

This leads directly to the uncertainty principle for operators, which is one of the central points of quantum theory that has always been subject of many discussions from the moment of its formulation. For the gaussian wavefunction (3.5) centered at $x_0 = 0$, the variances of the particles position and momentum are

$$\langle(\Delta\hat{x})^2\rangle = \frac{1}{(2\pi)^{1/2}\sigma}\int_{-\infty}^{\infty}\exp\left(-\frac{x^2}{2\sigma^2}\right)x^2dx = \sigma^2 \tag{3.9}$$

and

$$\langle(\Delta\hat{p})^2\rangle = \mathrm{h}\sigma\sqrt{\frac{2}{\pi}}\int_{-\infty}^{\infty}\exp(-k^2\sigma^2)k^2dk = \frac{\mathrm{h}^2}{4\sigma^2} \tag{3.10}$$

respectively.

One-Dimensional System

Take a one-dimensional system with wavefunction $\psi(x)$ in its spatial representation, that is defined in the entire interval $(-\infty, \infty)$. Suppose that $\langle\hat{x}\rangle = 0$ and $\langle\hat{p}\rangle = 0$ hold. This assumption is not necessary but taking arbitrary values for the expectation values of position and momentum only makes the calculation more complicated without increasing the understanding. The relation

$$\int_{-\infty}^{\infty}\left|\alpha x\psi + \frac{d\psi}{dx}\right|^2 dx$$
$$= \left(\int_{-\infty}^{\infty}\left(\alpha^2x^2|\psi|^2 + x\frac{d\psi^*}{dx}\psi + x\psi^*\frac{d\psi}{dx} + \frac{d\psi^*}{dx}\frac{d\psi}{dx}\right)dx \geq 0\right) \tag{3.11}$$

is evident since the square modulus is positive in $(-\infty, \infty)$. Here α is an arbitrary real constant. Talking into account (3.1), (3.3) and (3.7) and performing integration by parts we obtain

$$\alpha^2\langle(\Delta\hat{x})^2\rangle - \alpha + \frac{1}{\hbar}\langle(\Delta\hat{p})^2\rangle \geq 0 \tag{3.12}$$

This inequality is to hold for arbitrary real α . Thus,

$$\langle(\Delta\hat{x})^2\rangle\langle(\Delta\hat{p})^2\rangle \geq \frac{\hbar^2}{4} \tag{3.13}$$

must hold. (3.13) is called the Heisenberg uncertainty principle. Heisenberg derived it in 1925, and it gave rise to the creation of a new physical discipline called quantum mechanics. The simple significance is that the product of the variances of position and momentum has a minimum, which is $\hbar^2/4$. Since it clearly derives from the commutator relations its physical interpretation is that position and momentum may not be determined by measurement with arbitrary precision. Note that (3.13) only gives a lower limit. Now, turning back to the example of a gaussian wavefunction we see that it is a function, we see that it is a function fulfilling (3.13) exactly with the "="-sign. Thus we call it a state of "minimal uncertainty".

3.1.2 The Schrödinger Equation

The quantitative description of wavefunctions was given by Erwin Schrödinger in 1924. An example of a classical wave equation is a partial differential equation of second order in space and time. The corpuscular nature of the light field as discovered by Einstein relates the energy E of the photon to the angular frequency of the electromagnetic wave by $E = \hbar\omega$, while its momentum is given by $\boldsymbol{p} = \hbar\boldsymbol{k}$. The assumption of De Broglie for those relations to hold also for particles together with $E = \hbar^2k^2/(2m)$ led Schrödinger to take an equation of first order in time. This equation of motion must read

$$i\hbar\frac{\partial}{\partial t}\Psi(\boldsymbol{x}, t) = -\frac{\hbar^2}{2m}\nabla^2\Psi(\boldsymbol{x}, t) \tag{3.14}$$

which is a partial differential equation of first order in time and second order in space. Note the fact that an imaginary coefficient of the time derivative turns (3.14) into a wave equation. The spatial derivative of second order with its coefficient $\hat{T} = -\hbar^2/(2m)\nabla^2$ is, according to the discussion about the momentum operator in 3.1.1, the operator of the kinetic energy in a spatial representation. The equation of motion for the Schrödinger field Ψ in a potential is given by

$$i\hbar\frac{\partial}{\partial t}\Psi(\boldsymbol{x}, t) = (\hat{T} + U(\boldsymbol{x}))\Psi(\boldsymbol{x}, t) = \hat{H}\Psi(\boldsymbol{x}, t) \tag{3.15}$$

$\hat{H} = \hat{T} + U(\boldsymbol{x})$ is called the Hamilton operator of a closed system if $U(\boldsymbol{x})$ does not depend explicitly on the time variable t. This means that the solution of the system described by (3.15) is invariant under translations in time. Wavefunctions for which the energy has a specific value are called stationary states. We denote them as Ψ_n and call them eigenfunctions of the Hamilton operator. All of them follow the equation $\hat{H}\Psi_n = E_n\Psi_n$, which is called the eigenvalue equation for the Hamilton operator. The total of eigenfunctions and eigenvalues is called the spectrum of the Hamilton operator. In this case we integrate (3.15) in time, which gives

$$\Psi_n = \exp\left(-\frac{i}{\hbar}E_n t\right)\psi_n(\boldsymbol{x}) \tag{3.16}$$

where the function ψ_n depends only on the coordinates. (3.16) gives the time dependence of the wavefunction, while the spatial dependence and the energy of the eigenstate must be found solving the respective eigenvalue problem resulting inserting (3.16) into (3.15)

$$\hat{H}\psi_n(\boldsymbol{x}) = E_n\psi_n(\boldsymbol{x}) \tag{3.17}$$

Quantum Numbers

The numbers n counting the different energies are called quantum numbers. The stationary state with the lowest energy we call the ground state. We recall some technical aspects from linear algebra:

- For every discrete eigenvalue problem, there might be multiple eigenvalues which results in different eigenvectors for the same eigenvalue.

- With the knowledge of the whole energy spectrum, i.e., all the eigenvalues and eigenvectors that form a complete basis set, we may represent an arbitrary vector as an expansion into the eigenvectors.

$$\Psi(x, t) = \sum_n a_n \exp\left(-\frac{i}{\hbar} E_n t\right) \psi_n(x) \tag{3.18}$$

- The eigen-vectors are orthogonal ($\langle\psi_n|\psi_m\rangle = \delta_{nm}$). This means that the "scalar product" as defined for the wavefunctions through (3.1) is zero for $n \neq m$ and one for $n = m$.

The square moduli $|a_n|^2$ denote the probability of finding the system in a special eigenstate. The spatial probability distribution is given by $|\Psi_n|^2 = |\psi_n|^2$, it does not depend on time, which also holds for the expectation values of operators that do not depend explicitly on time. All those quantities are called conserved quantities. All conserved quantities commute with the Hamilton operator, i.e., we may measure them at the same time we measure the energy of the system. Nevertheless, they may not always give the same result for their expectation values and seem not to be uniquely defined in the case where there exist multiple energy eigenvalues. Thus we need to know more about the system than only the energy of the respective state or, in other words, in some situations the energy spectrum does not contain enough information about the system to be described uniquely.

When talking about conserved quantities, we may ask for the probability flux through the surface Γ of a finite volume Ω, i.e., the continuity equation for the probability density in the same way as it exists for the charge density in electrodynamics. For this purpose, we integrate the probability density (3.2) on the entire volume Ω and take its time derivative:

$$\begin{aligned} \frac{\partial}{\partial t}\int_\Omega (|\Psi|^2 dV) &= \int_\Omega \left(\frac{\partial\psi^*}{\partial t}\psi + \psi^*\frac{\partial\psi}{\partial t}\right) dV \\ &= \frac{i}{\hbar}\int_\Omega ((\Psi\hat{H}\Psi^* - \Psi^*\hat{H}\Psi)dV) \end{aligned} \tag{3.19}$$

where we used (3.15) and the fact that for the Hamilton operator $\hat{H} = \hat{H}^*$ holds. We use

$$\hat{H} = -\frac{\hbar}{2m}\nabla^2 + U(x) \tag{3.20}$$

and

$$\Psi\nabla^2\Psi^* - \Psi^*\nabla^2\Psi = \nabla(\Psi\nabla\Psi^* - \Psi^*\nabla\Psi) \tag{3.21}$$

which yields a continuity equation of the form

$$\frac{d}{dt}\int_{\Omega}|\Psi|^2 dV = -\int_{\Gamma}\nabla j dV \tag{3.22}$$

where Γ is the surface of the region Ω. Here $\mathbf{j}$ is the current density vector, which is identified by

$$j = \frac{i\hbar}{2m}(\Psi\nabla\Psi^* - \Psi^*\nabla\Psi) = \frac{1}{2m}(\Psi\hat{p}^*\Psi^* - \Psi^*\hat{p}\Psi) \tag{3.23}$$

Box 3.1. The Gauss Theorem.

The Gauss Theorem. Consider a vector field $A(r)$ and a volume element V with a closed surface S_V. The normal vector n of the surface points outward the volume element. Then we have the following relation between the surface integral and the volume integral of the vector field

$$\oint_{S_V} A ds = \int_V \nabla A dV \tag{B 3.1.1}$$

This is called the Gauss theorem. Due to this equation the properties of the field inside V -especially its sources- are related to its properties at the surface – especially its flux.

This is also true for situations where the volume is a non simply connected region, i.e., there may be holes in V.

In components we formulate the theorem for arbitrary dimensions D as follows

$$\oint_{S_V} ds_i\ldots = \int_V dV\frac{\partial}{\partial x_i}\ldots \tag{B 3.1.2}$$

where the dots are to be replaced by a vector field, or even a scalar field, while $i = 1, 2, \ldots, D$. This means that the surface integral is related to the volume integral of the derivative of the field.

Current Density

This is the current density of probability or simply the current density. For a given electronic system its expectation value multiplied by the unit charge corresponds to a measurable electrical current density.

Gauss theorem

Applying the Gauss theorem, (see Box 3.1), we see that

$$\frac{d}{dt}\int|\Psi|^2 dV = -\int j dS \qquad (3.24)$$

holds. (3.24) has a very simple interpretation: the change of probability density with time in the volume Ω is caused by the flux through the surface Γ.

3.2 Free and Bound Electrons, Dimensionality Effects

Another more sophisticated point is the fact that the energy spectrum may show a both discrete and a continuous part. This leads us to the following question: How does the spectrum depend on the imposed boundary conditions?

To determine finally the functional form of the wavefunction, we need information about these boundary conditions that the electronic system has to fulfill. The resulting spectrum of observables will be discrete, continuous or mixed and allows us to talk about its dimensionality [3.1], [3.2].

3.2.1 Finite and Infinite Potential Boxes

The One-Dimensional Potential Box

The simplest case to study appears to be the one-dimensional potential box with finite potential. The physical interpretation of its solution and boundary conditions, and the transition to infinite potential walls, are very instructive. Given the potential

$$U(x) = \begin{cases} 0, & x \in [0, a] \\ U_0, & x \notin [0, a] \end{cases} \qquad (3.25)$$

the spectrum of the Hamiltonian will have both discrete and continuous eigenvalues. The one-dimensional time independent Schrödinger equations read

$$\frac{d^2}{dx^2}\Psi + \frac{2m}{\hbar^2}E\Psi = 0 \text{ , for } x \in [0, a] \tag{3.26a}$$

$$\frac{d^2}{dx^2}\Psi + \frac{2m}{\hbar^2}(E - U_0)\Psi = 0 \text{ , for } x \notin [0, a] \tag{3.26b}$$

The Schrödinger equation, as given in (3.26a) and (3.26b), are differential equations of second order. To find their solutions we must provide additional information, i.e., conditions to hold in the physical domain Ω or on its boundary.

- The probability density must be smooth at $x = 0$ and $x = a$, and so must be the wavefunction. This is because there are no sources or sinks for the probability density, i.e., there is neither particle creation nor destruction going on, i.e., (3.24) holds:
  $\lim_{\varepsilon \to 0}(\Psi(+\varepsilon) - \Psi(-\varepsilon)) = 0$ and $\lim_{\varepsilon \to 0}(\Psi(a+\varepsilon) - \Psi(a-\varepsilon)) = 0$
- For the same reason the flux of probability density must also be continuous at the interface and so

$$\lim_{\varepsilon \to 0}\left(\frac{d}{dx}\Psi(+\varepsilon) - \frac{d}{dx}\Psi(-\varepsilon)\right) = 0 \text{ and} \tag{3.27a}$$

$$\lim_{\varepsilon \to 0}\left(\frac{d}{dx}\Psi(a+\varepsilon) - \frac{d}{dx}\Psi(a-\varepsilon)\right) = 0 \text{ .} \tag{3.27b}$$

The trial function for the solution of (3.26a) and (3.26b) is $\psi = const. \cdot \exp(\lambda x)$. Inserting this, we obtain for λ

$$\lambda = \pm ik = \pm\sqrt{\frac{-2mE}{\hbar^2}} \text{ , for } x \in [0, a] \tag{3.28a}$$

$$\lambda = \kappa = \pm\sqrt{\frac{2m(U_0 - E)}{\hbar^2}} \text{ , for } x \notin [0, a] \tag{3.28b}$$

Bound Electrons, Free Electrons

There are harmonic solutions for $E > U_0$, because the radicands in (3.28a) and (3.28b) are always negative in this case and thus the solution for λ is imaginary. Electrons in these states are called unbound electrons or free electrons. Their λ is the well known wavevector k and their wavefunctions do not necessarily vanish for $x \to \pm\infty$. We shall discuss their properties together with the free electron states. For now we are interested in the solutions with energy eigenvalues $E < U_0$, called bound states. Their wavefunction decays exponentially in regions I and III (see Figure 3.3). The plus sign in (3.28b) must hold for $x < 0$, while the

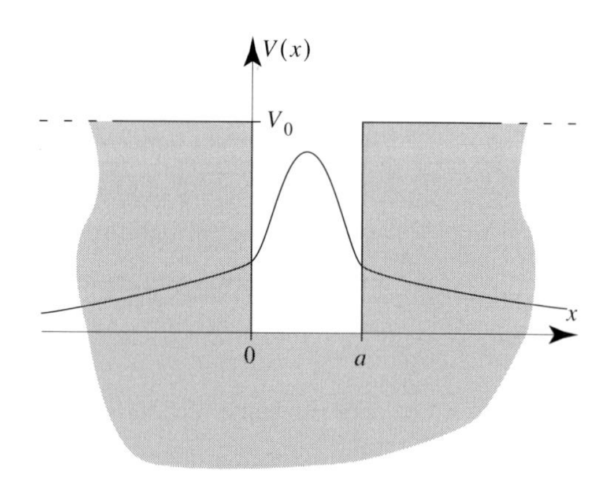

Figure 3.3. Finite box potential in one dimension.

minus sign must hold for $x > a$. In the regions I and III we call $|\lambda| = \kappa$. Inside the box we have a harmonic solution, with wavevector k. Thus we obtain

$$\psi_I(x) = A\exp(\kappa x) \quad (3.29a)$$

$$\psi_{II}(x) = B\exp(ikx) + C\exp(-ikx) \quad (3.29b)$$

$$\psi_{III}(x) = D\exp(-\kappa x) \quad (3.29c)$$

The boundary conditions read

$$\psi_I\big|_{x=0} = \psi_{II}\big|_{x=0} \tag{3.30a}$$

$$\psi_I'\big|_{x=0} = \psi_{II}'\big|_{x=0} \tag{3.30b}$$

$$\psi_{II}\big|_{x=a} = \psi_{III}\big|_{x=a} \tag{3.30c}$$

$$\psi_{II}'\big|_{x=a} = \psi_{III}'\big|_{x=a} \tag{3.30d}$$

Inserting (3.29a)-(3.29c) into (3.30a)-(3.30d), we obtain a linear system of equations to determine A-D:

$$\begin{bmatrix} 1 & -1 & -1 & 0 \\ 0 & e^{ika} & e^{-ika} & -e^{-\kappa a} \\ \kappa & -ik & ik & 0 \\ 0 & ike^{ika} & -ike^{-ika} & \kappa e^{-\kappa a} \end{bmatrix} \begin{bmatrix} A \\ B \\ C \\ D \end{bmatrix} = 0 \tag{3.31}$$

A solution can only be found if the determinant of the coefficient matrix in (3.31) is zero:

$$2\kappa k \cos(ka) + (\kappa^2 - k^2)\sin(ka) = 0 \tag{3.32}$$

With $ak = \xi$ we have $a\kappa = \sqrt{(2mU_0a^2)/\hbar^2 - \xi^2} = \sqrt{C^2 - \xi^2}$ and we may write (3.32) as

$$\frac{2\xi\sqrt{C^2 - \xi^2}}{C^2 - 2\xi^2} + \tan\xi = 0 \tag{3.33}$$

This equation must be solved either graphically, see Figure 3.4 and Figure 3.5, or by numerical methods. A numerical solution of the discretized model for the discussed eigenvalue scenario may be counterchecked in the special case with $U_0 \ll \hbar^2/(ma^2)$ where there is exactly one bound state found with $E_0 \approx U_0(1 - ma^2/(2\hbar^2))$.

Note that only in special limiting cases can an analytical solution be obtained easily. Usually, analytical methods end at a certain point and

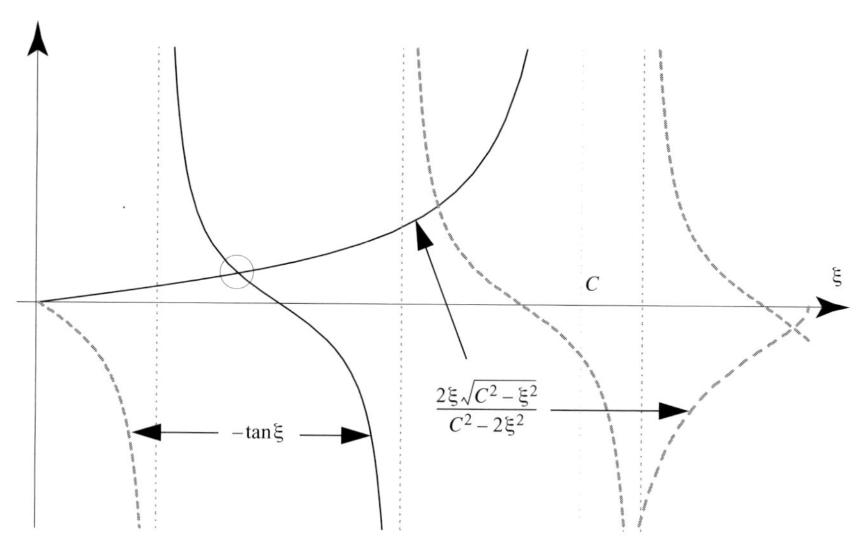

Figure 3.4. Graphical solution for equation (3.33). The intersection of the two expressions $-\tan\xi$ *and* $(2\xi\sqrt{C^2-\xi^2})/(C^2-2\xi^2)$, *indicated by the circle, is the first solution point. Also see Figure 3.4.*

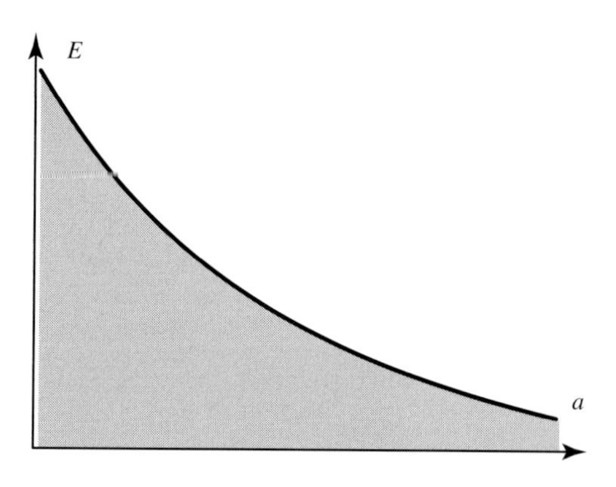

Figure 3.5. Change in energy E *as a function of well width* a, *for the first solution of equation (3.33), as given by the graphical solution (marked with a circle) in Figure 3.4.*

have to be succeeded by numerics. Since semiconductor structures are becoming more and more complicated, there is no way of dealing with the electronic structure calculations in an analytical way. The reader may now ask why all this rather theoretical material is discussed in detail here: simply because also the best simulation program is useless if there is no understanding of how to obtain a rough estimate of what the result has to be. We suggest implementing a simple numerical model for the calculation of 1D eigenvalue problems as is shown in Figure 3.8 for the harmonic oscillator potential.

Infinite Box Potential

We continue our discussion with the case where $U_0 \to \infty$. A simple physical argument tells us that for the energy of the system to be finite (the same holds for many other observable quantities), the wavefunction must vanish in the region where $U(x) \to \infty$, i.e., $\psi(x) = 0$ for $x \geq la$ and $x \leq 0$ and for all t. The trial solution for the wavefunction $\psi(x) = B\exp(ikx) + C\exp(-ikx)$ must fulfill the boundary condition $\psi(0) = 0$ and $\psi(a) = 0$. This leads to $\psi(x) = B\sin(kx)$, with $k = (n\pi)/a$ and n integer and positive starting at $n = 1$. Inserting this into the time–independent Schrödinger equation (3.17) gives us the energy spectrum of the infinite box potential:

$$E_n = \frac{\pi^2 \hbar}{2ma^2} n^2 \text{, with } n = 1, 2, 3, \ldots \tag{3.34}$$

The normalized wavefunction reads

$$\psi_n(x) = \sqrt{\frac{2}{a}} \sin\left(\frac{n\pi x}{a}\right) \tag{3.35}$$

We give another explanation to approach the infinite box potential. Suppose that $U_0 \gg \hbar^2/(ma^2)$ holds. Then there will be a part of the spectrum with low energies for which the wavefunction decays on a very short length scale outside the box, i.e., the electron is completely trapped inside the box.

Three–Dimensional Potential Boxes

The above results lead us immediately to the case of an electron in a three-dimensional potential box. Suppose the box is a cuboid region with $0 \leq x \leq a$, $0 \leq y \leq a$, and $0 \leq z \leq a$. Inside the box $U(x, y, z) = 0$, while outside $U(x, y, z) \to \infty$. The energy spectrum looks like

$$E_{n_x n_y n_z} = \frac{\pi^2 \hbar^2}{2ma^2}(n_x^2 + n_y^2 + n_z^2) \text{ , with } n_x^2, n_y^2, n_z^2 = 1, 2, 3, \ldots \quad (3.36)$$

Note that in this case multiply defined energy eigenvalues occur. For a given energy there are six ways of combining n_x, n_y, and n_z (presuming that $n_x \neq n_y \neq n_z$). A multiple energy eigenvalue is called degenerate when its multiplicity counts the number of states with the same energy. The quantum numbers n_x, n_y, and n_z indicate the momentum eigenvalues in the respective direction. To resolve the ambiguity of the energy indication of states the additional information needed is provided by the momentum. We shall encounter this problem several times. One may criticize that all that comes from the cuboid form of the potential box with equal edges of length a. This is correct, but in order to avoid degeneracy the quotients of the lengths of the edges must be irrational numbers. The normalized wavefunctions of the three-dimensional box reads

$$\Psi_{n_x n_y n_z} = \left(\frac{2}{a}\right)^{3/2} \sin\left(\frac{\pi n_x x}{a}\right) \cdot \sin\left(\frac{\pi n_y y}{a}\right) \cdot \sin\left(\frac{\pi n_z z}{a}\right) \quad (3.37)$$

The differences between one– and three–dimensional finite box potentials are not obvious. One might think that adding a dimension simply requires additional quantum numbers. This is certainly the case, but there is a fundamental difference. It can be shown that for a one-dimensional quantum box potential in any case at least one bound state can be found. This is not so in higher dimensions. In terms of box parameters there has to be a certain minimal width l of the box for a given U_0 in order to have a solution for a bound state.

3.2.2 Continuous Spectra

In the preceding paragraphs an electron showing a continuous spectrum has been described and therefore it is called unbound. Its wavefunction does not vanish for $x, y, z \to \infty$. Nevertheless, its value remains finite. We have hesitated up to now to introduce the concept of a free electron because of this fact. For wavefunctions with discrete eigenvalues that follow $\lim_{x, y, z \to \infty} \Psi = 0$ there is a well defined mathematical description, are defined in a so–called Hilbert space. This simply means that they must be square integrable. The number of basis vectors Ψ_n spanning the Hilbert space of the respective problem need not to be finite. Electrons with a continuous spectrum do not fit such a description because their wave-function is not square-integrable. It was Dirac who expanded the Hilbert space picture by normalizing the continuous states to the delta function. Remember the expansion of an arbitrary wavefunction in terms of the discrete eigenvectors of the Hamiltonian. We now make the transition where we let n *discrete* $\to$ **k continuous** and write (3.18) as

$$\Psi(x) = \int \Phi_k \Psi_k(x) dk \tag{3.38}$$

where now $a_n \to \phi_k$ (we omit the time variable t). The coefficient $|\phi_k|^2$ is interpreted as the probability density to find the system in the state indexed by k. Thus $\int |\phi_k|^2 dk = 1 = \int |\Psi_k|^2 dk$ must hold if the wave-function is normalized and the continuous basis functions are normalized $\int |\Psi_k|^2 dk = 1$ as well. We write

$$\int \Psi^*(x)\Psi(x)dx = \int |\Phi_k|^2 dk = \int \Phi_k^* \int \Psi_k^*(x)\Psi(x) dx dk \tag{3.39}$$

which gives us $\phi_k = \int \Psi_k^*(x)\Psi(x)dx$ and further with (3.38)

$$\phi_k = \int \Phi_{k'} \int \Psi_{k'}(x)\Psi_k^*(x)(dx)dk' \tag{3.40}$$

In order for this to hold we must require that

$$\int \Psi_{k'}(x)\Psi_k^*(x)dx = \delta(k - k') \tag{3.41}$$

with

$$\int \Phi_{k'} \delta(k - k') dk'$$

gives the definition of the delta function $\delta(k - k')$ which is zero for all $k - k' \neq 0$ and infinite at $k - k' = 0$ such that $\int \delta(k - k')dk = 1$ is the condition for normalizing the eigenfunctions of a continuous spectrum. We have chosen the expansion coefficients as ϕ_k on purpose in order to show that the situation of the gaussian wavefunction expanded in plane waves is exactly such a case. The infinitely extended momentum eigenfunctions describe ideal situations having an exact value for the momentum of the electron. According to the Heisenberg principle, allowing the transition for the momentum variance to vanish, this yields an infinite variance of its position such that their product is larger than $\hbar^2/4$. This can only be a hand–waving argument to visualize the situation. Nevertheless, it gives a feeling for the consistency problems arising. The gaussian wavefunction, in contrast, is an excellent example of a more realistic situation. Its interpretation is that of a moving electron having an expectation value of momentum and position with their respective variances in accordance with the Heisenberg principle. It is a more realistic situation because the arrival of electrons at a certain position may be measured and it coincides with an event of the measuring instrument. It is not possible here to go deeper into this subject but nevertheless we want to draw readers' attention to the literature on the problem of measurement in quantum mechanics.

The Continuous Spectrum of the Quantum Box

Suppose $E > U_0$ in (3.28a) and (3.28b). Then the solutions in the three regions I, II and III read

$$\psi_I(x) = a_1 \exp(ik_I x) + a_2 \exp(-ik_I x) \tag{3.42a}$$

$$\psi_{II}(x) = b_1 \exp(ik_{II} x) + b_2 \exp(-ik_{II} x) \tag{3.42b}$$

$$\psi_{III}(x) = c_1 \exp(ik_{III} x) + c_2 \exp(-ik_{III} x) \tag{3.42c}$$

where we have $k_{I,\,III} = \sqrt{2m(E-U_0)}/\hbar$ and $k_{II} = \sqrt{2mE}/\hbar$. In Figure 3.3 a wavefunction for a free electron is indicated. The larger kinetic energy in the box region yields a larger wavevector k and thus a wavefunction oscillating faster in space.

3.2.3 Periodic Boundary Conditions

A practical approach to set boundary conditions would be to have an infinite box potential with a large extension from $x = 0$ to $x = L$ forcing the wavefunction to vanish on the boundaries. This seems to be artificial: because of the vanishing wavefunction it is hard to imagine how electrons get into or out of the sample. There is another way of introducing boundary conditions for free electrons in one dimension that requires the wavefunction to be periodic, i.e., $\psi(0) = \psi(L)$. This allows us to have plane wave solutions for the Schrödinger equation, that are normalized to the length L:

$$\psi(x) = (L)^{-1/2}\exp(ikx) \tag{3.43}$$

Applying periodic boundary conditions the wavevector is restricted to

$$k = \frac{2\pi n}{L} \tag{3.44}$$

Taking $L = N \cdot a$, i.e., in multiples of the atomic spacings, there are N discrete values for k to fulfill the periodic boundary condition.

3.2.4 Potential Barriers and Tunneling

Let us discuss the situation inverse to that given in Figure 3.3, i.e., a potential of the form

$$U(x) = \begin{cases} U_0, & x \in [0, a] \\ 0, & x \notin [0, a] \end{cases} \tag{3.45}$$

(see also Figure 3.6). We insert the potential (3.45) into the Hamiltonian. Again we choose $\psi = \text{const.} \cdot \exp(\lambda x)$ as the trial solution. Thus we obtain

$$\lambda = \pm i k_I = \pm\sqrt{\frac{-2mE}{\hbar^2}}\text{, for } x \notin [0, a] \tag{3.46a}$$

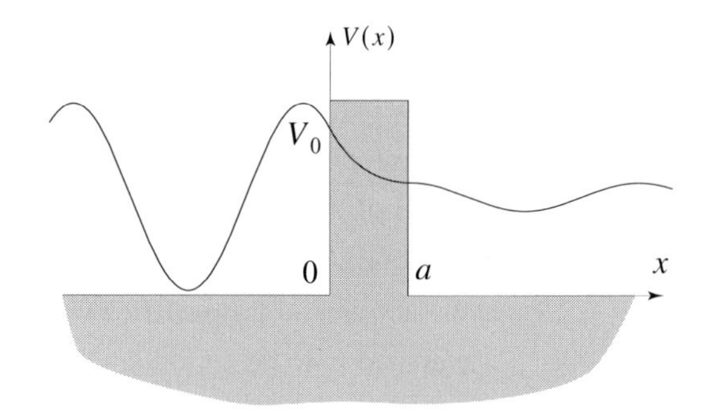

Figure 3.6. Barrier potential and sketched real part of a harmonic wavefunction hitting the barrier with energy $E = V_0/2$, i.e., smaller than V_0.

$$\lambda = \pm k_{II} = \pm\sqrt{\frac{2m(U_0 - E)}{\hbar^2}}, \text{ for } x \in [0, a] \tag{3.46b}$$

There are no bound states in this case because the region where harmonic solutions exist is infinite. The states with energies $E > U_0$ show the inverse behavior as for the box potential (3.25). This is indicated in Figure 3.6. Remember that there exist solutions that decay exponentially inside the barrier and extend to infinity on either side of the barrier for $E < U_0$.

We analyze this situation further. Consider a free electron impinging the barrier from region I to have a definite wavevector k_I. Thus our trial functions read

$$\psi_I(x) = a_1 \exp(ik_I x) + a_2 \exp(-ik_I x) \tag{3.47a}$$

$$\psi_{II}(x) = b_1 \exp(k_{II} x) + b_2 \exp(-k_{II} x) \tag{3.47b}$$

$$\psi_{III}(x) = c_1 \exp(ik_I x) + c_2 \exp(-ik_I x) \tag{3.47c}$$

We assume the impinging wave to have unit amplitude ($a_1 = 1$) and further that there is no wave travelling from region III towards the barrier ($c_2 = 0$). Note that $k_I = k_{III}$. On either side of the barrier the conditions for the continuity of the wavefunction ψ and its spatial derivative

$d\psi/dx$ must be fulfilled. This yields a system of four equations that determine the four remaining coefficients a_2, b_1, b_2, and c_1. The solutions reads

$$a_2 = (\exp(ik_{II}a) - \exp(-ik_{II}a))(k_I^2 - k_{II}^2)D^{-1} \quad (3.48a)$$

$$b_1 = 2\exp(i(k_I - k_{II})a)k_I(k_I + k_{II})D^{-1} \quad (3.48b)$$

$$b_2 = 2\exp(i(k_I + k_{II})a)k_I(k_{II} - k_I)D^{-1} \quad (3.48c)$$

$$c_1 = 4k_I k_{II} D^{-1} \quad (3.48d)$$

with $D = \exp(ik_I a)[\exp(ik_{II}a)(k_I - k_{II})^2 - \exp(-ik_{II}a)(k_I + k_{II})^2]$ .In Figure 3.6 the real parts of the different superposing wavefunctions are sketched for values $E > U_0$ and $E < U_0$. There are several remarkable features to discuss:

Reflection, Transmission and Tunneling

- c_1 is non-zero for $E < U_0$. In this case the particle behaves rather wave-like. This is very similar to a light-wave going through materials with different refractive index. There is always a reflected and a transmitted part, i.e., $a_2 \neq 0$ and $c_1 \neq 0$.
- For $E > U_0$ the situation changes. We find resonances for the transmission at $k_{II}a = n \cdot \pi$ with n an integer. There is no reflection in this case, whereas there is always a transmitted part, i.e., $c_1 \neq 0$.

Reflection and transmission are not uniquely defined by the amplitudes of the partial waves alone. Therefore, we define the reflexivity R as the quotient of reflected and incoming current density; the quotient of transmitted and incoming current density we call the transmitivity T:

$$R = |a_2|^2/|a_1|^2 = |a_2|^2 \quad (3.49a)$$

$$T = |c_1|^2/|a_1|^2 = |c_1|^2 \quad (3.49b)$$

In the case where there is neither creation nor destruction of particles, $R + T = 1$ must hold. In Figure 3.7 T is shown with increasing energy E of the incoming wave. There are well defined resonances where the

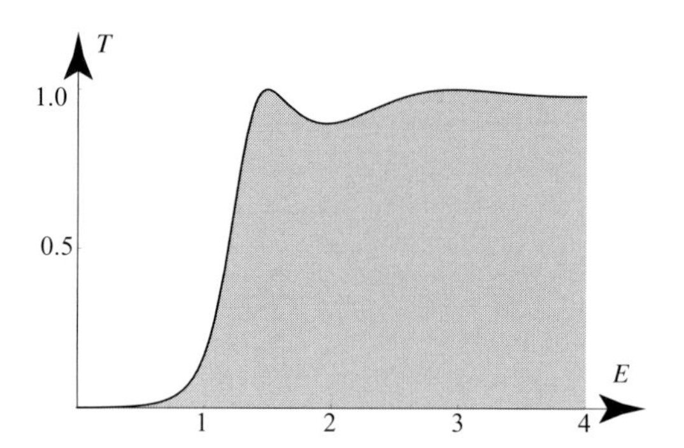

Figure 3.7. Transmission probability T versus E. E is measured in multiples of V_0.

whole incoming wave is transmitted. T also shows pronounced minima for $E \geq V_0$, which one does not expect in a classical behavior.

3.2.5 The Harmonic Oscillator

In Section 2.4.1 the normal modes of vibration for a lattice were introduced. It has been shown that in a first approximation the Hamilton function of the total lattice vibration is a sum of non-interacting harmonic oscillators. The quantization of the lattice vibrations has been presumed in 2.4.1 to explain the discrete nature of phonons. The intercommunity with electrons crops up due to the form of the potential energy.

Harmonic Oscillator Potential

$U = m\omega^2 q^2/2$ is the potential energy of a linear harmonic oscillator in one dimension. The Hamiltonian for a particle moving in such a potential reads

$$\hat{H} = \hat{T} + \hat{U} = \frac{\hat{p}^2}{2m} + \frac{m}{2}\omega^2 q^2 \tag{3.50}$$

The potential energy U goes to infinity as $q \to \pm\infty$. Therefore, the particles' wavefunction must go to zero for $q \to \pm\infty$. The resulting spectrum of the Hamiltonian will be discrete, with energy eigenvalues E_n and the respective eigenfunctions $\Psi_n = \exp(-(i/\hbar)E_n t)\psi_n(q)$. The variable q denotes the amplitude of the oscillator spring or the position of the electron in the harmonic potential. This is why both phenomena may be described in the same way.

We describe the oscillator in the q-representation. This means that the momentum operator is given by $\hat{p} = -i\hbar\, d/dq$, and the commutator is $\hat{p}\hat{q} - \hat{q}\hat{p} = -i\hbar$. The time–independent Schrödinger equation reads

$$\left\{\left(-\frac{\hbar^2}{2m}\right)\frac{d^2}{dq^2} + \frac{m}{2}\omega^2 q^2\right\}\psi(q) = E\psi(q) \tag{3.51}$$

With the coordinate transformation $q = \sqrt{\hbar/(m\omega)}\xi$ (3.51) reads

$$\frac{\hbar\omega}{2}\left\{-\frac{d^2}{d\xi^2} + \xi^2\right\}\psi(\xi) = E\psi(\xi) \tag{3.52}$$

The operator in braces in (3.52) has the form $a^2 - b^2 = (a-b)(a+b)$ and thus may be written as

$$\frac{\hbar\omega}{2}\left\{-\frac{d^2}{d\xi^2} + \xi^2\right\} = \hbar\omega \underbrace{\frac{1}{\sqrt{2}}\left\{-\frac{d}{d\xi} + \xi\right\}}_{b^+} \underbrace{\frac{1}{\sqrt{2}}\left\{\frac{d}{d\xi} + \xi\right\}}_{b} + \frac{\hbar\omega}{2} \tag{3.53}$$

With the definitions of the operators b^+ and b as given by the underbraces in (3.53) and shifting the zero point of the energy according to $E' = E - \hbar\omega/2$, the Schrödinger equation reads

$$\hbar\omega\, b^+ b\, \psi = E'\psi \tag{3.54}$$

This simple form allows us to calculate the wavefunctions very easily. We calculate the commutator of b^+ and b, which gives

$$[b, b^+] = bb^+ - b^+ b = 1 \tag{3.55}$$

Take the ground–state function ψ_o that has the lowest energy eigenvalue E_o'. Let the operator b act on both sides of (3.54)

$$\hbar\omega\; bb^{+}b\;\psi_o = E_o' b\psi_o \tag{3.56}$$

which gives with (3.55)

$$\hbar\omega\;(1 + b^{+}b)\; b\psi_o = E_o' b\psi_o \tag{3.57}$$

or

$$\hbar\omega\; b^{+}b\; b\psi_o = (E_o' - \hbar\omega) b\psi_o \tag{3.58}$$

(3.58) means that there is a ground state $\psi_o' = b\psi_o$ with an even smaller energy than E_o'. This is contradictory to the presupposition that ψ_o be the ground state. Even if the first choice of ψ_o had not been the right ground state, repeated execution of the operation in (3.56) would lead us to an energy eigenvalue $E_o' \to -\infty$. This cannot be, since the potential energy is bound, with its minimum at $U = 0$. On the other hand, (3.56) is a valid equation with a valid operation. Thus the only solution is that $b\psi_o = 0$. Inserting the definition of the operator b given in (3.53) yields

$$\left\{\frac{d}{d\xi} + \xi\right\}\psi_o(\xi) = 0 \tag{3.59}$$

with the solution

Ground–State Oscillator Wavefunction

$$\psi_o(\xi) = \frac{1}{(\pi)^{1/4}} \cdot \exp\left(-\frac{\xi^2}{2}\right) \tag{3.60}$$

$b\psi_o = 0$ implies that the l.h.s. of (3.54) is zero for the ground–state wavefunction and thus we have $E_o' = 0$. This gives us the ground state energy $E_o = (\hbar\omega/2)$. We perform the same operation as in (3.56), now letting b^{+} act on the Schrödinger equation for the ground state, which gives

$$\hbar\omega\; b^{+}b^{+}b\;\psi_o = E_o' b^{+}\psi_o \tag{3.61}$$

Considering the commutator for the operators b^+ and b we obtain

$$\hbar\omega\, b^+ b\, b^+ \psi_o = (E_o' + \hbar\omega) b^+ \psi_o \tag{3.62}$$

Obviously $\psi_1 = b^+\psi_o$ is an eigenvector of the operator $\hbar\omega\, b^+ b$ with eigen-value $E_o' + \hbar\omega$.

Number Operator

Repeating this procedure n times we obtain the n-th eigen-vector ψ_n with eigen-value $E_n' = n\hbar\omega$ (remember $E_o' = 0$). Thus we have $E_n = (n + 1/2)\hbar\omega$. The quantum number n characterizes the oscillator and is the eigenvalue of the quantum number operator $b^+ b$. The normalized wavefunction for an arbitrary excited state with quantum number n reads

Excited State Wave-Function

$$\psi_n = \frac{1}{\sqrt{n!}}(b^+)^n \psi_o = |n\rangle = \frac{1}{\sqrt{n!}}(b^+)^n |0\rangle \tag{3.63}$$

where $|n\rangle$ indicates the Dirac notation of a quantum state with n excited oscillator quanta and $|0\rangle$ describes the oscillator vacuum with no quanta present. Hence we have, from b^+, the number of oscillator quanta incremented by one, and applying on $|n\rangle$ results in

$$b^+|n\rangle = \sqrt{n+1}\,|n+1\rangle \tag{3.64}$$

while applying b yields

$$b|n\rangle = \sqrt{n}\,|n-1\rangle \tag{3.65}$$

Bosons

Note that the oscillator quantum number can be arbitrarily large for a given state. This is the case for particles that follow Bose statistics (see Chapter 5) and therefore these particles are called bosons.

Minimum Uncertainty

The quantum mechanical harmonic oscillator states apply for the description of various different phenomena and not only in semiconductors. In addition, it has a very important property because the ground–state wavefunction fulfills the equality in (3.13), i.e., the wavefunction is a quantum state of minimum uncertainty. This is easily shown by calculating the variances as (3.9) and (3.10). We recall the transformation $q = \sqrt{\hbar/(m\omega)}\xi$, thus we have the momentum operator given by $\hat{p} = -i\hbar\, d/dq = \sqrt{(m\omega)/\hbar}\, d/d\xi$. Since the potential is symmetric about $q = 0$ we have $\langle\hat{q}\rangle = 0$ and $\langle\hat{p}\rangle = 0$. We calculate $\langle\hat{q}^2\rangle$ and $\langle\hat{p}^2\rangle$:

$$\langle\hat{q}^2\rangle = \frac{\hbar}{m\omega}\frac{1}{(\pi)^{1/2}}\int_{\infty}^{\infty}\exp\left(-\frac{\xi^2}{2}\right)\xi^2\exp\left(-\frac{\xi^2}{2}\right)d\xi = \frac{1}{2}\frac{\hbar}{m\omega} \tag{3.66a}$$

$$\langle\hat{p}^2\rangle = -\hbar m\omega\frac{1}{(\pi)^{1/2}}\int_{\infty}^{\infty}\exp\left(-\frac{\xi^2}{2}\right)\frac{d^2}{d\xi^2}\exp\left(-\frac{\xi^2}{2}\right)d\xi = \frac{\hbar m\omega}{2} \tag{3.66b}$$

and thus their product yields

$$\langle(\Delta\hat{x})^2\rangle\langle(\Delta\hat{p})^2\rangle = \langle\hat{p}^2\rangle\langle\hat{q}^2\rangle = \frac{\hbar^2}{4} \tag{3.67}$$

A numerical treatment of the harmonic oscillator problem together with the plots of eigenfunctions is given in Figure 3.8.

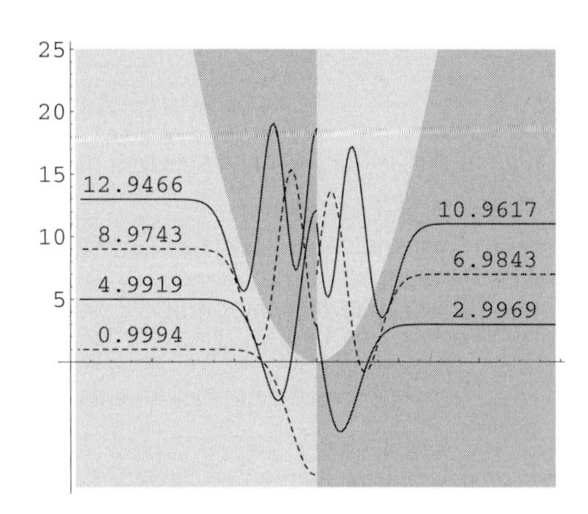

Figure 3.8. A harmonic potential is indicated by the gray shaded background. The black lines to the left show the symmetric eigenmodes of this potential, aligned with their respective energy eigenvalues. The black lines to the right are the anti-symmetric eigenmodes, similarly aligned.

3.2.6 The Hydrogen Atom

The paradigm for the electronic states of a central symmetric potential is the hydrogen atom since it consists of one proton in the nucleus surrounded by one electron. It helps in understanding the principle of electronic states of atoms which is similar to the electronic states of defects and traps for electrons in the semiconductor or even helps to understand the formation of a band structure. We shall not go into details of the derivation but we shall have a closer look at the form of the electronic states.

Coulomb Potential

The Coulomb potential of the proton is spherically symmetric in three dimensions, which means that it depends only on the distance r of the symmetry center $U(\boldsymbol{r}) = -e^2/r$ (e is the unit charge). The best choice to tackle the problem is spherical coordinates, with the radius vector **r** given by its modulus r, the polar angle ϕ and the azimutal angle θ. The Hamilton operator thus reads

$$H = -\frac{\hbar^2}{2\mu}\nabla^2 - \frac{e^2}{r} \tag{3.68}$$

To solve the Schrödinger equation we must write the ∇^2 in spherical coordinates, which reads

$$\nabla^2 = \frac{1}{r^2}\frac{\partial}{\partial r}\left(r^2\frac{\partial}{\partial r}\right) + \frac{1}{r^2}\left(\frac{1}{\sin\theta}\frac{\partial}{\partial\theta}\sin\theta\frac{\partial}{\partial\theta} + \frac{1}{\sin^2\theta}\frac{\partial^2}{\partial\phi^2}\right) \tag{3.69}$$

Inserting (3.69) together with (3.68) and solving the respective Schrödinger equation is beyond the scope of this book. We restrict the discussion to an interpretation of the electronic wavefunction and an illustration of the atomic orbitals.

The common approach to calculate the bound electronic states of the hydrogen atom (energy $E < 0$) starts with a product trial function for the wavefunction $\psi(\boldsymbol{r}) = R(r)Y(\theta, \phi)$. Then the Schrödinger equation separates into two eigenvalue problems, one for the radial motion ($R(r)$) and one for the angular motion ($Y(\theta, \phi)$). The solutions of the angular part are the spherical harmonic functions $Y_l^m(\theta, \phi)$ (see Table 2.2). The radial functions are $R(r) = R_{n,l}(r)/r$, where the $R_{n,l}(r)$ follow a differential equation of Laguerre type, and are shown in Figure 3.9• for dif-

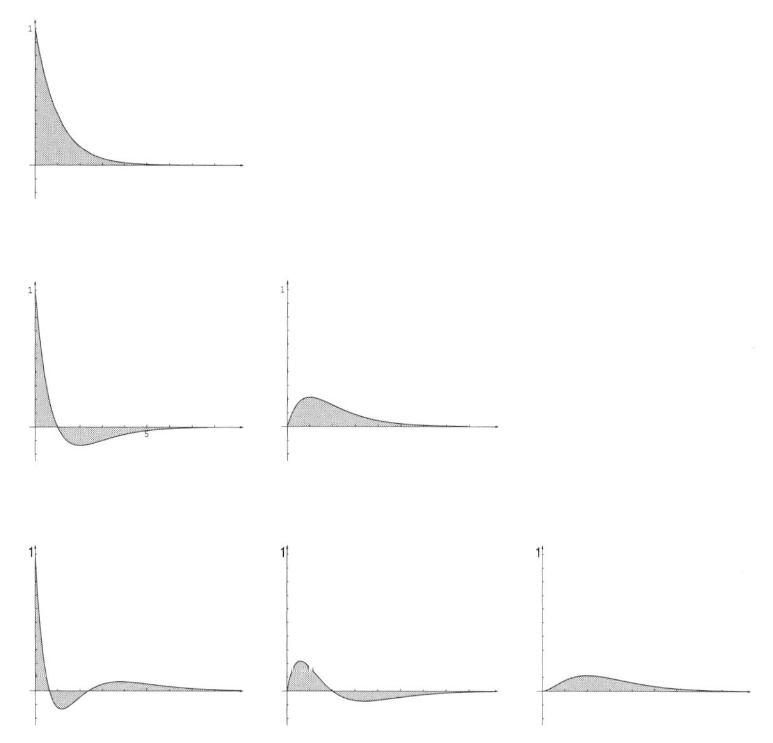

Figure 3.9. The radial wavefunction $R_{n,l}(r)$ for different n and l.

ferent (n, l). The wavefunction $\psi_{n,l,m}(r, \theta, \phi)$ (see Table 3.1) has three integer parameters:

- n : the quantum number of radial motion,
- $l = 0, 1, \ldots, n-1$: the first quantum number of angular motion,
- $m = -1, \ldots, 1$: the second quantum number angular motion (note that m is an integer, not to be confused with the mass of the electron).

Table 3.1. The wavefunction $\psi_{n,l,m}(r, \theta, \phi)$.

n	l	m	$\psi_{n,l,m}(r, \theta, \phi)$			
1	0	0	$\frac{\gamma_2^{3/2}}{\sqrt{\pi}} \cdot$		$\exp(-r)$	
2	0	0	$\frac{\gamma_2^{3/2}}{\sqrt{\pi}} \cdot$	$(1 - r/2) \cdot$	$\exp(-r/2)$	
2	1	0	$\frac{\gamma_2^{3/2}}{\sqrt{\pi}} \cdot$	$r \cdot$	$\exp(-r/2)$	$\cos\theta$
2	1	±1	$\frac{\gamma_2^{3/2}}{\sqrt{2\pi}}$	r	$\exp(-r/2)$	$\sin\theta e^{\pm i\phi}$
3	0	0	$\frac{\gamma_2^{3/2}}{3\sqrt{\pi}}$	$(2 - 2r + 2r^2)$	$\exp(-r/3)$	
3	1	0	$\frac{\sqrt{2}\gamma_2^{3/2}}{\sqrt{3\pi}}$	$(2 - r/3)r$	$\exp(-r/3)$	$\cos\theta$
3	1	±1	$\frac{\gamma_2^{3/2}}{\sqrt{3\pi}}$	$(2 - r/3)r$	$\exp(-r/3)$	$\sin\theta e^{\pm i\phi}$
3	2	0	$\frac{\gamma_2^{3/2}}{3\sqrt{2\pi}}$	r^2	$\exp(-r/3)$	$(3\cos^2\theta - 1)$
3	2	±1	$\frac{\gamma_2^{3/2}}{\sqrt{3\pi}}$	r^2	$\exp(-r/3)$	$\cos\theta \sin\theta e^{\pm i\phi}$
3	2	±2	$\frac{\gamma_2^{3/2}}{2\sqrt{3\pi}}$	r^2	$\exp(-r/3)$	$\sin^2\theta e^{\pm 2i\phi}$

Degenerate Spectrum

Each quantum state represents an atomic energy level $E_{n,l,m}$ that is typical for the hydrogen atom, i.e., typical for the potential energy given in the Hamiltonian (3.68). There are nevertheless sets of quantum numbers, i.e., well distinguishable wavefunctions, that have the same energy eigenvalue. This multiple eigenvalue phenomenon is well known from linear algebra, and is called degeneracy. All energies of the hydrogen spectrum with equal quantum number n are degenerate. Note that the Pauli principle is still valid, because the wavefunctions are different. This degeneracy can be lifted by adding a potential that acts on the respective wavefunctions differently for different quantum numbers. In fact, nature provides us with such additional interaction potentials and an exact measurement of the spectrum a hydrogen atom does not show theses degeneracies. The conclusion is that our model is too simple. Nevertheless, it explains the principles that the electronic system of atoms follows. One major incompleteness is that for many-electron systems the Coulomb interaction between the electrons must be taken into account. This makes the Schrödinger equation highly non-linear and thus other techniques including numerics must be used. Moreover, if we want to understand the formation of crystal symmetry by just putting atoms together, this interaction is responsible for the details in the electronic band structure.

Scattering

So far we have only dealt with the bound states of the atom and an electron occupying one of them. There is no dynamics in this picture, because there is no process represented by an interacting potential that might change this static situation. Suppose a free electron collides with the atom, then there are several possibilities:

- the electron gets *trapped* and subsequently occupies a quantum state of the atom, it emits a photon carrying away the energy difference between the free state and the bound state;
- it is scattered and moves on with a different wave-vector, there is a momentum transfer and an energy transfer between the electron and the atom;
- the transferred energy excites a bound electron already occupying a quantum state, the free electron moves on with a lower kinetic energy, the bound electron either occupies an excited state higher in energy, or even leaves behind an ionized atom moving freely;

- the excited electron bounces back to its ground state, emitting a photon.

All those processes require the transitions of electrons between different electronic states. The scenario described above is what happens in a gas discharge lamp. The analogous process can be found in the conduction band of a semiconductor where an electron is hitting a defect atom that has a spectrum of localized electronic states below the conduction bandedge. The energy scale is of course one order of magnitude lower and there are additional degrees of freedom to which the excess energy of the electron may be transferred, e.g., phonons.

The quantum mechanical description of the electron as given above focused on the stationary states of the electrons. Now the time dependence of the Schrödinger equation has must be exploited.

3.2.7 Transitions Between Electronic States

The superposition of the time–dependent solutions of the Schrödinger equation (3.16) gives the general form of the wavefunction with arbitrary initial conditions

$$\Psi^0(\boldsymbol{x}, t) = \sum_n c_n \exp(-i\omega_n t)\psi_n(\boldsymbol{x}) \tag{3.70}$$

where $\omega_n = E_n/\hbar$ and the $\psi_n(\boldsymbol{x})$ in (3.70) are the solution of the stationary eigen-value problem

$$\hat{H}_0\psi_n = E_n\psi_n \tag{3.71}$$

The coefficients c_n are determined by the initial condition, i.e., $c_n = \langle\psi_n(\boldsymbol{x})|\Psi^0(\boldsymbol{x}, 0)\rangle$. The superscript 0 of the wavefunction and the subscript 0 of the Hamiltonian indicate the unperturbed stationary problem.

Time-Dependent Perturbation

Now we add a *small* time–dependent perturbation potential to the Hamiltonian. Think of an electromagnetic wave hitting the hydrogen atom, or an additional electrostatic potential switched on instantaneously. Before the electromagnetic wave arrives or the potential is switched on, the stationary description defines the state of the system very well. Thus it is a good approximation to write the perturbed solution as a superposition of eigen-function belonging to the stationary unperturbed problem with time-dependent coefficients

$$\Psi(\boldsymbol{x}, t) = \sum_n c_n(t)\exp(-i\omega_n t)\psi_n(\boldsymbol{x}) \tag{3.72}$$

To determine the $c_n(t)$ we insert (3.72) in the Schrödinger equation including the perturbation in the Hamiltonian

$$i\hbar\frac{\partial}{\partial t}\Psi(\boldsymbol{x}, t) = (\hat{H}_0 + \hat{V})\Psi(\boldsymbol{x}, t) \tag{3.73}$$

which gives

$$\underbrace{\sum_n c_n(t)E_n\exp(-i\omega_n t)\psi_n}\ + i\hbar\sum_n\frac{dc_n(t)}{dt}\exp(-i\omega_n t)\psi_n = \underbrace{\sum_n \hat{H}_0 c_n(t)\exp(-i\omega_n t)\psi_n}\ + \sum_n \hat{V}c_n(t)\exp(-i\omega_n t)\psi_n \tag{3.74}$$

The first term on the l.h.s. and the first term on the r.h.s. of (3.74) cancel because of (3.71). Multiplying (3.74) by ψ_m and integrating over all space give

$$i\hbar\frac{dc_m(t)}{dt}\exp(-i\omega_m t) = \sum_n V_{mn}c_n(t)\exp(-i\omega_n t), \tag{3.75}$$

where $V_{mn} = \langle\psi_m|\hat{V}\psi_n\rangle$ and $\omega_{mn} = \omega_m - \omega_n = (E_m - E_n)/\hbar$ (note that $\langle\psi_m|\psi_n\rangle = \delta_{mn}$). We obtain a set of differential equations for the c_m:

$$i\hbar\frac{dc_m(t)}{dt} = \sum_n V_{mn}c_n(t)\exp(-i\omega_{mn}t) \tag{3.76}$$

For the following we assume that the perturbation is switched on instantaneously at $t = 0$ and is subsequently constant. Then the c_m are given at $t = 0$ by the initial condition. We choose $c_k(0) = 1$ and $c_n(0) = 0$ for $n \neq k$. For very short times after $t = 0$ the situation has not changed very much so that we insert the initial condition in (3.76):

$$i\hbar\frac{dc_m(t)}{dt} = V_{mk}c_k(0)\exp(-i\omega_{mk}t) = V_{mk}\exp(-i\omega_{mk}t) \tag{3.77}$$

(3.77) can be easily integrated to give

$$\begin{aligned} c_m(t) &= -\frac{i}{\hbar}\int_0^t V_{mk}\exp(-i\omega_{mk}t')dt' \\ &= -\frac{1}{\hbar\omega_{mk}}V_{mk}(\exp(-i\omega_{mk}t) - 1) \end{aligned} \tag{3.78}$$

The change in the probability of finding the system in a state m is given by $|c_m(t)|^2$. This is the squared modulus of (3.78), which reads

$$|c_m(t)|^2 = \frac{4}{\hbar^2}|V_{mk}|^2\frac{\sin(\omega_{mk}t/2)}{\omega_{mk}^2} \tag{3.79}$$

Transition Rate

(3.79) is the probability density of finding the system at time t in ψ_m if it has been found in ψ_k at time $t = 0$. The transition rate from k to m is given by

$$\Gamma_{mk} = \frac{|c_m(t)|^2}{t} \tag{3.80}$$

In the long time limit

$$\lim_{t \to \infty}\frac{\sin(\omega_{mk}t/2)}{\omega_{mk}^2} \to \frac{\pi t}{2}\delta(\omega_{mk}) \tag{3.81}$$

When the impact of the switched on interaction is over we obtain

$$\Gamma_{mk} = \frac{2\pi}{\hbar^2}|V_{mk}|^2\delta(\omega_{mk}) \tag{3.82}$$

(3.82) is a very important result, called the *Fermi golden rule*. The transition rate from state k to m is determined by the squared matrix element of the V_{mk} of the perturbing potential, where the delta-function $\delta(\omega_{mk})$ ensures that the energy is conserved.

The approach of constant perturbation in this section can be extended to potentials which vary arbitrarily in time. We will not go into detail but rather explain the result. Let $V(\omega, t) = V^0 \exp(\pm\omega t)$ be a perturbation varying with a fixed frequency ω and an amplitude V^0. Then the transition rate becomes

$$\Gamma_{mk} = \frac{2\pi}{\hbar^2}|V_{mk}|^2\delta(\omega_{mk}\pm\omega) \tag{3.83}$$

The difference between (3.83) and (3.82) is that the energies E_m and E_k of the initial and final states differ now by $\hbar\omega$, which is the energy of the special perturbation mode with frequency ω. We already know that electrons couple to periodically oscillating phenomena such as phonons and electromagnetic waves. Hence, (3.83) will be the basis for describing the scattering events that electrons in a conduction band of a semiconductor experience.

3.2.8 Fermion number operators and number states

In the same manner as for the harmonic oscillator we represent the quantum state of a fermion by means of a creation operator c^+ and a destruction operator c using the Dirac notation. Defining a fermion vacuum $|0\rangle$ we have

$$c_\lambda^+|0\rangle = |\lambda\rangle \tag{3.84}$$

for an electron placed in specific state λ. For the reverse operation we have

$$c_\lambda|\lambda\rangle = |0\rangle \tag{3.85}$$

where an electron has been moved out of the state λ. We must take into account that only one fermion may occupy a quantum state that is uniquely defined by its quantum numbers λ. Therefore, trying to put another electron in λ must yield a zero eigenvalue

$$(c_\lambda^+)^2|\lambda\rangle = 0 \tag{3.86}$$

and the same must hold for twice trying to remove an electron from λ

$$(c_\lambda)^2|\lambda\rangle = 0 \tag{3.87}$$

The action of the operator $c_\lambda^+ c_\lambda + c_\lambda c_\lambda^+$ on an arbitrary state $|\beta\rangle$ yields

$$(c_\lambda^+ c_\lambda + c_\lambda c_\lambda^+)|\beta\rangle = \left(\begin{array}{l}(1+0) \text{ for } \beta=\lambda \\ (0+1) \text{ for } \beta \neq \lambda\end{array}\right. = 1|\beta\rangle \tag{3.88}$$

which means that, regardless of whether the state λ is occupied or unoccupied, the resulting eigenvalue is 1. This leads us to the definition of the anti-commutator product

Anti-commutator

$$\{c_\lambda^+, c_\lambda\} = c_\lambda^+ c_\lambda + c_\lambda c_\lambda^+ = 1 \tag{3.89}$$

We use curly brackets to indicate the anti-commutation relation. Note the difference in the sign when compared to the commutator of oscillator creation and destruction operators (3.55). $c_\lambda^+ c_\lambda = n$ is still the correct number operator. In the case of fermions, the number operator's eigenvalues are either 1 or 0 and indicate whether a particle in the respective state is present or not.

3.3 Periodic Potentials in Crystal

Lattice periodic structures create a very special situation for the electrons. Thus the properties of electrons in such kinds of potentials are different from those of a free electron [3.3]–[3.5].

3.3.1 The Bloch Functions

Symmetry

A perfect crystal has a given periodicity. The special property of the electronic potential V is that it has exactly the same periodicity, i.e.

$$V(\boldsymbol{x}+\boldsymbol{l}) = V(\boldsymbol{x}) \tag{3.90}$$

where **l** is a arbitrary lattice vector as discussed in Chapter 2. We write (3.90) as $T(\boldsymbol{l})V(\boldsymbol{x}) = V(\boldsymbol{x}+\boldsymbol{l}) = V(\boldsymbol{x})$, where we used the translation operator $T(\boldsymbol{l})$, that transforms a function to its value at the place shifted by the lattice vector **l** from the input. Operating on the product $\hat{H}(\boldsymbol{x})\Psi(\boldsymbol{x})$ it is shifted by **l**. Since the Hamiltonian is invariant under translation, we have

$$T(\boldsymbol{l})\hat{H}(\boldsymbol{x})\Psi(\boldsymbol{x}) = \hat{H}(\boldsymbol{x})\Psi(\boldsymbol{x}+\boldsymbol{l}) = \hat{H}(\boldsymbol{x})T(\boldsymbol{l})\Psi(\boldsymbol{x}) \tag{3.91}$$

which means $T(\boldsymbol{l})\hat{H}(\boldsymbol{x}) - \hat{H}(\boldsymbol{x})T(\boldsymbol{l}) = 0$, i.e., T and $\hat{H}$ commute. This implies that $\Psi(\boldsymbol{x})$ is also an eigenfunction of T, which means $T\Psi(\boldsymbol{x}) = \lambda\Psi(\boldsymbol{x})$. Applying the translation operator n times yields $T^n\Psi(\boldsymbol{x}) = \lambda^n\Psi(\boldsymbol{x})$. Since the wavefunction must be bound $|\lambda| \le 1$. Suppose $\lambda < 1$ and replace **l** by –**l**, than again the wavefunction would not be bound in the inverse direction, thus only $|\lambda| = 1$ remains as a solution. Let us write $\lambda = \exp(i\boldsymbol{kl})$ and therefore $\Psi(\boldsymbol{x}+\boldsymbol{l}) = \exp(i\boldsymbol{kl})\Psi(\boldsymbol{x})$. To fulfil this we write the wavefunction as

$$\Psi(\boldsymbol{x}) = \exp(i\boldsymbol{kx})u_k(\boldsymbol{x}) \tag{3.92}$$

where $u_k(\boldsymbol{x})$ is a lattice periodic function

$$u_k(\boldsymbol{x}+\boldsymbol{l}) = u_k(\boldsymbol{x}) \tag{3.93}$$

Equation (3.92) together with (3.93) is called the Bloch theorem. Thus electronic wavefunctions in a perfect crystal lattice are plane waves modulated by a lattice periodic function. We immediately recognize that the problem left is to determine the form of $u_k(\boldsymbol{x})$. Therefore, we insert (3.92) into the stationary Schrödinger equation, which gives

$$\left[\frac{\hbar^2}{2m}(\boldsymbol{k}^2 - 2i\boldsymbol{k}\nabla - \nabla^2) + V(\boldsymbol{x})\right]u_k(\boldsymbol{x}) = E_{k,\,j}u_k(\boldsymbol{x}) \tag{3.94}$$

We see that $u_k(\boldsymbol{x})$ and $E_{k,\,j}$ depend on the plane wave-vector **k**. In addition, (3.94) allows for given **k** a series of eigen-values accounted for by the index j.

3.3.2 Formation of Band Structure

To construct a simple model semiconductor we take a one-dimensional potential composed of barriers repeated periodically with the periodicity length $L = a + b$, e.g.

$$V(x) = \begin{cases} V_0, & x \in [-b, 0] \\ 0, & x \in [0, a] \end{cases} \tag{3.95}$$

as sketched in Figure 3.10. We insert (3.95) in (3.94) and solve the one-

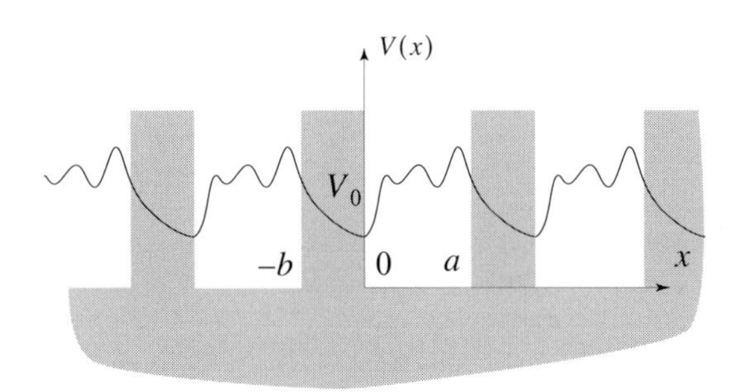

Figure 3.10. Bloch waves in a periodic crystal lattice potential.

dimensional eigenvalue problem with the ansatz

$$u(x) = \begin{cases} (Ae^{-i(k-\kappa)x} + Be^{-i(k+\kappa)x}), \text{ for } 0<x<a \\ (Ce^{-(ik-\lambda)x} + De^{-(ik+\lambda)x}), \text{ for } -b<x<0 \end{cases} \tag{3.96}$$

where $\kappa = \sqrt{2mE}/\hbar$ and $\lambda = \sqrt{2m(V_0 - E)}/\hbar$. To solve for the coefficients A, B, C and D in (3.96) we apply the boundary. The wavefunctions and their derivatives must be equal at $x = 0$ ($u_1(0) = u_2(0)$ and $u'_1(0) = u'_2(0)$), and the wavefunctions and their derivatives at $x = -b$ must equal those at $x = a$ ($u_1(a) = u_2(-b)$ and $u'_1(a) = u'_2(-b)$). This gives us a homogeneous system of four linear equations, which has only non-trivial solutions if its determinant is zero, which yields

$$\frac{\lambda^2 - \kappa^2}{2\kappa\lambda}\sinh(\lambda b)\sin(\kappa a) + \cosh(\lambda b)\cos(\kappa a) = \cos(kL) \tag{3.97}$$

We see that already for the simple periodic potential barrier model (3.97) cannot be solved analytically. A graphical solution is given in Figure 3.11.

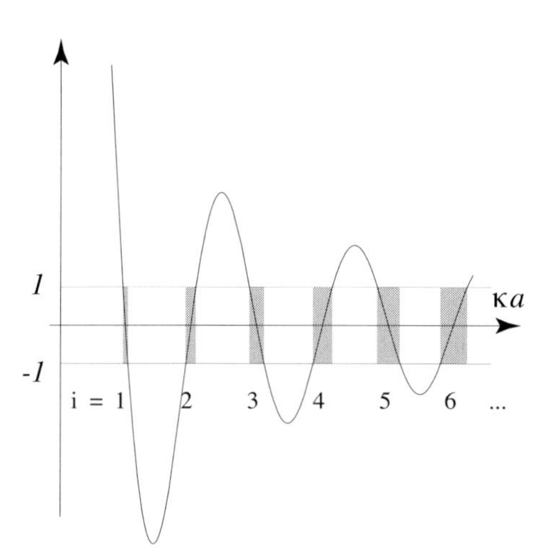

Figure 3.11. Graphical solution for equation (3.97) that shows the emergence of allowable energy bands.

Compare this result with the case of a single potential box given in Section 3.2.1. The major difference is that in the periodic case there are regions of κa where a whole continuum of solutions is possible, whereas in the single well case we have only discrete values.

Moving Atoms Together

The transition between the two situations can be easily seen by making b in (3.97) very large. In this way we recover (3.32) from (3.97). This procedure corresponds to moving atoms away from each other. Another effect is given by adding to a single well potential a second identical potential box at a distance b, then a third one and so on. The case of two potential boxes at distance b gives rise to the question of what happens with the lowest energy levels that each box contributes to the total system. This is shown in Figure 3.12. The levels split in energy symmetri-

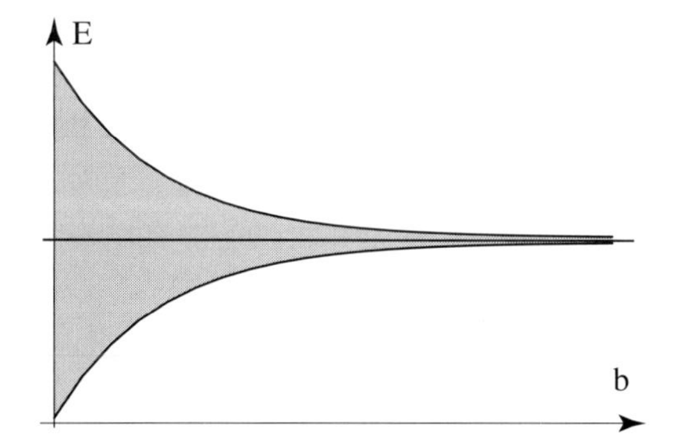

Figure 3.12. For two potential boxes of distance b apart, the energy levels tend to split with an amount that increases with decreasing b.

cally around the solution for $b \to \infty$. Imagine that we fill one electron in this system and it will occupy the lower of the two levels, then the total energy is less than for an electron sitting in one box with the other at an infinite distance. Since systems tend to occupy the lowest possible energy, this means that the boxes experience binding, i.e., attraction due to quantum mechanical properties of the system. If we think of the boxes as two atoms, we can observe the formation of molecules. This is not the whole truth because positively charged atomic bodies will repel each other. Hence there is binding only if the total energy of electrostatic repulsion and quantum mechanical attraction has a minimum for a finite b.

3.3.3 Types of Band Structures

Valence and Conduction Bands

The graphical solution of (3.97) as given in Figure 3.11 provides us with the possible electronic states that can be occupied by electrons. In a semiconductor the Fermi energy lies in the gap between two bands, i.e., there is an uppermost completely filled band called the valence band. The following unoccupied band is called the conduction band. As we shall see in Section 5.4, this band contains a certain number of electrons depending on the temperature of the electronic system.

Si, Ge, GaAs

The wave-vector k entering the r.h.s. of (3.97) so far was taken in its limiting cases $k = 0$ and $k = \pi/L$, resulting in the upper and lower bounds of the bands by the fact that $\cos 0 = 1$ and $\cos \pi = -1$. We know ask for the structure of the bands in between those two limits for arbitrary $k \in [0, \pi/L]$. From Section 2.2 we know that this corresponds to the center and the edge of the Brillouin zone. The crystal structure is not isotropic and therefore we shall have different band structures in the respective crystal directions. This leaves us with a function $E_i(\boldsymbol{k})$, where i is the band index in Figure 3.11 and $\boldsymbol{k}$ is now a real three-dimensional vector. In Figure 3.13 the typical band structure for cubic crystals is shown in different crystallographic directions. There are several things to observe in Figure 3.13:

- after passing the edge of the Brillouin zone they periodically repeat;
- the valence band, i.e., the uppermost occupied band is separated by an energy gap E_g from the conduction band (the valence bands are highlighted);
- in general, the minimum energy of the conduction band is not necessarily found at the same k-value as the maximum of the valence band (points B or D are lower in energy than point C). This is called an indirect band gap as is the case for Si and Ge. GaAs shows a direct

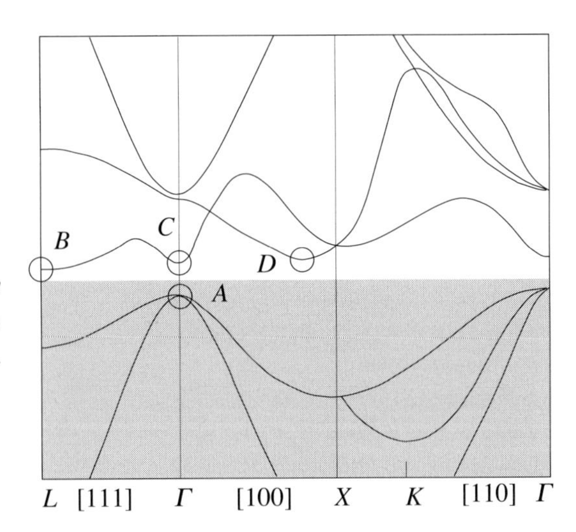

Figure 3.13. Band structures of cubic crystals. The circles denote important extremes of the bands. A is the valence band energy at the center of the Brillouin zone [000]. B is the conduction band energy at the Brillouin zone edge in the [111] crystal direction. C is the conduction band energy at the center of the Brillouin zone [000]. D is the conduction band minimum in the [100] crystal direction. The energy values of these points determine if we have a direct or indirect band gap.

band gap, i.e., the minimum of conduction band energy is found at the same k value as the maximum of the valence band energy;

- there are multiple valence bands found, the so called light-hole band and heavy-hole band.

Real Band Structures

Figure 3.13 is a schematic diagram of how the band structures appear. In fact, nobody can tell either by experiment or calculation exactly what the upper edge of the conduction band looks like. The calculations need assumptions that might not be enough to determine the band structure in the whole Brillouin zone. Most of the calculations have to be corrected with the experimental data available. A standard technique in calculating band structures is the Linear Combination of Atomic Orbitals (LCAO), also called the tight binding method, see Box 3.1. This method was sketched in Section 3.3.2 using very simple atomic orbitals. In principle, the atomic structure must be known, as discussed in its basic principles for the hydrogen atom in Section 3.2.6. The LCAO method is the starting point for more sophisticated methods used today. Nevertheless, there is always a correction of input parameters by comparison with experiment.

Box 3.1. The tight binding LCAO method [3.6]–[3.8].

Tight Binding. If we make the bold assumption that the wavefunctions of crystal atoms are only slightly perturbed from their free state, and we severely limit the interaction between atoms to those of nearest neighbors only, and we only consider the most essential of the "free" atom's orbitals, then a particularly straightforward calculation of the crystalline band structure becomes possible that correctly predicts the bands of the tightly-bound valence electrons.

Hamiltonian. The crystal electron's Hamiltonian is written as $H = p^2/2m + \sum_k V_k(\boldsymbol{r})$. We assume that the basis states are the $3s$ and $3p$ states. Since each of these four states (s, p_x, p_y, p_z) can occur for each of the two Si sites in the unit cell, the Hamiltonian will be an 8×8 matrix.

Basis states. The crystal basis states are

$$\phi_l^{(m)} = \frac{1}{\sqrt{N}} \sum_j e^{i(\boldsymbol{k} \cdot \boldsymbol{R}_j^{(m)})} \psi_l(\boldsymbol{r} - \boldsymbol{R}_j^{(m)}) \quad \text{(B 3.1.1)}$$

for each of the two types of atoms, i.e., $m = 1, 2$.

Hamiltonian Matrix. For the diamond structure only four parameters are unique when we evaluate the matrix elements by $H_{ml} = \langle \phi_m | H | \phi_m \rangle$, i.e.

$$\left.\begin{matrix} V_{ss} & V_{sx} \\ V_{xx} & V_{xy} \end{matrix}\right\} \quad \text{(B 3.1.2)}$$

We also obtain the phase parameters

$$\left.\begin{aligned} g_1 &= \frac{1}{4}(e^{id_1 \cdot k} + e^{id_2 \cdot k} + e^{id_3 \cdot k} + e^{id_4 \cdot k}) \\ g_2 &= \frac{1}{4}(e^{id_1 \cdot k} + e^{id_2 \cdot k} - e^{id_3 \cdot k} - e^{id_4 \cdot k}) \\ g_3 &= \frac{1}{4}(e^{id_1 \cdot k} - e^{id_2 \cdot k} + e^{id_3 \cdot k} - e^{id_4 \cdot k}) \\ g_4 &= \frac{1}{4}(e^{id_1 \cdot k} - e^{id_2 \cdot k} - e^{id_3 \cdot k} + e^{id_4 \cdot k}) \end{aligned}\right. \quad \text{(B 3.1.3)}$$

and the factors

$$\left.\begin{matrix} G_1 = V_{ss} g_1 & G_2 = V_{sx} g_2 \\ G_3 = V_{sx} g_3 & G_4 = V_{sx} g_4 \\ G_5 = V_{xx} g_1 & G_6 = V_{xy} g_2 \\ G_7 = V_{xy} g_3 & G_8 = V_{xy} g_4 \end{matrix}\right\} \quad \text{(B 3.1.4)}$$

The Hamiltonian matrix now becomes

$$H = \begin{bmatrix} E_s & G_1 & 0 & 0 & 0 & G_2 & G_3 & G_4 \\ G_1{}^* & E_s & -G_2 & -G_3 & -G_4 & 0 & 0 & 0 \\ 0 & -G_2{}^* & E_p & 0 & 0 & G_5 & G_8 & G_6 \\ 0 & -G_3{}^* & 0 & E_p & 0 & G_8 & G_5 & G_7 \\ 0 & -G_4{}^* & 0 & 0 & E_p & G_6 & G_7 & G_5 \\ G_2{}^* & 0 & G_5{}^* & G_8{}^* & G_6{}^* & E_p & 0 & 0 \\ G_3{}^* & 0 & G_8{}^* & G_5{}^* & G_7{}^* & 0 & E_p & 0 \\ G_4{}^* & 0 & G_6{}^* & G_7{}^* & G_5{}^* & 0 & 0 & E_p \end{bmatrix} \quad \text{(B 3.1.5)}$$

Computed Band Structure. We can now evaluate the Hamiltonian matrix using numerical values for the parameters (see [3.6]). If the eigenspectrum is plotted along the lines of high symmetry of the Brillouin zone we easily obtain the familiar band structure shown in Figure B3.1.1. This

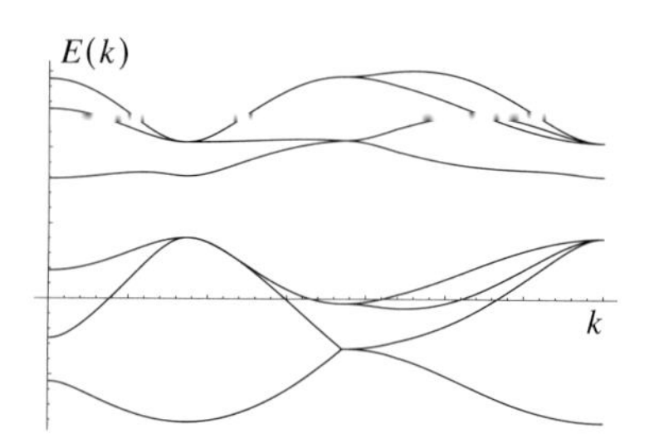

Figure B3.1.1. The LCAO approximation for the band structure of silicon. Note that the conduction bands are quite wrong, for the figure predicts an almost direct band gap.

should be compared to the actual band structure of silicon shown in Figure 3.13.

The main features of a band structure as discussed in Figure 3.13 are experimentally well known and are therefore the corner pillars that any calculation must reproduce.

3.3.4 Effective Mass Approximation

Tensor

A closer look to the band extremes indicates that electrons close to those points are well described with a quadratic energy dispersion as for the free electron if we approximate the local behavior around these points by a parabolic band structure. The center of the Brillouin zone in the conduction band as shown in Figure 3.13 shows this quadratic dependence of the energy E on the wave vector $\boldsymbol{k}$. The energy of a free electron is given as $E = \hbar^2 k^2/(2m)$. Hence we can approximate the electronic behavior at point C in Figure 3.13 by taking the second derivative of the energy with respect to $\boldsymbol{k}$

$$\frac{\partial^2}{\partial k_i \partial k_j} E(\boldsymbol{k}) = \hbar^2 \frac{1}{m^*_{ij}} \tag{3.98}$$

(3.98) allows us to use the dispersion relation $E(\boldsymbol{k}) = (\hbar^2 k_i k_j)/2m^*_{ij}$ near the conduction band minimum. The mass calculated by (3.98) in general is not equal to the free electron mass. Moreover, it is a tensor–like quantity that depends on the direction in k-space where the respective derivative is taken. This is why it is called the effective mass. Therefore, in this sense the electron is a quasi–particle, as will be discussed in Section 5.2.4, that only behaves like a free electron but with a changed mass. The importance of this approximation will become clear with the fact that most electrons are to be found at the band edge minimum at $T = 300K$. Therefore, it will be the effective mass that enters all relations of electronic transport.

Anisotropy

The definition given in (3.98) immediately suggests that there must be some remainders from the crystal anisotropic structure inside the electronic mass. Indeed, the masses in different semiconductors depend strongly on the crystal direction or the direction in the reciprocal lattice. In silicon, e.g., the conduction band minimum is at point D (see Figure 3.13). This leaves us with six equivalent valleys in the [100] crystal direction as we know from Section 2.2.1. In this direction the electronic effective mass is $m_L = 0.91m_0$ while in the perpendicular two directions we obtain $m_T = 0.19m_0$, with the free electron mass $m_0 = 9.1 \times 10^{-27}kg$. This will be of importance when it comes to applying an electric field to induce a current, then the crystal direction determines the conductivity (see Chapter 6).

3.4 Summary for Chapter 3

The chapter started with the free electron, which introduced wavefunctions and the Schrödinger equation. We saw that physical observations are the interpretation of a resultant wavefunction. This wavefunction has undergone a variety of wavelike interferences on its way through space. Next, the electron was bound to the potential of the hydrogen nucleus. This made visible the bonding orbitals that we met in Chapter 2, the atomic potential, and the electronic levels of an atom. Placing many atoms in a regular crystal lattice, we obtained a periodic potential. Even with a simple box potential as a primitive model for the atomic potential, we already saw the emergence of a band structure. This model also produced a forbidden band, and band splitting as interatomic distance was varied. Once we obtained the band structure of silicon, we saw that only a small region in k-space, right around the minimum of the conduction band and the maximum of the valence band, is where the important action takes place. In these regions, the carrier bands can be assumed parabolic, again a harmonic model. From the second derivatives of the bands we obtain the effective mass of the carriers, a property we will need to compute the flow of electrons through the crystal.

3.5 References for Chapter 3

3.1 Calude Cohen-Tannoudji, Bernard Diu, Franck Laloë, *Quantum Mechanics*, Vols. 1 and 2, John Wiley & Sons, New York (1977)

3.2 David K. Ferry, *Quantum Mechanics*, IOP Publishing, Bristol (1995)

3.3 Neil W. Ashcroft, N. David Mermin, *Solid State Physics*, Saunders College Publishing, Philadelphia (1988)

3.4 Franklin F. Y. Yang, *Introduction to Solid State Electronics*, 2nd Ed., North-Holland, Amsterdam (1989)

3.5 Otfried Madelung, *Introduction to Solid-State Theory*, Springer-Verlag, Heidelberg (1981)

3.6 Peter Y. Yu, Manuel Cordona, *Fundamentals of Semiconductors*, Springer Verlag, Berlin (1996)

3.7 Manuel Cordona, Fred H. Pollak, *Energy-Band Structure of Germanium and Silicon: The $k \cdot p$ Method*, Phys. Rev. 142(2) (1966) 530–543

3.8 Dr. Jony J. Hudson, Private communication (j.j.hudson@susx.ac.uk)

Chapter 4 The Electromagnetic System

Electrodynamics has had an unprecedented technological impact on our everyday lives. The phenomena that collectively belong to the field span many orders of magnitude, and include long wavelength radio signals, millimeter wavelength microwaves at airfields and in the kitchen, light all the way between infrared, visible and ultraviolet wavelengths, and higher still all the way to harmful ionizing radiation. As long as we consider the free propagation of electromagnetic waves, one theory covers it all—a remarkable discovery.

James Clarke Maxwell (1831–1879) culminated the search for a unified electromagnetic theory that could explain all the effects of "electricity" in one formalism. In truth, Maxwell's formalism was correct but cumbersome. Oliver Heaviside (1850–1925), a pioneer in his field, a great admirer of Maxwell and a champion of the use of vectors, first formulated the electrodynamic equations as we know them now. In parallel, Heinrich Herz (1857–1894) did the same, and for a brief period in history the one or the other name was associated with the equations. It was only in following the usage established by Albert Einstein (1879–1955), in his seminal work on the photoelectric effect, that we now call the governing electrodynamic equations the Maxwell equations. Einstein went on to unify space-time electromagnetic theory in his work on relativity, resulting in a single expression for the Maxwell equations.

At interaction dimensions of the order of atomic spacings and smaller, we have to also include the rules of quantum mechanics. We consider both viewpoints.

Chapter Goal

Our goal for this chapter is first to obtain a complete description of classical electrodynamics, and then to extend this model of radiation to a quantum viewpoint.

Chapter Roadmap

Our road map is as follows: to state the Maxwell equations, quantify the concepts leading to electro-quasi-statics and magneto-quasi-statics, and to their completely static counterparts. Next, we take a closer look at light, which is that part of the electromagnetic spectrum that ranges from the near infrared all the way through to the near ultraviolet, by treating both its wave-like and particle-like characteristics. There will be practically no "optics" here, and the interaction between light and matter will appear only later in the book.

4.1 Basic Equations of Electrodynamics

The Maxwell equations are remarkably structured. If it were not for the absence of magnetic "charges" in nature, we would be able to exchange the roles of the electric field $\boldsymbol{E}$ and the magnetic field $\boldsymbol{H}$, as well as the electric displacement $\boldsymbol{D}$ and the magnetic induction $\boldsymbol{B}$. We first write the Maxwell equations in differential form; each equation must be satisfied in every point of space:

The Faraday Law

$$\nabla\times\boldsymbol{E} = -\frac{\partial\boldsymbol{B}}{\partial t} \tag{4.1a}$$

The Ampere Law

$$\nabla\times\boldsymbol{H} = \boldsymbol{J}+\frac{\partial\boldsymbol{D}}{\partial t} \tag{4.1b}$$

The Gauss Law for Electric Fields

$$\nabla\bullet\boldsymbol{D} = \rho \tag{4.1c}$$

The Gauss Law for Magnetic Fields

$$\nabla\bullet\boldsymbol{B} = 0 \tag{4.1d}$$

Each of the equations can be integrated over space, and after applying some vector identities we obtain the integral representations of the Maxwell equations:

The Faraday Law

$$\oint_C \boldsymbol{E}\bullet d\boldsymbol{l} = -\frac{\partial}{\partial t}\int_S \boldsymbol{B}\bullet d\boldsymbol{S} \tag{4.2a}$$

The Ampere Law

$$\oint_C \boldsymbol{H}\bullet d\boldsymbol{l} = \int_S \boldsymbol{J}\bullet d\boldsymbol{S}+\frac{\partial}{\partial t}\int_S \boldsymbol{D}\bullet d\boldsymbol{S} \tag{4.2b}$$

The Gauss Law for Electric Fields

$$\oint_S \boldsymbol{D}\bullet d\boldsymbol{S} = \int_V q\,dV = Q \tag{4.2c}$$

The Gauss Law for Magnetic Fields

$$\oint_S \boldsymbol{B}\bullet d\boldsymbol{S} = 0 \tag{4.2d}$$

The Faraday law (4.1a) describes the electric field $\boldsymbol{E}$ that is generated by a time-varying magnetic induction $\boldsymbol{B}$. Note that the electric field will, in general, not be spatially uniform. In particular, it tells us that the electric field vector is perpendicular to the magnetic induction vector, because of the curl operator on the left-hand side of (4.1a). This becomes clear when we look at, for example, the x-component:

$$\left(\frac{\partial E_z}{\partial y}-\frac{\partial E_y}{\partial z}\right) = -\frac{\partial B_x}{\partial t} \tag{4.3}$$

Equation (4.2a) has a particularly simple interpretation. The line integral $\oint_C \boldsymbol{E} \bullet d\boldsymbol{l} = U_{\mathrm{Ind}}$ is the induced electric potential, which is equal to minus the rate of change of flux $-\dot{\phi} = -\partial/\partial t(\int_S \boldsymbol{B} \bullet d\boldsymbol{S})$ that passes through the surface enclosed by the loop, or

$$U_{\mathrm{Ind}} = -\dot{\phi} \tag{4.4}$$

The Ampere law (4.1b) describes how a magnetic field is generated in response to both a charge current density $\boldsymbol{J}$ and a time-varying electric displacement $\boldsymbol{D}$. By the same argument as before, the magnetic field vector is perpendicular to the current density vector and the electric displacement vector, and it will, in general, also not be spatially uniform.

The Gauss laws (4.1c) and (4.1d) confirm the experimental fact that only charge monopoles, and no magnetic monopoles, are observed in nature. In addition, the electrostatic law states that the electric displacement field generated in a closed region of space is only due to the enclosed charges. Inversely, if a region of space contains no charges, and hence the right-hand side of (4.2c) is zero, we would measure no net electric displacement: the divergence is of $\boldsymbol{D}$ zero.

Three constitutive equations are necessary to complete the set. They link the various fields to each other and describe the influence of the material on the propagation of the fields. The electric field $\boldsymbol{E}$ drives the charge current $\boldsymbol{J}$, and is hampered in doing so by the conductivity σ of the material:

$$\boldsymbol{J} = \sigma \boldsymbol{E} \tag{4.5}$$

In vacuum, Equation (4.5) of course falls away, and in semiconductors it is a gross simplification of the actual nonlinear charge conduction mechanism taking place, which we discuss in more detail in Chapter 7. The magnetic flux density $\boldsymbol{B}$ is driven by the magnetic field $\boldsymbol{H}$, and is hampered thereby by the permeability μ of the material:

$$\boldsymbol{B} = \mu \boldsymbol{H} = \mu_0 \mu_r \boldsymbol{H} \tag{4.6}$$

In vacuum, and for the semiconductor silicon, the permeability $\mu = \mu_0 = 1/c^2\varepsilon_0$. Note that, by convention, $\mu_0 = 4\pi \times 10^{-7}\ Js^2C^{-2}m^{-1}$, which immediately fixes the value for ε_0. In vacuum the permittivity $\varepsilon = \varepsilon_0 = 8.854 \times 10^{-12}\ AsV^{-1}m^{-1}$. For magnetically active materials the dependence of the relative permeability on the magnetic field is in fact highly nonlinear and frequency dependent. The electric field $\boldsymbol{E}$ also drives the electric displacement $\boldsymbol{D}$, and is hampered thereby by the dielectric permittivity ε of the material

$$\boldsymbol{D} = \varepsilon\boldsymbol{E} = \varepsilon_0\varepsilon_r\boldsymbol{E} \tag{4.7}$$

For solids, the relative permittivity ε_r is a function of the spatial distribution of atomic charge, as well as the charge's mobility. In fact, its value is strongly frequency dependent. A more detailed discussion of the cause of permittivity can be found in Section 7.2.2. (Solving the Maxwell equations is outlined in *Box 4.1)*.

Taking the curl ($\nabla\times$) of the Faraday law (4.1a) and the time derivative ($\partial/\partial t$) of the Ampére law (4.1b), we obtain

$$\nabla\times\nabla\times\boldsymbol{E} = -\nabla\times\frac{\partial\boldsymbol{B}}{\partial t} = -\nabla\times\frac{\partial}{\partial t}\mu\boldsymbol{H} \tag{4.8a}$$

$$\nabla\times\frac{\partial\boldsymbol{H}}{\partial t} = \frac{\partial\boldsymbol{J}}{\partial t} + \frac{\partial^2\boldsymbol{D}}{\partial t^2} = \frac{\partial}{\partial t}\sigma\boldsymbol{E} + \frac{\partial^2}{\partial t^2}\varepsilon\boldsymbol{E} \tag{4.8b}$$

Assuming vacuum conditions, so that $\sigma = 0$, $\mu = \mu_0$, and $\varepsilon = \varepsilon_0$, and requiring $\rho = 0$, we obtain

$$\nabla\times(\nabla\times\boldsymbol{E}) = -\mu_0\nabla\times\frac{\partial\boldsymbol{H}}{\partial t} = -\mu_0\varepsilon_0\frac{\partial^2\boldsymbol{E}}{\partial t^2} \tag{4.9}$$

Using the identity $\nabla\times(\nabla\times\boldsymbol{E}) = \nabla(\nabla\bullet\boldsymbol{E}) - \nabla^2\boldsymbol{E}$, and the Gauss law (4.1c), we obtain the second order partial differential Helmholtz equation

$$\nabla^2\boldsymbol{E} - \mu_0\varepsilon_0\ddot{\boldsymbol{E}} = 0 \tag{4.10}$$

Box 4.1. The Finite Difference Time-Domain Method.

To solve the Maxwell equations in the time domain, computer programs are used that discretize the space and time coordinates, most frequently using the finite difference (FD) method. In 1966, K. S. Yee [4.2] invented a discretization scheme that still dominates the field, for he was able to satisfy exactly all the Maxwell equations in one swoop, at the same time obtaining a numerically stable scheme for the explicit time integration. The trick lies in the so-called Yee stencil, see Figure B4.1.1.

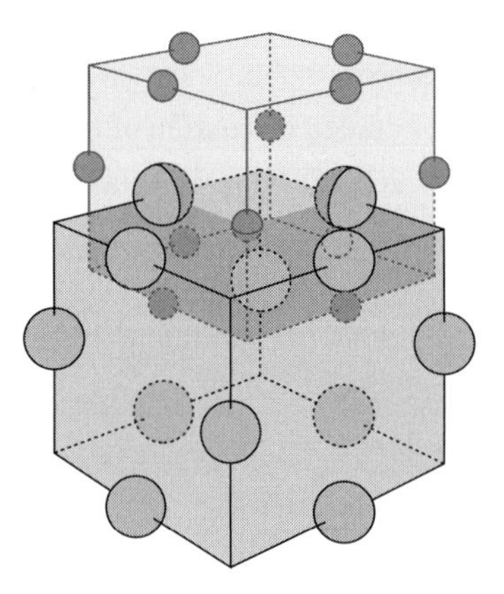

Figure B4.1.1 Yee finite difference time domain (FDTD) stencil for a primitive cell.

The values of the field components are only known at the positions shown by the balls, and correspond to the coordinate direction to which the cell edge lies parallel. The large balls correspond to positions where we evaluate the electric field components, and the small balls to positions where we evaluate the magnetic field components, see Figure B4.1.2. The algorithm alternately updates the $\boldsymbol{E}$ and $\boldsymbol{H}$ fields at time-step intervals of Δt, forming a so-called leap-frog method. (The offset grids of the two fields ensure that the simplest FD update formula is accurate to second order in space and time.) The Yee cell also guarantees that the Faraday and Ampere laws are automatically satisfied at each point in the grid through the update formulas. The two exemplary equations below correspond to the Yee-cell loops in Figure B4.1.2:

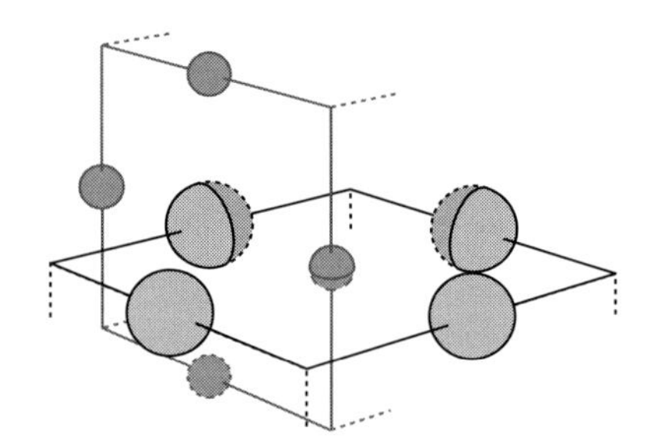

Figure B4.1.2 The top electric and left-front magnetic cell faces of Figure B4.1.2

$$\frac{\partial H_z}{\partial t} = \frac{1}{\mu}\left(\frac{\partial E_x}{\partial y} - \frac{\partial E_y}{\partial x}\right) \tag{B 4.1.1}$$

$$\frac{\partial E_x}{\partial t} = \frac{1}{\varepsilon}\left(\frac{\partial H_z}{\partial y} - \frac{\partial H_y}{\partial z} - \sigma E_x\right) \tag{B 4.1.2}$$

Discretization results in

$$\frac{\Delta H_z\big|_{i,j,k}^{\Delta n}}{\Delta t} = \frac{1}{\mu_{i,j,k}}\left[\Delta E_x\big|_{i,\Delta j,k}^{n} / (\Delta y) - \Delta E_y\big|_{\Delta i,j,k}^{n} / \Delta x\right] \tag{B 4.1.3}$$

$$\frac{\Delta E_x\big|_{i,j,k}^{\Delta n}}{\Delta t} = \frac{1}{\varepsilon_{i,j,k}}\left[\Delta H_z\big|_{i,\Delta j,k}^{n} / \Delta y - \Delta H_y\big|_{i,j,\Delta k}^{n} / \Delta z - \sigma_{i,j,k} E_x\big|_{i,j,k}^{n}\right] \tag{B 4.1.4}$$

Note the space-saving formalism

$$\Delta E_x\big|_{i,\Delta j,k}^{n} = E_x\big|_{i,j+\frac{1}{2},k}^{n} - E_x\big|_{i,j-\frac{1}{2},k}^{n} \tag{B 4.1.5}$$

which is a wave equation for $\boldsymbol{E}$ with periodic solutions of the following type

$$\boldsymbol{E} = \boldsymbol{E}_0 g(\omega t - \boldsymbol{k} \bullet \boldsymbol{r}) \tag{4.11}$$

where the second derivatives of g are assumed to exist. Notice that (4.11) requires g to be periodic but does not specify the exact form. The trigonometric sin, cosine and the exponential functions fit the specifications exactly.

4.1.1 Time-Dependent Potentials

Up to now the formulation for electromagnetics has been in terms of the electric and magnetic field variables. We can go one step further by introducing potentials for the field variables. If a vector field is divergence free, i.e., when $\nabla \bullet \boldsymbol{B} = 0$ holds, then there exists a vector $\boldsymbol{A}$ such that

$$\boldsymbol{B} = \nabla \times \boldsymbol{A} \tag{4.12}$$

$\boldsymbol{A}$ is the vector potential. Insert this definition for $\boldsymbol{B}$ into the Faraday law (4.1a):

$$\nabla \times \boldsymbol{E} = -\nabla \times \frac{\partial \boldsymbol{A}}{\partial t} \Leftrightarrow \nabla \times \left(\boldsymbol{E} + \frac{\partial \boldsymbol{A}}{\partial t}\right) = 0 \tag{4.13}$$

Whenever the curl of a quantity is zero (we say that it is irrotational or rotation free), it can be written as the gradient of a scalar potential, say Φ, which provides us with a definition for the electric field entirely in terms of the vector and scalar potentials

$$\boldsymbol{E} + \frac{\partial \boldsymbol{A}}{\partial t} = -\nabla \Phi \Leftrightarrow \boldsymbol{E} = -\left(\frac{\partial \boldsymbol{A}}{\partial t} + \nabla \Phi\right) \tag{4.14}$$

The two Maxwell equations that we started with are automatically satisfied by the potentials, and the remaining two now become (after a number of manipulation steps)

$$\nabla^2 \boldsymbol{A} - \mu\varepsilon\frac{\partial^2 \boldsymbol{A}}{\partial t^2} - \nabla\left(\nabla \bullet \boldsymbol{A} + \mu\varepsilon\frac{\partial\Phi}{\partial t}\right) = -\mu\boldsymbol{J} \tag{4.15a}$$

$$\frac{\partial}{\partial t}\nabla \bullet \boldsymbol{A} + \nabla^2\Phi = -\frac{\rho}{\varepsilon} \tag{4.15b}$$

We now consider the case for a vacuum; recall that $\mu_0\varepsilon_0 = 1/c^2$. Because of the relation $\boldsymbol{B} = \nabla\times\boldsymbol{A}$, $\boldsymbol{B}$ is unchanged through the addition of the gradient of an arbitrary scalar function Λ, because $\nabla\times(\boldsymbol{A} + \nabla\Lambda) = \nabla\times\boldsymbol{A}$. However, through equation (4.14), which under the same substitution for $\boldsymbol{A}$ results in

$$\boldsymbol{E} = -\frac{\partial \boldsymbol{A}}{\partial t} - \nabla\left(\frac{\partial\Lambda}{\partial t} + \Phi\right) \tag{4.16}$$

we see that the scalar potential must be transformed according to $\Phi \Rightarrow \Phi - \partial\Lambda/\partial t$ to achieve invariance of the electric field. Now choosing Λ such that

$$\nabla^2\Lambda + \mu\varepsilon\frac{\partial^2 \Lambda}{\partial t^2} = 0 \tag{4.17}$$

implies that

$$\nabla \bullet \boldsymbol{A} + \frac{1}{c^2}\frac{\partial\Phi}{\partial t} = 0 \tag{4.18}$$

which completely de-couples the remaining two vacuum Maxwell equations to give

$$\nabla^2\Phi - \frac{1}{c^2}\frac{\partial^2 \Phi}{\partial t^2} = -\frac{\rho}{\varepsilon} \tag{4.19a}$$

$$\nabla^2 \boldsymbol{A} - \frac{1}{c^2}\frac{\partial^2 \boldsymbol{A}}{\partial t^2} = -\mu\boldsymbol{J} \tag{4.19b}$$

These two equations represent four separate scalar wave equations of identical form, for which one can develop a (see the discussion in [4.1]) time-dependent fundamental solution or Green function. The Green function is nothing else than the analytical solution of the wave equation to a unit (read Dirac delta function) right-hand side. We define the Kronecker delta function positioned at s' as

$$\delta(s' - s) = \begin{cases} 1 & s = s' \\ 0 & \text{otherwise} \end{cases} \tag{4.20}$$

so that it looks like a unit pulse function at s'. For the wave equation, the Green function

$$G^{(\pm)}(r, t, r', t') = \frac{\delta\left(t' - \left[t \mp \frac{[r - r']}{c}\right]\right)}{|r - r'|} \tag{4.21}$$

is called retarded/advanced because it describes the effect of a unit load at time t' and position r' on another point located at r, at a later/earlier time t. Equations (4.19a) and (4.19b) are linear, and therefore permit superposition of solutions. By superposition of elemental right-hand sides, the Green functions can be used to build up a complete response to a right-hand side build–up of a spatially and temporally distributed charge and/or current density.

4.1.2 Quasi-Static and Static Electric and Magnetic Fields

Many effects of interest to us are dominated either by the electric field or by the magnetic field, and do not require us to consider the complete coupling of both. We follow the discussion in [4.5], which is highly recommended for additional reading. We start by normalizing the electromagnetic quantities using unit-carrying scale factors and unit-less scalar or vector variables (indicated by under-bars). We choose the scaling of length, time and the electric field:

$$x = \lambda \underline{x} \qquad t = \tau \underline{t}$$
$$E = \Sigma \underline{E} \tag{4.22}$$

Using the Maxwell equations (4.1a)–(4.1d) and the constitutive equations (4.5)–(4.6), we apply dimensional analysis to derive the scale factors of all other quantities in terms of the first three

$$[D] \equiv [\varepsilon E] \Rightarrow D = \varepsilon_o \Sigma \underline{D} \tag{4.23a}$$

$$[\nabla \bullet D] \equiv [\rho] \Rightarrow \rho = \frac{\varepsilon_o \Sigma}{\lambda} \underline{\rho} \tag{4.23b}$$

$$[J] \equiv \left[\frac{\partial D}{\partial t}\right] \equiv [H] \Rightarrow J = \frac{\varepsilon_o \Sigma}{\tau} \underline{J}$$
$$\Rightarrow H = \frac{\varepsilon_o \Sigma \lambda}{\tau} \underline{H} \tag{4.23c}$$

$$[\nabla \times E] \equiv \left[\frac{\partial B}{\partial t}\right] \Rightarrow B = \frac{\tau \varepsilon_o \Sigma}{\lambda} \underline{B} \tag{4.23d}$$

For the material properties, ε_o, μ_o and a typical conductivity σ_o are the scale factors. We now insert the normalizations above into the Maxwell equations:

$$\nabla_{\underline{x}} \times \underline{E} = \frac{\mu_o \varepsilon_o \lambda^2}{\tau^2} \frac{\partial}{\partial \underline{t}} \underline{H} = -\frac{\tau_{em}^2}{\tau^2} \frac{\partial}{\partial \underline{t}} \underline{H} = -\beta^2 \frac{\partial}{\partial \underline{t}} \underline{H} \tag{4.24a}$$

$$\nabla_{\underline{x}} \times \underline{H} = \frac{\tau \sigma_o}{\varepsilon_o} \underline{\sigma} \underline{E} + \frac{\partial}{\partial \underline{t}} \underline{E} = \frac{\tau}{\tau_e} \underline{\sigma} \underline{E} + \frac{\partial}{\partial \underline{t}} \underline{E} \tag{4.24b}$$

$$\nabla_{\underline{x}} \bullet \varepsilon_r \underline{E} = \underline{\rho} \tag{4.24c}$$

$$\nabla_{\underline{x}} \bullet \mu_r \underline{H} = 0 \tag{4.24d}$$

In (4.24b) we have defined the parameter

Charge Relaxation Time

$$\tau_e = \frac{\varepsilon_0}{\sigma_0} \tag{4.25}$$

which is the typical time for electric-field dominated charge signals to settle. The charge relaxation equation

$$\nabla \bullet \sigma \boldsymbol{E} = -\frac{\partial}{\partial t}\nabla \bullet \varepsilon \boldsymbol{E} \text{ or } \nabla \bullet \boldsymbol{J} = -\dot{q} \tag{4.26}$$

which can be obtained by manipulating the Maxwell equation set (see also Table 4.1), is a diffusion equation governing the conservation of free charge, and describes the way charge imbalance settles. Such processes are dominated by τ_e. An electromagnetic wave traversing our system is characterized by the time

$$\tau_{em} = \sqrt{\mu_0 \varepsilon_0}\lambda = \frac{\lambda}{c} = \sqrt{\tau_m \tau_e} \tag{4.27}$$

which we have obtained by straightforward association. In addition, in (4.24a) we have defined the factor

Time-Rate Parameter

$$\beta = \left(\frac{\tau_{em}}{\tau}\right)^2 \tag{4.28}$$

which characterizes the importance of the electromagnetic signal's transit time to times of interest in our system.

If we had based our normalization on the magnetic field instead on the electric field, we would have obtained the parameter

Magnetic Diffusion Time

$$\tau_m = \mu_0 \sigma_0 \lambda^2 \tag{4.29}$$

which is the typical time for magnetic field dominated signals to diffuse through the system. This time dominates the magnetic diffusion equation:

$$\frac{1}{\mu\sigma}\nabla^2 \boldsymbol{B} = \frac{\partial \boldsymbol{B}}{\partial t} \tag{4.30}$$

which in turn is obtained by a straightforward manipulation of the magneto-quasi-static equations (neglecting material transport) given in Table

4.1. What it tells us is that the magnetic induction field effectively diffuses into a material.

Together, the four characteristic time constants can be used to decide on which effects to effectively ignore. Thus, if $\beta \ll 1$, we can certainly ignore the wave-like effects and concentrate on diffusion-like formulations, since the electromagnetic wave passes our system faster than it can respond. Going one step further, and considering the case where

Electro-Quasi-Statics

$$\tau_m < \tau_{em} < \tau_e \text{ and } \beta \ll 1 \tag{4.31}$$

we can formulate the electrical equations as if the magnetic phenomena were instantaneous. For

Magneto-Quasi-Statics

$$\tau_e < \tau_{em} < \tau_m \text{ and } \beta \ll 1 \tag{4.32}$$

we can assume that charge relaxation effects are instantaneous. Note that quasi-statics by no means imply steady-state phenomena, which we treat next, but merely address the extent of dynamic coupling between the constituent charges, magnetic fields and the electromagnetic waves that excite our system. In other words, for the dynamic equation the other field appears effectively static because its time constant is small. The resulting equation sets for the quasi-static approximations are summarized in Table 4.1.

Table 4.1. The quasi-static equations of electrodynamics. Adapted from [4.4].

	Equations	
Magneto-quasi-statics	$\nabla\times\boldsymbol{H} = \boldsymbol{J}$	$\nabla\times\boldsymbol{E} = -\dfrac{\partial\boldsymbol{B}}{\partial t}$
	$\nabla\bullet\boldsymbol{B} = 0$	$\boldsymbol{B} = \mu\boldsymbol{H} = \mu_0(\boldsymbol{H}+\boldsymbol{M})$
	$\nabla\bullet\boldsymbol{J} = 0$	

Table 4.1. The quasi-static equations of electrodynamics. Adapted from [4.4].

	Equations	
Electro-quasi-statics	$\nabla\times\boldsymbol{E} = 0$	$\nabla\times\boldsymbol{H} = \boldsymbol{J} + \frac{\partial\boldsymbol{D}}{\partial t}$
	$\nabla\bullet\boldsymbol{D} = \rho$	$\boldsymbol{D} = \varepsilon\boldsymbol{E} = \varepsilon_0\boldsymbol{E} + \boldsymbol{P}$
	$\nabla\bullet\boldsymbol{J} = -\frac{\partial\rho}{\partial t}$	

Poisson Equation of Electrostatics

When we consider the electro-quasi-static case with steady-state conditions for the electric displacement and charge, i.e. $\boldsymbol{J} = \partial\boldsymbol{D}/\partial t = 0$, so that $\nabla\times\boldsymbol{H} = 0$, we can completely ignore all coupling of electric with magnetic phenomena and only consider the effects of an electric charge density ρ distributed in space setting up an electric displacement field $\boldsymbol{D}$ according to $\nabla\bullet\boldsymbol{D} = \rho$. When we integrate equation (4.1c) over an arbitrary control volume (with outward surface normal vector $\boldsymbol{n}$) that contains some total charge Q, then we obtain

$$\oint_{\Omega}(\nabla\bullet\boldsymbol{D})d\Omega = \oint_{\partial\Omega}(\boldsymbol{D}\bullet\boldsymbol{n})d\partial\Omega = \oint_{\Omega}\rho d\Omega = Q \tag{4.33}$$

The second and fourth terms provide a simple fact: the amount of $\boldsymbol{D}$ leaving the volume perpendicular through its enveloping surface is equal to the charge contained in the volume. If all terms are stationary in time, then equation (4.1c) or (4.33) describes electrostatics. To perform electrostatic calculations we usually go one step further by introducing the electrostatic potential ψ by noting that

$$\boldsymbol{D} = \varepsilon\boldsymbol{E} = -\varepsilon\nabla\psi \tag{4.34}$$

When we insert equation (4.34) into equation (4.1c) we obtain

$$\nabla\bullet(\varepsilon\nabla\psi) = \rho \tag{4.35}$$

Equation (4.35) is amenable to numerical solution by so-called fast-Poisson solvers, computer programs that exploit the following fundamental property of the Poisson equation. If we consider a region of homogeneous ε, then inside this region equation (4.35) becomes a Poisson equation $\nabla^2\psi = \rho/\varepsilon$. We consider the situation where ρ/ε represents a point charge at position $\boldsymbol{r}$ which we represent by a Dirac delta function:

$$\frac{\rho}{\varepsilon} = \delta(\boldsymbol{r}' - \boldsymbol{r}) \tag{4.36}$$

Green Function

For this case, we obtain the so-called Green function for the Poisson equation:

$$\bar{\psi} = \frac{1}{4\pi}\frac{1}{(\boldsymbol{r}' - \boldsymbol{r})} \tag{4.37}$$

that satisfies equation (4.35) exactly. Since we are, in principle, able to represent any spatial charge distribution as a linear superposition of individual point charges, we can obtain the associated potential by a linear superposition of the appropriate fundamental solutions via equation (4.37). In reality this is not very practical, as we would require large sums of terms at each point of interest in space: for n charges and m evaluation points we would require of the order of $n \times m$ evaluations.

Multipole Expansion

An important simplification technique is the so-called multipole expansion. The idea is remarkable. Consider a group of charges in space within an enclosing sphere of radius r_g. Clearly, if we are far enough from the group of charges, i.e., $r \gg r_g$, the values of the Green functions of the individual charges will not indicate strongly the spatial separation of the charges, merely their number and charge polarity. Algebraically, we can represent the combined Green function of a group of charges using a multipole expansion. The power of the method is that a group of groups of charges can again be represented in this manner. What this ultimately means is that a single sum over the Green function of many charges can be made much more efficient. We first spatially partition the charges into a hierarchy of clusters, then, starting inside the first groups in the hierarchy, we form the multipole expansion coefficients. These are now translated through the hierarchy so that, at the very worst, it is possible to perform only of the order of $n(\log n)$ vs. n evaluations for a system of n charges. Remarkably, further efforts have shown that the number of evaluations can be reduced to the order of the number of charges n in the system.

Magnetostatics

In analogy with the electrostatic case, we now consider magneto-quasistatic situations where, in addition, $\partial \boldsymbol{B}/\partial t = 0$, so that again magnetic and electrostatic phenomena are completely de-coupled, but stationary magnetic phenomena dominate. We consider the equation $\nabla \times \boldsymbol{H} = \boldsymbol{J}$. From the equation for the vector potential $\boldsymbol{B} = \nabla \times \boldsymbol{A}$, and the constitutive equation $\boldsymbol{H} = \boldsymbol{B}/\mu$, we obtain

$$\nabla \times \left(\frac{1}{\mu} \nabla \times \boldsymbol{A}\right) = \boldsymbol{J} \tag{4.38}$$

The vector identity $\nabla \times (\nabla \times \boldsymbol{A}) = \nabla(\nabla \bullet \boldsymbol{A}) - \nabla^2 \boldsymbol{A}$, together with $\nabla \bullet \boldsymbol{A} = 0$, transforms (4.38) to

$$-\nabla^2 \boldsymbol{A} = \mu \boldsymbol{J} \tag{4.39}$$

which now represents a separate Poisson equation for each component of the vector potential, i.e.

$$\begin{bmatrix} -\nabla^2 A_x \\ -\nabla^2 A_y \\ -\nabla^2 A_z \end{bmatrix} = \mu \begin{bmatrix} J_x \\ J_y \\ J_z \end{bmatrix} \tag{4.40}$$

This is very convenient, for we may now use the same methods to solve (4.39) as for the electrostatic equation (4.35).

4.2 Basic Description of Light

Around the beginning of the 20th century, new experimental evidence indicated that light, when interacting with a solid material, seems to behave also as a "particle" – now called a photon. Up to that stage, the wave nature model of light had sufficed, and could be cleverly used to explain most phenomena observed. Each model of course has major technological significance. The difference between the models becomes clear when we consider what happens to a light wave when it has to have a finite energy.

4.2.1 The Harmonic Electromagnetic Plane Wave

We can consider the harmonic electromagnetic plane wave in a vacuum that satisfies Equation (4.10) as a basic component with which to build up more detailed descriptions. Thus, following Equation (4.11), we select

$$\boldsymbol{E} = \boldsymbol{E}_0 \exp[-i(\omega t - \boldsymbol{k} \bullet \boldsymbol{r})] \tag{4.41}$$

We first insert (4.41) into (4.1c), noting that $\rho = 0$, to give

$$\nabla \bullet \boldsymbol{E} = 0 = i\boldsymbol{E}_0 \bullet \boldsymbol{k} \exp[-i(\omega t - \boldsymbol{k} \bullet \boldsymbol{r})] \text{ or } \boldsymbol{E}_0 \bullet \boldsymbol{k} = 0 \tag{4.42}$$

We next use (4.1b), noting that, since $\sigma = 0$ that $\boldsymbol{J} = 0$, to obtain

$$\boldsymbol{H} = \frac{1}{\omega\mu_0} \boldsymbol{k} \times \boldsymbol{E}_0 \exp[-i(\omega t - \boldsymbol{k} \bullet \boldsymbol{r})] \text{ or } \boldsymbol{H}_0 = \boldsymbol{k} \times \boldsymbol{E}_0 \tag{4.43}$$

We see that $\boldsymbol{E}_0$, $\boldsymbol{H}_0$ and $\boldsymbol{k}$ are cyclically perpendicular to each other. Thus the mutually orthogonal vectors $\boldsymbol{B}$ and $\boldsymbol{E}$ always lie in a plane perpendicular to the direction of propagation $\boldsymbol{k}$. Without sacrificing generality for the plane wave case, we can assume now that the wave propagates along the x-axis, the $\boldsymbol{E}$-field lies parallel to the y-axis and the $\boldsymbol{B}$-field lies parallel to the z-axis, hence

$$\begin{aligned} E_y &= E_0 \cos(\omega t - kx) \qquad & B_z &= B_0 \cos(\omega t - kx - \alpha) \\ E_x &= E_z = 0 & B_x &= B_z = 0 \end{aligned} \tag{4.44}$$

The argument of the cosine functions is called the phase of the wave components, and is a maximum if the argument is an integer multiple of 2π. Assuming that we fix a point in space, say $x = x_0$, then the wave varies in *time* as it passes our location with a period of

$$T = 2\pi/\omega \tag{4.45}$$

In exactly the same way we can fix a point in time, say $t = t_0$, to find that the wave varies in *space* as it passes our time point with a *spatial* period or wavelength of

$$\lambda = 2\pi/|k| \tag{4.46}$$

The speed of propagation of the peak of the wave is found from the cosine argument again. We can write that, between two wave peaks m cycles apart,

$$\omega t - kx = 2\pi m \tag{4.47}$$

Looking at one wave peak, so that $m = 0$, and on dividing equation (4.47) by kt, we obtain

$$\frac{\omega}{k} = \frac{x}{t} = c \tag{4.48}$$

Phase Velocity

which is the *phase velocity* of the wave in a vacuum. Everywhere but in a vacuum will the phase velocity become dependent on the frequency of the wave. The factor α in equation (4.44) causes a relative phase shift between the electric and magnetic field components. If $\alpha = 0$, then the light is linearly polarized as shown in Figure 4.1 (a). If, however, $\alpha \neq 0$, then the light is circularly polarized as illustrated in Figure 4.1 (b). The sign of α determines whether the wave is polarized left/right (or clockwise/anti-clockwise).

An electromagnetic wave has an energy density of

$$v_E(\boldsymbol{x}, t) = \varepsilon_0 \boldsymbol{E} \bullet \boldsymbol{E} \tag{4.49}$$

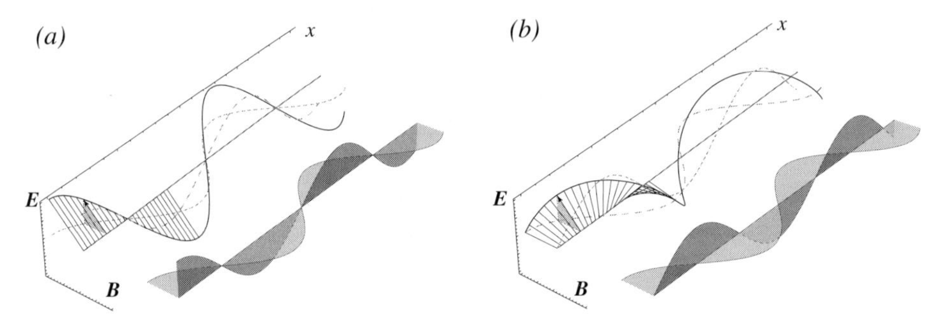

Figure 4.1. (a) A linearly polarized planar electromagnetic wave at a given instant in time t_0. (b) A circularly polarized planar electromagnetic wave at a given instant in time t_0. The $\boldsymbol{B}$ and $\boldsymbol{E}$ field components are confined to planes perpendicular to each other and to the direction of wave propagation. A differing phase causes the rotation of the field vectors. For both figures, the plot in the foreground shows the $\boldsymbol{B}$ and $\boldsymbol{E}$ field strengths, and the background plot the resultant vector $\boldsymbol{R} = \boldsymbol{B} + \boldsymbol{E}$, clearly indicating the planar and circular nature of the wave.

which, when integrated over a period of the wave, gives an average value of

$$\bar{\nu}_E = \frac{\varepsilon_0 E_0^2}{2} \tag{4.50}$$

4.2.2 The Electromagnetic Gaussian Wave Packet

The energy density of a plane wave as found in equation (4.50) implies that the wave has an energy proportional to the space to which the wave is limited, which is potentially unbounded. A wave with finite energy, which we require based on observations of radiation, is possible only if the wave and hence its energy are localized in space. A mathematically convenient way to achieve this is by forming a wave packet centered around a wavevector $\boldsymbol{k}_0$. A wave packet has an envelope that defines the maximum amplitude of a cosine wave. For a 1D wave the electric field component is defined by

$$E(x,t) = E_0 \exp\left[-\frac{\sigma_k^2}{2}(ct-x)^2\right]\cos(\omega_0 t - k_0 x) \tag{4.51}$$

and is illustrated in Figure 4.2. The center of this wave packet travels at the speed of light c to the right, and has been defined so that its spatial extension and its wavevector extension fulfil the relation

$$\Delta x \Delta k = \left(\frac{1}{\sigma_k}\right)(\sigma_k) = 1 \tag{4.52}$$

so that, if the packet is spread in space, it will have a precise frequency, and if it is concentrated in space, its spectrum will be spread. The energy density of the wave packet is now finite, and can be analytically computed as

$$v_E = \frac{\varepsilon_0}{2} E_0^2 \exp[-\sigma_k^2 (ct-x)^2] \tag{4.53}$$

This is simply a constant shape Gaussian that moves to the right with the speed of light c, see Figure 4.2. The Gaussian does not change its shape

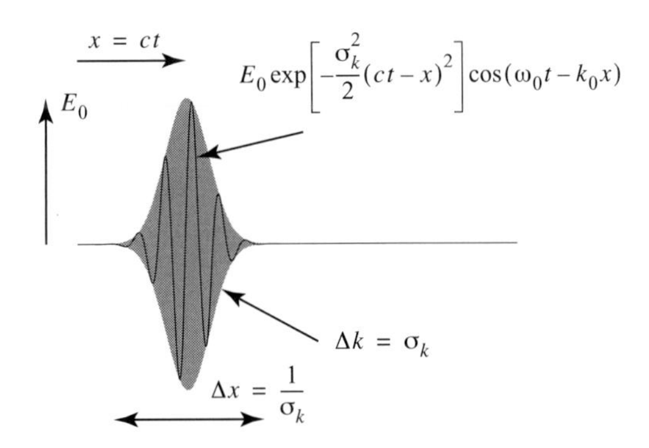

Figure 4.2. The general features of a Gaussian wave packet.

because of the linear dispersion relation $\omega = ck$. For example, an electronic wave packet, for $\omega(k) \neq \text{const}$, would change its shape, because $\omega(k)$ is not linear.

4.2.3 Light as Particles: Photons

Convenient as wave packets are, they do not describe photons yet, and what is missing of course is a quantum-mechanical approach to light. Since photons act as located particles, for example in the photoelectric effect, they are not representable by a space-filling wave packet, yet they are observed to interfere as waves.

Discussion

As we have shown in more detail in Chapter 3, the quantum-mechanical solution to this apparent contradiction is to make the wave packet represent the probability density of the particle. Before we look at the details, it is instructive to see what is resolved by this new representation. Since the probability is wave-like, it obeys wave mechanics, and hence can produce wave–like interference with the probability waves of other photons. If a photon is detected (it is "observed"), it is with a process that requires the photon to be particle-like. At this instant of detection, the resultant probability of the photon is evaluated by the detector, and the photon reveals its position. On the way to the detector, the photon does not reveal its position. Indeed, there is no way to tell how it got from A to B apart from disturbing it on its way. What quantum mechanics does elegantly is to let the photon's *probability* "propagate" with the speed of light, and interact with equipment in a determined way. Each measurement becomes an evaluation of the evolved probability distribution of the photon.

Probability Density

To describe a photon that is consistent with measurements and observations, we use a wave packet with a Gaussian spectral function and a total energy of $2\pi h\omega$:

$$f(k) = \frac{1}{\sqrt{2\pi}\sigma_k}\exp\left[-\frac{(k-k_0)^2}{2\sigma_k^2}\right] \tag{4.54}$$

The probability density of the photon is then proportional to $|f(k)|^2$, with the proportionality constant chosen so that the probability of finding the photon anywhere is equal to one. The probability of finding a photon in the interval $k - \Delta k/2$ and $k + \Delta k/2$ is then

$$P(k) = 2\sqrt{\pi}\sigma_k |f(k)|^2 \tag{4.55}$$

Macroscopic Electro-magnetics

We still must create an intuitive link between electromagnetics and photons. The "trick" lies in the fact that electromagnetic radiation in vacuum can be brought into the exact same form as a harmonic oscillator (the harmonic oscillator is extensively discussed in Section 3.2.5). In quantum mechanics, the harmonic oscillator quantizes the energy as

$$E = \left(n + \frac{1}{2}\right)\hbar\omega,\ n = 0, 1, 2, \ldots \tag{4.56}$$

The unbounded vacuum does not place any limitation on ω, which may vary arbitrarily. We could view light in this context as an electromagnetic wave whose energy is restricted to integer multiples of $\hbar\omega$ starting at the zero-point energy of $\hbar\omega/2$. Here a photon corresponds to an energy step, which for visible light with a wavelength of 500 nm is the very small quantity $\hbar\omega = 1.66 \times 10^{-18}$ J, so that for typical optical ray transmission applications the fundamental step size is so insignificantly small that the energy appears as a continuous variable.

Another possibility is to consider an electromagnetic light wave as a superposition of photon-sized electromagnetic wave packets that are in phase with each other. Each photon carries a fundamental quantum of energy, $\hbar\omega$, and the superposition of a large quantity of such photons appears as a macroscopic electromagnetic wave.

Cavities

We now consider what happens when electromagnetic radiation is trapped in a 1D cavity of length L with perfectly conducting and perfectly reflecting walls. The electric field has a vanishing component parallel to a conducting surface. Clearly, the cosine functions are good candidates for this case as long as $k = n_j\pi/L$, where $n_j = 1, 2, 3, \ldots$, and so we use

$$\boldsymbol{E}(x,t) = \boldsymbol{E}_0 \cos[\omega t - n_j \pi x / L] \tag{4.57}$$

Inserting equation (4.57) into the wave equation (4.10) we obtain

$$\left[(n_j\pi/L)^2 - \left(\frac{\omega}{c}\right)^2\right](\boldsymbol{E}_0 \cos[\omega t - n_j \pi x / L]) = 0 \tag{4.58}$$

which has non-trivial solutions for

$$\omega_j = \frac{n_j \pi c}{L}, n_j = 1, 2, 3, \ldots \tag{4.59}$$

Thus the frequency ω_j of the trapped wave is quantized. This contrasts with the unbounded wave case, where the frequency can take on any value.

4.3 Waveguides

The salient behavior of small waveguides, even down to the micrometer dimensions typical for features on a silicon or gallium arsenide microchip, can be treated by only considering the Maxwell equations. Waveguides represent a very important technological tool, by forming the interconnects for communication and sensor equipment based on millimeter, micrometer and optical wavelength electromagnetic waves. Knowledge of the design of waveguides is an important asset of a microsystem engineer's "toolbox". In this context, an optical fiber is a waveguide, and so is a co-axial cable. Integrated waveguides use differences in diffractive constants to form a "channel" that guides the electromagnetic wave.

The idea of a waveguide is to support optimally the propagation of a wave *along* its axis, and to prevent the wave from escaping (or dissipating) in the transverse direction along the way. Usually, waveguides consist of long straight stretches and shorter curved segments. Clearly, a curved segment requires the most general 3D treatment, because the wave will be deflected from its straight path by the waveguide walls along the curve. For a straight segment we can simplify the analysis by considering simple harmonic propagation along the axis of the waveguide, thereby "separating" the variables somewhat. Thus we start the analysis by assuming that a wave of the form (4.10) that propagates in the x-direction through an isotropic medium:

$$\begin{aligned} \boldsymbol{E}(t) &= \boldsymbol{E}_t(y, z)\exp[-i(\omega t - kx)] \\ \boldsymbol{H}(t) &= \boldsymbol{H}_t(y, z)\exp[-i(\omega t - kx)] \end{aligned} \tag{4.60}$$

Note that $\boldsymbol{E}_t(y, z)$ and $\boldsymbol{H}_t(y, z)$ are real values. For this case, harmonicity implies that $\partial/\partial t \equiv -i\omega$ and axial propagation of a planar distribution implies that $\nabla \equiv (ik, \partial/\partial y, \partial/\partial z) = (ik, \nabla_t)$, so that $\nabla_t \equiv (\partial/\partial y, \partial/\partial z)$. We now check to see what this assumption induces by inserting (4.60) into the four Maxwell equations (4.1a)-(4.1d) as corrected for dielectric materials:

$$\nabla_t \times \boldsymbol{E}_t + ik\boldsymbol{e}_x \times \boldsymbol{E}_t = -i\omega\boldsymbol{B}_t \text{ gives } \nabla_t \times \boldsymbol{E}_t = 0\,,$$

$$\boldsymbol{e}_x \times \boldsymbol{E}_t = -\frac{\omega}{k}\boldsymbol{B}_t \tag{4.61a}$$

$$\nabla_t \times \boldsymbol{H}_t + ik\boldsymbol{e}_x \times \boldsymbol{H}_t = -i\omega\boldsymbol{D}_t \text{ gives } \nabla_t \times \boldsymbol{H}_t = 0\,,$$

$$\boldsymbol{e}_x \times \boldsymbol{H}_t = -\frac{\omega}{k}\boldsymbol{D}_t \tag{4.61b}$$

$$\varepsilon\left(\frac{\partial \boldsymbol{E}_{ty}}{\partial y} + \frac{\partial \boldsymbol{E}_{tz}}{\partial z}\right) = \varepsilon(\nabla_t \bullet \boldsymbol{E}_t) = 0 \text{ implies that } \nabla_t \bullet \boldsymbol{D}_t = 0 \tag{4.61c}$$

$$\mu\left(\frac{\partial \boldsymbol{H}_{ty}}{\partial y} + \frac{\partial \boldsymbol{H}_{tz}}{\partial z}\right) = \mu(\nabla_t \bullet \boldsymbol{H}_t) = 0 \text{ implies that } \nabla_t \bullet \boldsymbol{B}_t = 0 \tag{4.61d}$$

Additionally, we have assumed that ε and μ vary so slightly with space that the term involving its gradient can be dropped. We see that the magnetic induction is perpendicular to the electric field, and the magnetic field is perpendicular to the electric displacement. Furthermore, we see that the transverse field distributions $\boldsymbol{E}_t$ and $\boldsymbol{H}_t$ are rotation free and divergence free. Finally, we insert (4.60) into (4.10), and cancel the common exponential factor to obtain

$$\nabla_t^2 \boldsymbol{E}_t - [k^2 - k_0^2 \varepsilon]\boldsymbol{E}_t = \nabla_t^2 \boldsymbol{E}_t - \gamma \boldsymbol{E}_t = 0 \tag{4.62}$$

where $k_0 = \omega / c$. We have thus obtained an eigenmode equation for the planar transverse electric field vector $\boldsymbol{E}_t(y, z)$. In general, solving (4.62) will yield the eigenpairs $(\gamma^{(n)}, \boldsymbol{E}_t^{(n)})$ that represent the allowable solutions. From each $\gamma^{(n)} = [k^{2(n)} + k_0^2 \varepsilon]$ we obtain the allowable $\pm k^{(n)}$. We make two observations:

- Since we obtain both a positive and a negative value for $k^{(n)}$, this implies that the solution generates waves that propagate in the positive and negative directions along the waveguide.
- Equation (4.62) has the exact form of the stationary Schrödinger equation. The role of the dielectric constant takes on that of the potential in the Schrödinger equation, and we could use solution methods already developed for solving potential well problems. We now use this fact to investigate a classical case.

4.3.1 Example: The Homogeneous Glass Fiber

We consider a circular cross-section glass fiber waveguide with a quadratically varying refractive index $n^2 = A - br^2$, with A and b constant [4.7]. Since $\varepsilon = n^2$, this means that we have a spatially linear function for the dielectric constant ε. In cylindrical coordinates (r, θ), equation (4.62) now becomes

$$\nabla_t^2 \boldsymbol{E}_t - [k^2 - k_0^2 \varepsilon]\boldsymbol{E}_t = \frac{\partial^2 \boldsymbol{E}_t}{\partial r^2} + \frac{1}{r}\frac{\partial \boldsymbol{E}_t}{\partial r} + \frac{1}{r^2}\frac{\partial^2 \boldsymbol{E}_t}{\partial \theta^2} - [k^2 - k_0^2 \varepsilon]\boldsymbol{E}_t = 0 \tag{4.63}$$

The geometry is axially symmetric. We therefore assume that we can decouple the radial and angular solution components using $\boldsymbol{E}_t(r, \theta) = R(r)\Theta(\theta)$. Inserting this into (4.63) we obtain

$$\frac{r^2}{R}\frac{\partial^2 R}{\partial r^2} + \frac{r}{R}\frac{\partial R}{\partial r} - r^2[k^2 - k_0^2\varepsilon(r)] = p^2 = -\frac{1}{\Theta}\frac{\partial^2 \Theta}{\partial \theta^2} \tag{4.64}$$

Since radial and angular components are separated, we can consider the radial terms alone

$$\frac{\partial^2 R}{\partial r^2} + \frac{1}{r}\frac{\partial}{\partial r}(R) - R\left[k^2 - k_0^2\varepsilon(r) - \frac{p^2}{r^2}\right] = 0 \tag{4.65}$$

where p^2 is the constant linking the radial and angular equations. We make two further substitutions, setting $U = k^2 - k_0^2 A$, $\alpha = k_0 b$, to obtain

$$\frac{\partial^2 R}{\partial r^2} + \frac{1}{r}\frac{\partial}{\partial r}(R) - R\left[U - (\alpha r)^2 - \frac{p^2}{r^2}\right] = 0 \tag{4.66}$$

This is exactly the form of the 2D harmonic oscillator (in cylindrical coordinates) with total energy U and potential $V = (\alpha r)^2$, the well-known solutions of which are of the form

$$R(r) = \exp(-\alpha r^2/2) \sum_{j = m_0}^{m} a_j r^j \tag{4.67}$$

and where the summation limits come from the boundary conditions.

4.4 Summary for Chapter 4

We encounter electromagnetic radiation as light from the sun and other sources, as radio and television signals, mobile telephone communications, fibre-optic laser light rays, and of course in the form of background radiation from deep space. The very successful model for the propagation of radiation is the set of wave equations of Maxwell. Solid state interactions expose the particle nature of light, where a quantum-mechanical view becomes necessary. In electronic circuits, light can be guided around the surface by half-open "fibres" that we call waveguides. We have included them here, but leave all other interactions between photons and matter for Chapter 7.

4.5 References for Chapter 4

4.1 J. D. Jackson, *Classical Electrodynamics*, Wiley, New York (1975)

4.2 K. S. Yee, *Numerical Solution of Initial Boundary Value Problems Involving Maxwell's Equations in Isotropic Media*, IEEE Trans. Antennas and Propagation, vol. 14 (1966) 302-307

4.3 A. Tavlove, *Computational Electrodynamics – The Finite-Difference Time-Domain Method*, Artech House, Boston (1995)

4.4 H. H. Woodson, J. R. Melcher, *Electromechanical Dynamics*, Robert E. Krieger Publishing Company, Florida, Reprint Edition (1985)

4.5 J. Melcher, *Continuum Electromechanics*, MIT Press, Cambridge, Massachusetts (1981)

4.6 S. Wolfram, *The Mathematica Book*, Cambridge University Press, 4th Ed., (1999)

4.7 S. G. Lipson, H. S. Lipson, D. S. Tannhauser, *Optik*, Springer Verlag, Berlin, (1997)

Chapter 5 Statistics

The number of particles that occur in a semiconductor is extraordinarily large. From chemistry we got to know the mol as a unit, which counts off chunks of materials weighing in at a couple of grams, yet containing 1.6×10^{23} particles (this is of course Avogadro's number). If we wish to calculate properties down at the individual particle level, as indeed the molecular dynamics people do, then we can consider only a very small amount of matter. One way out is to consider the statistical nature of the particles, and this works well if we have very many of them, as is indeed the case for most realistic physical systems.

Chapter Goal The goal of this chapter is to introduce the concepts of statistics and statistical mechanics necessary to understand the variety of pseudo particles that we encounter in the study of the solid state.

Chapter Roadmap Our road map is to first introduce the ensembles. These groups of particles are convenient models from thermodynamics, and provide a basis for a consistent theory. Next we consider ways to count particles and particle states. This leads to the famous particle distributions. Armed with these we can describe the behavior of many-particle systems in a very compact manner. The rest of the chapter shows how this is done for bosons (named after Bose, and include phonons and photons), and for fermions (named after Enrico Fermi, and include electrons and holes).

5.1 Systems and Ensembles

In representing a system A which is composed of a very large number N of identical subsystems, there is a trade-off between the available resources (i.e., computer memory, computation time) and the accuracy that we can achieve for the representation. If N is so large that the representation of the exact state, including all subsystems, by far exceeds all available resources, then the only way to represent the state of A is by means of statistical statements, i.e., probability distributions for the respective system properties. In this chapter we develop the basic statistical techniques required to represent the 'particle' systems of our solid state semiconductor. A thorough treatment of the basics may be found in textbooks [5.1] and [5.2]. The application to semiconductors is dealt with in [5.3].

Our discussion starts with the microcanonical ensemble, which considers a large number of identical isolated systems. Each system resides in the smallest possible energy range that we can possibly consider. This system delivers us with a definition for the density of states. The microcanonical ensemble is extended in range to the next level, the canonical ensemble, where we only allow heat exchange with a very large reservoir. This model delivers the canonical partition function. Stepping up, we consider the grand ensemble, where we allow particle exchange with the environment. From this model we obtain a definition for the chemical potential, the driving force for particle exchange, as well as for the grand partition function. With these results, we move on to statistically describe particle counts. We specialize the statistics for different types of particle systems as they are found in microsystem devices: photons, electrons and holes, quasiparticles. We end the chapter with applications of the statistical results, and show how they characterize the particles and quasiparticles that we encounter in describing the behavior of semiconductor-based devices.

5.1.1 Microcanonical Ensemble

An isolated system A (see Figure 5.1a) in equilibrium, with a fixed num-

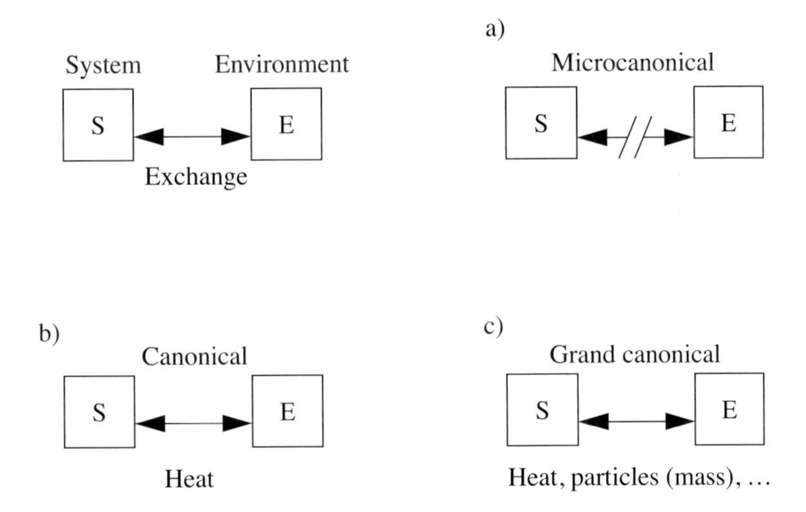

Figure 5.1. Characterization of systems according to the way they interact with the environment.

ber N of components, which resides in a fixed volume V, assumes different microscopic implementations in the small energy interval $E < E_i < E + \delta E$. The index i counts the different implementations. A large number of identical such systems A is called a microcanonical ensemble. An implementation i is characterized by its energy E_i and a detailed description of the state of each of the N subsystems. We can think of the N components as particles. Each particle j then has a given vector momentum $\boldsymbol{p}_j$ and vector position $\boldsymbol{q}_j$; each vector in turn has three spatial components. With N particles this forms a $6N$-dimensional phase space to represent the different implementations i. We can now construct a hypersurface in the $6N$-dimensional space that corresponds to the energy E_i.

In an interval δE that is small in comparison to the total energy of the system, an external observer cannot distinguish between the different implementations i. Even if there was a slight difference in the energy it would not be resolvable by the accuracy limit of possible measurements. Every implementation i therefore has an equal probability P_i to be realized, because from a macroscopical point of view the implementations are indistinguishable.

$$P_i = \begin{cases} c \text{ for } (E < E_i < E + \delta E) \\ 0 \text{ otherwise} \end{cases} \tag{5.1}$$

Here c is a constant which comes from the normalization that requires that $\Sigma P_i = 1$, where the sum counts every state that A is allowed to assume in the given energy interval. This means that in an ensemble that has much more members than the maximum number of implementations each implementation is realized by the same number of member systems.

The volume $\Gamma(E)$ that the system occupies in the $6N$-dimensional phase space, i.e., the number of possible implementations in the small energy interval $(E < E_i < E + \delta E)$, is given by

$$\Gamma(E) = \int_{(E < E_i < E + \delta E)} \mathrm{d}^{3N}q\,\mathrm{d}^{3N}p \tag{5.2}$$

which evaluates to

$$\Gamma(E) = V(E + \delta E) - V(E) \tag{5.3}$$

where $V(E)$ is the volume of the system occupying in the energy range $(0 < E_i < E)$. For the case where $\delta E \ll E$, we expand (5.3) into a linear Taylor series to obtain

$$\Gamma(E) = D(E)\delta E \tag{5.4}$$

which defines the density of states

Density of States

$$D(E) = \partial V(E)/\partial E \tag{5.5}$$

This rather abstract definition becomes lucid when applying to an electron in a cube with a cubic crystal structure with lattice constant a. The edge of the cube has a length $L = n \cdot a$. Applying periodic boundary conditions the k-vector comes in discrete portions $\mathbf{k}= (2\pi n_x, 2\pi n_y, 2\pi n_z)/L$ as shown in (3.44). Given the energy E we calculate the number of states that are within the range of $(0, E)$, i.e., the number of combinations of (n_x, n_y, n_z) that results each in an energy lower than E, or in other words the sum over all possible $\mathbf{k}$. We have $n_x = (k_x L)/(2\pi)$ etc.We write the sum over all $\mathbf{k}$ as the integral over k-space over all states with energies lower than E in spherical coordinates

$$\sum_{k} \rightarrow \left(\frac{L}{2\pi}\right)^3 \int_{(0,E)} \mathrm{d}k_x \mathrm{d}k_y \mathrm{d}k_z = \left(\frac{L}{2\pi}\right)^3 \int_0^{k(E)} 4\pi k^2 dk \tag{5.6}$$

(5.6) divided by the sample volume L^3 gives the volume $V(k(E))$ the states occupy in k-space. The energy is given by $E = (\hbar^2 k^2)/(2m)$ and thus we have $k = \sqrt{(2mE)/\hbar^2}$, where k is the absolute value of the vector $\mathbf{k}$. Furthermore we have $dk = m \cdot dE/(\hbar^2 k)$. We insert this into (5.6) in order to calculate the volume with respect to a given energy E

$$V(E) = \frac{1}{4\pi^2}\left(\frac{2m}{\hbar}\right)^{3/2} \int_0^E \sqrt{E'} dE' \tag{5.7}$$

With the expansion procedure as shown in (5.3) we obtain for the density of states

$$D(E) = \frac{1}{4\pi^2}\left(\frac{2m}{\hbar}\right)^{3/2} \sqrt{E} \tag{5.8}$$

Taking into account that (5.8) is valid for spin up and down particles it must be multiplied by a factor 2.

5.1.2 Canonical Ensemble

Let us now consider a small system A which stays in contact with a very large heat reservoir R (see Figure 5.1b). We only allow A to exchange heat with R (think of a tin-can filled with liquid that is immersed in a lake). In the following we assume that $A \ll R$. The total energy of the system, given by $E_{tot} = E_A + E_R$, is conserved. If A can be found in a well defined state A_i with energy E_i, then the energy of the heat reservoir is given by $E_R = E_{tot} - E_i$. The probability P_i to find exactly an implementation A_i, is proportional to the number of possible implementations $\Omega(E_R)$ for the heat reservoir R, each of which has energy E_R

$$P_i = c \cdot \Omega(E_{tot} - E_i) \tag{5.9}$$

where c is the normalization factor. The prerequisite $A \ll R$ immediately implies that $E_i \ll E_R$. We next expand the logarithm of Ω by a Taylor series about $E_i = 0$, (or about E_R) to obtain a good approximation

$$\ln\Omega(E_{tot} - E_i) = \ln\Omega(E_{tot}) - \left[\frac{\partial \ln\Omega}{\partial E_R}\right]_{E_i = 0} E_i + \dots \tag{5.10}$$

Now imagine that, for one specific E_i, the reservoir only has to fulfil the condition that $E_R = E_{tot} - E_i$. Due to its huge size compared to A, the number of implementations of R is obviously very large and also increases strongly with increasing energy. Inserting (5.10) in (5.9) yields $P_i = c \cdot \exp{-\beta E_i}$ and performing the normalization we obtain

$$P_i = \frac{\exp(-\beta E_i)}{\sum_k \exp(-\beta E_k)} \tag{5.11}$$

Temperature

where $[\partial \ln\Omega / \partial E_R]_{E_i = 0} = \beta$ is independent of the energy E_i. The parameter β specifies the average energy per degree of freedom of the system A. This average energy is the first moment of the probability distribution $\bar{E} = \Sigma_i(E_iP_i)$. The factor β, which we can also write as $\beta = (k_BT)^{-1}$, is the inverse of the thermal energy. Here T is the temperature and k_B is the Boltzmann constant. Note that the sum in the denominator of (5.11) accounts for the normalization of P_i. Equation (5.11) is called the canonical probability distribution.

Average Energy

If we write $Z = \sum_k \exp(-\beta E_k)$ for the denominator of (5.11), then the average energy is given by

$$\bar{E} = -\frac{1}{Z}\frac{\partial Z}{\partial\beta} = -\frac{\partial \ln Z}{\partial\beta} \tag{5.12}$$

Canonical Partition Function

We call Z the canonical partition function. Suppose that we know the total number of particles for a specific implementation k to be the sum $N_k = \Sigma_i n_i$, where n_i is the number of particles with energy E_i such that $E_k = \Sigma_i n_i E_i$ holds. Then we may write the partition function as

$$Z = \sum_k \exp(-\beta[\Sigma_i n_i E_i]_k),\ \text{with the constraint} \Sigma_i n_i = N_k \tag{5.13}$$

Average Number of Particles

The bracket $[\Sigma_i n_i E_i]_k$ denotes one specific implementation k of E_k. This gives us direct access to the average number of particles $\bar{n}_j$ in a specific state j. For a canonical distribution this is given by

$$\bar{n}_j = \frac{\sum_k n_j \exp(-\beta[\Sigma_i n_i E_i]_k)}{\sum_k \exp(-\beta[\Sigma_i n_i E_i]_k)} = -\frac{1}{\beta Z}\frac{\partial Z}{\partial E_j} = -\frac{1}{\beta}\frac{\partial \ln Z}{\partial E_j} \tag{5.14}$$

Sum Over States

The canonical partition function Z is sometimes also called the sum over states. We have already seen that Z is a central term in statistical mechanics, from which many other system properties may be derived. Let us consider the energies of all realizations k that the system A may assume for a small energy interval $[E, E + \delta E]$. In this case the probability of finding A with the energy E is given by the sum over all implementations k

$$P(E) = \sum_k P_k = c\Omega(E)\exp(-\beta E) \tag{5.15}$$

Continuous Probability

with k such that $E_k \in [E, E + \delta E]$. Equation (5.15) contains only the energy E for the same reasons that are valid for the microcanonical ensemble: the energies cannot be distinguished by measurement. Since $\Omega(E)$ is the number of implementations for the heat reservoir R with energy E and is, due to the large size of R, a continuous function, the probability P becomes a continuous function of energy.

5.1.3 Grand Canonical Ensemble

For a grand canonical ensemble the system A with a given number of particles N_k is allowed to exchange particles with its reservoir R (see Figure 5.1). The total number of particles is conserved so that $N_{tot} = N_R + N_i$. Following the same arguments as for the canonical ensemble, the number of particles N is included in the argument list of Ω. The concept of a reservoir now also implies that, in addition to $E_i \ll E_R$, we have that $N_{tot} \gg N_i$. We expand Ω into a Taylor series around $E_i = 0$ and $N_i = 0$ which yields

$$\begin{aligned} &\ln\Omega(E_{tot} - E_i, N_{tot} - N_i) \\ &= \ln\Omega(E_{tot}, N_{tot}) - \left[\frac{\partial \ln\Omega}{\partial E_R}\right]_{E_i = 0} E_i - \left[\frac{\partial \ln\Omega}{\partial N_R}\right]_{N_i = 0} N_i \ldots \end{aligned} \tag{5.16}$$

An additional parameter arises of course from the derivative with respect to the particle number

$$\left[\frac{\partial \ln\Omega}{\partial N_R}\right]_0 = \alpha \tag{5.17}$$

This parameter α as we shall see later will be interpreted as the chemical potential. The probability to have a specific implementation then assumes the form

$$P_i \sim \exp(-\beta E_i - \alpha N_i) \tag{5.18}$$

and is called the grand canonical distribution. For a system which may occupy different energy states E_k the total energy of a specific implementation is given by $E_i = \Sigma_k n_k E_k$, where the number of particles is $N_i = \Sigma_k n_k$ fixed. Here n_k is the number of subsystems or particles with energy E_k. To calculate the normalization for (5.18) we have to sum over all possible numbers of particles N'

$$Q = \sum_{N'} Z(N')\exp(-\alpha N') \tag{5.19}$$

Grand Partition Function

Q is called the grand partition function. $Z(N')$ is the canonical ensemble partition function. As was observed for the partition function Z, we shall see that Q is a key quantity for a system. Let us analyze Q in more detail. It consists of a product of two functions $Z(N')$ and $\exp(-\alpha N')$, where the first term increases with increasing N', and the second term decreases with increasing N'. As a product they result in a sharp maximum at the equilibrium particle number N. For the case of large enough N and a sufficiently sharp peak with width ΔN, we may write that $Q = Z(N)\exp(-\alpha N)\Delta N$. This implies that we replace the sum of products in (5.19) by the argument function's value at the equilibrium particle number N times the width of the argument function. Hence we obtain $\ln Q = \ln Z(N) - \alpha N + \ln \Delta N$. For $\Delta N \ll N$, the last term in may be dropped. This gives

$$\ln Q = \ln Z(N) - \alpha N \tag{5.20}$$

and for the grand partition function we obtain that

$$Q = \sum_k \exp(-\beta[\Sigma_i n_i E_i]_k)\exp(-\alpha N_k) \tag{5.21}$$

In (5.21) we see that the sum over different implementations of the product $\exp(-\beta[\Sigma_i n_i E_i]_k)\exp(-\alpha N_k)$ is nothing else than the product of sums over each energy level E_i, and hence we may write that

$$Q = \prod_i \sum_{n_i} \exp(-(\beta E_i + \alpha) n_i) \tag{5.22}$$

The product in (5.22) accounts for all states i accessible to the system, while the sum over n_i depends on how many particles may occupy a state i.

5.2 *Particle Statistics: Counting Particles*

Different statistical properties of the system arise from the many ways that the sum in (5.13) and (5.22) has to be performed. Summation over all possible states i in turn implies summation over all possible values of n_i. We now consider these cases.

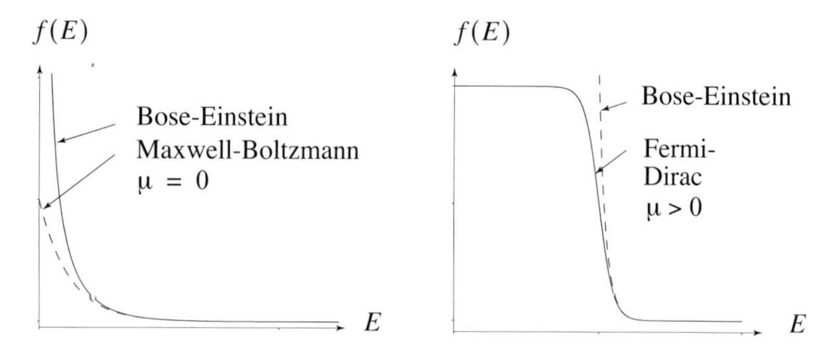

Figure 5.2. Maxwell-Boltzmann, Bose-Einstein, and Fermi-Dirac distributions.

5.2.1 Maxwell-Boltzmann Statistics

Suppose that there are N distinguishable and noninteracting particles in the system. Then, for a given implementation $\{n_1, n_2, \ldots\}$, n_j particles have the energy E_j, to which we must add the constraint that $\Sigma_k n_k = N$. According to (B 5.1.2) there are $N!/(n_1!n_2!\ldots)$ possibilities to distribute the N particles among all available states. This way of counting distinguishes the individual particles, but not their ordering once they

Box 5.1. Counting permutations, variations, and combinations.

The probability for a macroscopic event to be realized by a specific implementation, in the knowledge that none of the N individual implementations is to be preferred, is exactly 1/N. The number N of implementations depends on the rules on how implementations are counted:

1. The number of permutations of n distinguishable objects P^n, without duplication, is given by

$$N(P^n) = n! \qquad \text{(B 5.1.1)}$$

2. Repeating in **1.** the j-th element i_j times, with the constraint that $\sum_{j=1}^{r} i_j = n$, gives

$$N(P^n_{i_1 \dots i_r}) = \frac{n!}{i_1! \cdot \dots \cdot i_r!} \qquad \text{(B 5.1.2)}$$

3. Choosing k objects out of n different objects, without duplication, is called a variation V^n_k. The number of implementations for this type of variation is

$$N(V^n_k) = \frac{n!}{(n-k)!} \qquad \text{(B 5.1.3)}$$

4. Repeating objects in **3.** yields the variation $\bar{V}^n_k$. The number of implementation in this case is

$$N(\bar{V}^n_k) = n^k \qquad \text{(B 5.1.4)}$$

5. If the ordering in **3.** is not considered, this task is called the combination of k elements out of n objects without repetition C^n_k. Its number of implementations is

$$N(C^n_k) = \binom{n}{k}. \qquad \text{(B 5.1.5)}$$

6. Repeating objects in **5.** is called a combination of k elements out of n objects with repetition $\bar{C}^n_k$. Its number of implementations is given by

$$N(\bar{C}^n_k) = \binom{n+k-1}{k} \qquad \text{(B 5.1.6)}$$

accumulate in a specific state, so that we use the canonical ensemble as a model. Therefore, the partition function reads

$$Z = \sum_{n_1, n_2, \dots} \frac{N!}{n_1! n_2! \dots} e^{(-\beta(n_1 E_1 + n_2 E_2 + \dots))} = (e^{-\beta E_1} + e^{-\beta E_2} + \dots)^N \qquad (5.23)$$

The resulting average number of particles $\bar{n}_j$ in a specific state j is given by

$$\bar{n}_j = -\frac{1}{\beta}\frac{\partial \ln Z}{\partial E_j} = -\frac{N \exp(-\beta E_j)}{\sum_k \exp(-\beta E_k)} \qquad (5.24)$$

and is called the Maxwell-Boltzmann distribution after its inventors (see Figure 5.2). This, in fact is the result we already saw in (5.14) with the only difference that we gave an explicit rule how to distribute the different particles on the different states.

5.2.2 Bose-Einstein Statistics

If the total particle number N is tuned by a parameter α, then we have to use the grand partition function to derive the average number of particles per energy-level. For the case of Bose-Einstein (BE) statistics, arbitrary numbers n_k are allowed. This yields for (5.22)

$$Q = \prod_i \sum_{n_i}^{\infty} \exp(-(\beta E_i + \alpha) n_i) = \prod_i \frac{1}{1 - e^{-(\alpha + \beta E_i)}} \tag{5.25}$$

Thus the logarithm of (5.25) is

$$\ln Q = -\sum_i \ln(1 - e^{-(\alpha + \beta E_i)}) \tag{5.26}$$

and inserting this into (5.20), we obtain

$$\ln Z = \alpha N - \sum_i \ln(1 - e^{-(\alpha + \beta E_i)}) \tag{5.27}$$

The partition function already includes the constraint of the total number of particles and thus may be used to calculate the average number of particles $\bar{n}_i$ for a specific energy level E_i

$$\bar{n}_i = -\frac{1}{\beta}\frac{\partial \ln Z}{\partial E_i} = -\frac{1}{\beta}\left[-\frac{\beta e^{-(\alpha + \beta E_i)}}{1 - e^{-(\alpha + \beta E_i)}} + \frac{\partial \ln Z}{\partial \alpha}\frac{\partial \alpha}{\partial E_i}\right] = \frac{1}{e^{\alpha + \beta E_i} - 1} \tag{5.28}$$

The derivative $\partial \ln Z / \partial \alpha$ comes from the fact that α may depend on the position of the energy levels. The fact that, for a given N, the parameter α maximizes the grand partition function, results in $\partial \ln Z / \partial \alpha = 0$. Equation (5.28) is called the Bose-Einstein distribution.

Chemical Potential

The important relation

$$\alpha = \frac{\partial \ln Z}{\partial N} = -\frac{\mu}{kT} \tag{5.29}$$

defines the chemical potential μ as the change of $\ln Z$ with respect to N. We find that small changes $d\ln Z$ reveal the significant parameters of the system

$$d\ln Z = \frac{\partial \ln Z}{\partial N}dN + \frac{\partial \ln Z}{\partial \beta}d\beta = -\frac{\mu}{k_B T}dN - \bar{E}d\beta \tag{5.30}$$

5.2.3 Fermi-Dirac Statistics

The important difference for Fermi-Dirac statistics is that we constrain each energy level to be occupied by only one particle. This changes the sum in (5.25) to give

$$Q = \prod_i \sum_{n_i = 0}^{1} \exp(-(\beta E_i + \alpha)n_i) = \prod_i (1 + e^{-(\alpha + \beta E_i)}) \tag{5.31}$$

We now take logarithms

$$\ln Q = \sum_i \ln(1 + e^{-(\alpha + \beta E_i)}) \tag{5.32}$$

and thus obtain the average occupation number by inserting (5.32) into (5.20)

$$\bar{n}_i = -\frac{1}{\beta}\frac{\partial \ln Z}{\partial E_i} = \frac{1}{\beta}\frac{\beta e^{-(\alpha + \beta E_i)}}{1 + e^{-(\alpha + \beta E_i)}} = \frac{1}{e^{\alpha + \beta E_i} + 1} \tag{5.33}$$

called the Fermi-Dirac distribution function after its inventors. The definition of the chemical potential follows the same procedure as in (5.29).

Pauli Principle

The reason why, in the Fermi-Dirac distribution, each energy level may be occupied by at maximum one particle, is given by the Pauli principle. The Pauli principle does not allow two electrons to occupy the same quantum state. They must differ at least in one quantum number. This single fact gives rise to the orbital structure of atoms, and hence to the large variety of molecules that can be formed.

5.2.4 Quasi Particles and Statistics

The question as to which statistics should be applied to what is answered by quantum-mechanical considerations. All phenomena occurring in a semiconductor obey quantum mechanics. Often we are not aware of this fact, such as when we are dealing with a macroscopically-sized sample. For special cases, such as on a very small length scale, at very low temperatures, or in a coherent light field from a laser, the quantum nature of phenomena is directly observable. The lattice vibrational amplitudes show discrete changes, the light intensity as well, and the electric current is composed of single charge carriers.

This discrete nature of the phenomena is best described by so called quasi-particles. These are nothing else than the single quanta by which the amplitudes of the each observable quantity may increase or decrease by. It is just by way of talking about quantum phenomena that we introduce the term "phonon" for the quantum of a lattice vibration, "photon" for the quantum of electromagnetic oscillation, and "electron" or "hole" for quantum moving charged carrier. The most mechanically intuitive picture is perhaps that of a phonon, which is the coaction of all the individual ions forming a crystal lattice.

In a similar manner as for a photon, we have to imagine an electron in the conduction band of a semiconductor. The picture we obtain is the net result of a composed excitation of the bare free-electron, together with the crystal structure forming the conduction band and the other electrons in the semi-conductor (and of course strictly-speaking many more effects.) It is not the electron as a massive particle that makes it different from a phonon. Rather, it is the kind of interacting subsystems that form the resulting phenomenon. In this light the "-ons", the quasi-particle family members, lose their mystery and remain simple descriptions of a phenomenon.

5.3 Applications of the Bose-Einstein Distributions

Lattice Vibrations

The normal vibrations of the crystal lattice as described in Chapter 2 may be described as set of linear non-interacting harmonic oscillators when neglecting the nonlinear interaction between the lattice atoms. Each of these oscillators i has a specific frequency ω_i and each of the phonons contributes with an energy $h\omega_i$. With n_i phonons present frequency ω_i we have a contribution of $E_i = n_i h\omega_i$ to the energy contained in that lattice vibration mode.

Photons

For photons, the situation looks alike and the only constraint on the partition function is given by the parameter β, the temperature, which determines the average energy of the system. There is no prescribed number of particles for a single sate i in either cases of photons or phonons. Any number of photons or phonons may occupy a given energy level. For this case the partition function reads

$$Z = \sum_k \exp(-\beta[\Sigma_i n_i E_i]_k) = \prod_i \sum_{n_i = 0}^{\infty} \exp(-\beta E_i n_i) = \prod_i \frac{1}{1 - e^{-\beta E_i}} \tag{5.34}$$

From (5.34) we can calculate the average number of particles with energy E_i

$$\bar{n}_j = -\frac{1}{\beta}\frac{\partial \ln Z}{\partial E_j} = \frac{e^{-\beta E_j}}{1 - e^{-\beta E_j}} = \frac{1}{e^{\beta E_j} - 1} \tag{5.35}$$

Planck Distribution

which is the well known Planck distribution for photons. It gives the statistical occupation number of a specific energy level for a photonic system. Consider a piece of material at a specific temperature, i.e., with given β. Then (5.35) describes the intensity distribution of the irradiated electromagnetic spectrum with respect to the frequency (note that the energy of the electromagnetic wave is proportional to its frequency.) Of course this is a non-equilibrium situation. Strictly, we also have to take into account the power absorbed from the environment and calculate the balance in order to determine the heat radiation. Nevertheless, (5.35) reveals the statistical nature of heat radiation from a material.

The description of phonons via (5.35) is valid only with some restrictions. At low temperature only waves with large wavelengths compared to the interatomic spacing are excited. In this case the lattice can be treated as a continuum. At short wavelengths the spectrum of frequencies shows a maximum value ω_{max}, which is independent of the shape of the crystal. In general the spectrum is quite difficult and must be calculated or measured in detail to give realistic distributions at higher frequencies. Nevertheless in a wide range of applications (5.35) is a good approximation.

For the case of photons and phonons, the partition function does not depend on the constraint of the number of particles N. Therefore, the derivative $\ln Z$ with respect to N vanishes and thus photon or phonon statistics are special cases of the Bose-Einstein statistics with $\mu = 0$.

5.4 Electron Distribution Functions

The distribution functions for electrons in the periodic lattice of a semiconductor follow Fermi-Dirac statistics. To distribute a certain number of electrons in the conduction band of a semi-conductor, we have to know the shape of the band with respect to the momentum of the electron, as explained in Section 3.3.3.

5.4.1 Intrinsic Semiconductors

An intrisic semiconductor is a pure crystal, where the valence band is completely occupied and the conduction band is completely empty. This is the whole truth for a purely classical description. Such a classical model does not allow electrons to occupy the valence band. We know, however, that the temperature, which gives us the parameter β, enters into the term for the probability of finding an electron with an energy E above the conduction band edge energy E_C, and is given by

$$f_e(E) = \frac{1}{e^{\beta(E-\mu)} + 1} \tag{5.36}$$

An intrinsic semiconductor is supposed to be uncharged. Therefore, we find that with a certain probability $f_h(E) = 1 - f(E)$ an electron is missing in the valence band. This missing electron we call a hole or defect electron. It behaves like a carrier with opposite charge. If the number densities n_h of positive (holes) and n_e of negative (electrons) carriers are equal the semiconductor is charge free. This number density is given by the integral over all possible implementations in energy space

$$n_e = \int_{E_c}^{\infty} D_e(E) f_e(E) dE \tag{5.37a}$$

$$n_h = \int_{E_V}^{\infty} D_h(E) f_h(E) dE = \int_{E_V}^{\infty} D_h(E)(1 - f_e(E)) dE \tag{5.37b}$$

Because of the discussion that lead to (5.8) the density of states for the valence band (D_h) and for the conduction band (D_e) appears in the integral. The energy of the valence band edge E_V and the conduction band edge E_C is fixed. Their difference $E_G = E_C - E_V$, the band gap energy, is a typical material parameter. The only parameter left to fulfil $n_e = n_h$ is the chemical potential μ. In the intrinsic case this is usually several $k_B T$ below the conduction band, so that

$$f(E) = \exp\left(\frac{\mu - E}{k_B T}\right) \tag{5.38a}$$

$$1 - f(E) = \exp\left(\frac{E - \mu}{k_B T}\right) \tag{5.38b}$$

is a good approximation. This allows us to solve (5.37a) and (5.37b) which gives

$$n_e = D_C \exp\left(\frac{\mu - E_C}{k_B T}\right) \tag{5.39a}$$

$$n_h = D_V \exp\left(\frac{E_V - \mu}{k_B T}\right) \tag{5.39b}$$

with

$$D_C = \frac{1}{4}\left(\frac{2m_e k_B T}{\pi h}\right)^{3/2} \tag{5.40a}$$

$$D_V = \frac{1}{4}\left(\frac{2m_h k_B T}{\pi h}\right)^{3/2} \tag{5.40b}$$

(5.40a) and (5.40b) are called the effective density of states. As a rule of thumb we say that the approximations made calculating the density are only good if the densities are lower than the respective density of states. Charge neutrality now requires

$$(m_e)^{3/2} \exp\left(\frac{\mu - E_C}{k_B T}\right) = (m_h)^{3/2} \exp\left(\frac{E_V - \mu}{k_B T}\right) \tag{5.41}$$

For (5.41) to hold, it is

$$\mu_i = \frac{E_C + E_V}{2} + \frac{3k_B T}{4} \ln\left(\frac{m_h}{m_e}\right) \tag{5.42}$$

where the index i stands for “intrinsic”. We see that for equal electron and hole masses $m_e = m_h$ the chemical potential lies exactly half way between the conduction band edge and the valence band edge. Furthermore it will not show any temperature dependence. The intrinsic carrier concentration thus reads

$$n_e = n_h = n_i = 4.9 \cdot 10^{15} \left(\frac{m_e m_h}{m_0}\right)^{3/4} T^{3/2} \exp\left(-\frac{E_G}{k_B T}\right) \tag{5.43}$$

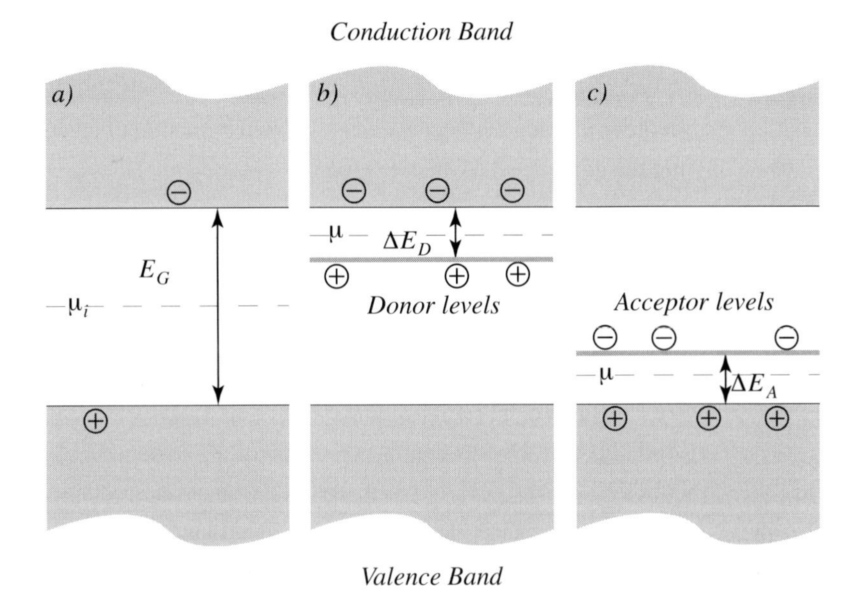

Figure 5.3. Band edges and chemical potentials for a) intrinsic, b) n-doped, and c) p-doped semiconductors.

We see that for a given T the n_i is highest for semiconductors with small band-gap. For given E_G it increases with temperature.

5.4.2 Extrinsic Semiconductors

A doped semiconductor is called extrinsic. Due to impurities or dopants that have energy levels in the band-gap near the conduction band (donors) or the valence band (acceptors), the carrier concentration is increased. Donors increase electron concentration, acceptors increase hole concentration (see Figure 5.3 b and c). For charge neutrality to hold

$$n_e + N_{A^-} = n_h + N_{D^+} \tag{5.44}$$

must be fulfilled. In (5.44) N_{A^-} denotes the ionized acceptor concentration and N_{D^+} denotes the ionized donor concentration (Note that in Figure 5.3 not all impurities are ionized). Let us focus on a purely n-type doping with total donor concentration N_D, and zero acceptor concentration N_A. In this case (5.44) turns into $n_e = n_h + N_{D^+}$.

In order to calculate N_{D^+} we must know the fraction of electron occupying the donor levels, which is given by

$$n_D = \frac{N_D}{1 + \exp\left(\frac{E_D - \mu}{k_B T}\right)} \tag{5.45}$$

(5.45) is calculated by applying integration in energy space of a Fermi-Dirac distribution (5.36) inside the band-gap. Usually there are no energy levels present. Due to doping we obtain at E_D a density of states given by $D(E) = \delta(E - E_D)$, where the delta function accounts for the fact that only the energy level E_D in the band-gap can be occupied. This gives $N_{D^+} = N_D - n_D$.

We are left with the task to calculate n_h. We do not know the chemical potential, but we know from (5.39a) and (5.39b) that, wherever it comes to lie

$$n_e n_h = D_C D_V \exp\left(\frac{-E_G}{k_B T}\right) = n_i^2 \tag{5.46}$$

holds. Thus we use $n_h = n_i^2 / n_e$ and obtain a second order equation for n_e reading

$$n_e^2 - n_e N_{D^+} - n_i^2 = 0 \tag{5.47}$$

Assuming that at the given temperature all donor atoms are ionized $N_D = N_{D^+}$ we obtain

$$n_e = \frac{N_D}{2}\left(1 + \sqrt{1 + \frac{4n_i^2}{N_D^2}}\right) \tag{5.48}$$

The value in (5.48) must equal that in (5.39a). This gives us the chemical potential

$$\mu = E_C + k_B T \ln\left(\frac{N_D}{2D_C}\left(1 + \sqrt{1 + \frac{4D_C D_V}{N_D^2}\exp\left(\frac{-E_G}{k_B T}\right)}\right)\right) \tag{5.49}$$

In the case of sufficiently low temperatures the ionized impurity concentration plays the leading part and we assume $n_h \ll N_{D^+}$ and thus write $n = N_{D^+}$. This limiting case has a chemical potential of

$$\mu = E_D + k_B T \ln\left(-\frac{1}{2} + \sqrt{\frac{1}{4} + \frac{N_D}{D_C}\exp\left(\frac{\Delta E_D}{k_B T}\right)}\right) \tag{5.50}$$

where ΔE_D was taken from Figure 5.3 b). In Figure 5.4 the chemical

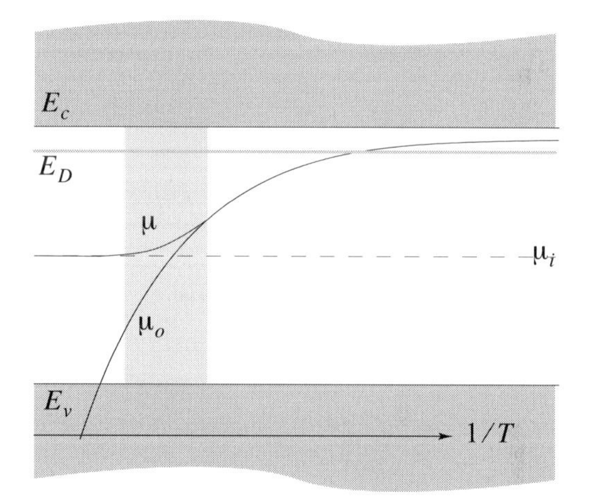

Figure 5.4. Chemical potential in three different temperature regimes.

potential in three different temperature regimes is shown. At high temperatures the intrinsic carrier density is much higher than the impurity concentration and the chemical potential shift to mid-gap. Then there follows a transition regime. For low temperatures the impurities are dominating and the chemical potential shifts between impurity level E_D and conduction band edge E_C.

5.5 Summary for Chapter 5

The concept of an ensemble allows us to tie many-particle model behavior to that of thermodynamics. Each subsequent ensemble introduces a further thermodynamic state variable, generalizing the concept. In order to handle the extraordinary large number of system states that this view presents, we need to evaluate the partition functions. For this, the particle statistics are introduced. One important result of this section is a method with which to obtain the density of states as a function of either k-space, or of energy. Armed with such an expression, we can describe the behavior of many-particle systems in a very compact manner. At the end of the chapter we have a good description for the population (the effective density of states) of charge carriers in the conduction and valence bands of silicon.

5.6 References for Chapter 5

5.1 F. Reif, *Fundamentals of Statistical and Thermal Physics*, McGraw-Hill (1965)

5.2 K Huang, *Statistical Mechanics*, John Wiley & Sons, New York (1987)

5.3 P.S.Kireev, *Semiconductor Physics*, MIR Publishers Moscow (1978)

Chapter 6 Transport Theory

The study of semi-classical time-dependent processes usually separates the constitutive from the balance equations. The idea is the following: constitutive equations describe the way in which a particular material can influence a particular transport process, usually by adding resistance. For example, the atoms of the silicon crystal represent scattering centers for electrons moving through. Chapter 2 to Chapter 4 discussed these constitutive relations in quite some detail without looking at either transport, or the interaction between transporting particles. The balance equations usually implement some postulate of mechanics—these are often called integrals of the motion. The balance equations demand the conservation of an extensive property, such as energy, or momentum, and so on. When we are dealing with an ensemble of particles, with a distribution among some states, and we do not want to track each particle individually, then we require a transport theory which is adjusted to this view. Since electrons have many features, and electron transport is so very important in device engineering, we focus much of our attention on describing the transport of electrons through silicon. In describing transport we introduce the concept of scattering. This term will obtain more meaning in Chapter 7, where we look at the interactions between the various subsystems.

Chapter Goal The goal of this chapter is to show how to describe transport phenomena and how to compute the transport of electrons through a semiconductor device.

Chapter Roadmap We start with the Boltzmann transport equation, describing its terms, and show how it should be modified for particular particle types. Zooming in, we first describe local equilibrium. Next, we see how local equilibrium and the global balance equations of thermodynamics are related. This leads us to a number of models for classical and semi-classical transport of charge carriers in silicon, including the very successful drift-diffusion equations. Finally, we look at numerical methods used to solve the transport equations.

6.1 The Semi-Classical Boltzmann Transport Equation

The semi-classical Boltzmann transport equation (BTE) gives a description of the electronic system in terms of single particle distribution functions $f(\boldsymbol{p}, \boldsymbol{x}, t)$, i.e., a time-dependent density in a six-dimensional single particle phase space. A single particle is identified by its three spatial coordinates and three momentum values. For the momentum of a free particle we know that $\boldsymbol{p} = \hbar\boldsymbol{k} = m\boldsymbol{v}$. In this case we could write the distribution function in terms of velocities $f(\boldsymbol{v}, \boldsymbol{x}, t)$. From the discussion in Section 3.3.3 we know that this simple relation does not hold in a semiconductor in general. Though we shall restrict our discussion to energies near the band extrema, where $\boldsymbol{p} = m^*\boldsymbol{v}$ holds, with the effective mass m^*, we prefer to take the wave vector $\boldsymbol{k}$ as the independent variable. This will remind us that in the case of electron we are dealing with wave phenomena as well as with particle properties. A more formal reason is that position and momentum are canonically conjugated variables. According to (5.6) the density of states is $1/(2\pi)^3$. Note that for two independent spin directions an electron may assume this value twice. Thus $(4\pi^3)^{-1}f(\boldsymbol{k}, \boldsymbol{x}, t)d^3\boldsymbol{k}d^3\boldsymbol{x}$ gives the number of particles which reside in a small volume element $d^3\boldsymbol{x}$ around the position vector $\boldsymbol{x}$ and in a small volume element $d^3\boldsymbol{k}$ around the wave vector $\boldsymbol{k}$. We think of this phase space simply as the vector space formed by the three positions and the three momenta of a carrier. $f(\boldsymbol{k}, \boldsymbol{x}, t)$ is a particle density in phase space similar to the charge density and total charge in (4.33). Performing the integration over the entire phase space Ω the system assumes

$$(4\pi^3)^{-1}\int_{\Omega} f(\boldsymbol{k}, \boldsymbol{x}, t)d^3\boldsymbol{k}d^3\boldsymbol{x} = N \tag{6.1}$$

gives us the number N of particles present in the system.The BTE is a law of motion for a single particle density to evolve in this space. A general treatment of the BTE can be found in 6.1 a more semiconductor specific description is given in [6.2] and [6.3]

$$\frac{\partial}{\partial t}f(\boldsymbol{k}, \boldsymbol{x}, t) + \dot{\boldsymbol{k}}\nabla_k f(\boldsymbol{k}, \boldsymbol{x}, t) + \dot{\boldsymbol{x}}\nabla_x f(\boldsymbol{k}, \boldsymbol{x}, t) = C(f(\boldsymbol{k}, \boldsymbol{x}, t)) \tag{6.2}$$

The l.h.s. of (6.2) is called the streaming motion term or convective term. The path a single particle follows we call a phase-space trajectory. Knowing this trajectory of a single particle with arbitrary initial condition the streaming motion term is determined. This will be discussed in the next section. The r.h.s. of (6.2) is called the collision term or scattering term and is discussed in 6.1.2.

6.1.1 The Streaming Motion

In the case where $C(f(\boldsymbol{k}, \boldsymbol{x}, t)) = 0$ there is only streaming motion along the individual particle trajectories. These trajectories are given once the temporal evolution of position $\boldsymbol{x}(t)$ and velocity $\boldsymbol{k}(t)$ of each particle are known. If there are no particles created or destroyed in a given volume element of the phase space the total time derivative of the phase-space density vanishes

$$\frac{d}{dt}f(\boldsymbol{k}, \boldsymbol{x}, t) = \frac{\partial}{\partial t}f(\boldsymbol{k}, \boldsymbol{x}, t) + \dot{\boldsymbol{k}}\nabla_k f(\boldsymbol{k}, \boldsymbol{x}, t) + \dot{\boldsymbol{x}}\nabla_x f(\boldsymbol{k}, \boldsymbol{x}, t) \tag{6.3}$$

where ∇_k and ∇_x denote the gradients with respect to the wave vector and the position. A volume in phase-space evolves in time according to the trajectories that its member particles follow. Assuming that $\dot{\boldsymbol{x}} = \boldsymbol{v} = \hbar \boldsymbol{k}/m$ and combine position and wave vector into a single vector $\xi = \{\boldsymbol{x}, \boldsymbol{k}\}$ in the six-dimensional phase-space, we rewrite (6.3) as a continuity equation like that for current continuity in Section 4.1.2

$$\frac{\partial}{\partial t}f(\xi(t), t) + \nabla_\xi(f(\xi(t), t)\dot{\xi}) = 0 \tag{6.4}$$

If we write $\dot{\boldsymbol{k}} = \boldsymbol{F}/\hbar$ considering the Newton law, where $\boldsymbol{F}$ does not depend on the wave vector, and take into account that $\boldsymbol{k}$ does not explicitly depend on $\boldsymbol{x}$, $\nabla_\xi \dot{\xi} = 0$ holds and (6.4) corresponds to (6.3). The term $f(\xi(t), t)\dot{\xi}$ we interpret as a probability current density in phase-space, i.e., the probability density that flows out of or into a six-dimensional phase-space volume element $d^3\boldsymbol{x}d^3\boldsymbol{k} = d^6\xi$.

For a constant external force $\boldsymbol{F}$ along the x-direction the deformation of a unit circle in a two dimensional phase space (x, k_x) is shown in Figure 6.1. Without any applied force the phase-space density changes

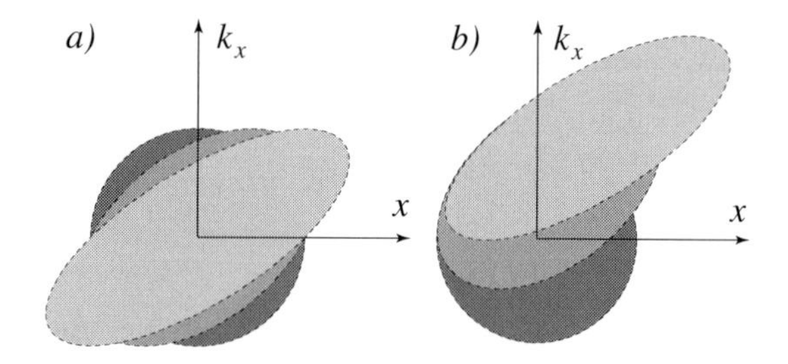

Figure 6.1. Phase-space density evolving according to (6.3), a) without applied force and b) with constant applied force in x-direction.

shape from the unit circle to an ellipse. With applied constant external force the initial circular density is shifted and distorted. Both cases conserve the total initial area, i.e., the particle density.

6.1.2 The Scattering Term

In the case where $C(f(\mathbf{k}, \mathbf{x}, t)) \neq 0$ there are particles removed from a position in the three dimensional wave vector subspace of the phase space and put to a new position. This process is nothing else than a momentum transfer in a scattering process. Suppose the scattering process itself takes place on a time-scale much shorter than that defined by the streaming motion and the time between two scattering processes. This leads to the assumption that on the latter time-scale all scattering processes happen instantaneously, i.e., it takes no time for the particle to switch from one velocity to another one due to the scattering process. This situation is shown in Figure 6.2 (a-c). Momentum is transferred in a

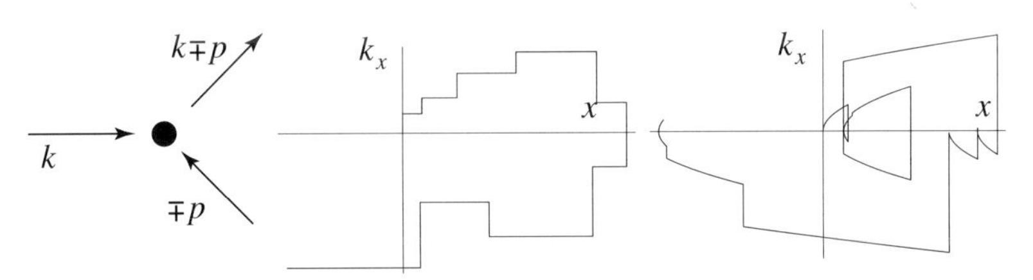

Figure 6.2. a) Momentum transfer in a scattering process, b) single particle trajectory without applied external force field, and c) trajectory in the presence of scattering and applied external field.

very short time from an external reservoir to the particle. Imagine small liquid molecules hitting an object immersed in the liquid or electrons being hit by phonons. In the case of absent external driving force this causes the single particle to follow a trajectory as sketched in Figure 6.2 (b) where the particle follows straight lines in space with constant velocity until it suffers the next scattering event. The situation changes with applied external field as depicted in Figure 6.2 (c). The trajectories between scattering events are no more straight lines. In this picture the approximation of instantaneous scattering events becomes clear: the change in the particle velocity or momentum due to scattering must always be unresolved on the scale shown compared with the acceleration that the particle suffers from an external field. The mathematical description of how the scattering process enters (6.2) is by means of transition probabilities, i.e.

$$C(f(\boldsymbol{k},\boldsymbol{x},t)) = \sum_{\boldsymbol{k}'}[W(\boldsymbol{k}'\rightarrow\boldsymbol{k})f(\boldsymbol{k}',\boldsymbol{x},t)-W(\boldsymbol{k}\rightarrow\boldsymbol{k}')f(\boldsymbol{k},\boldsymbol{x},t)] \quad (6.5)$$

$W(\boldsymbol{k}'\rightarrow\boldsymbol{k})$ stands for a transition rate of particles with wave vector $\boldsymbol{k}'$ being scattered into the state $\boldsymbol{k}$ and therefore called in-scattering term. $W(\boldsymbol{k}\rightarrow\boldsymbol{k}')$ stands for the transition rate of particles at $\boldsymbol{k}$ being scattered to $\boldsymbol{k}'$ which is accounted for as a loss term and thus we call it an out-scattering term. The transition rate W itself is nothing else than a number of events per second for the process $\boldsymbol{k}'\rightarrow\boldsymbol{k}$ or $\boldsymbol{k}\rightarrow\boldsymbol{k}'$. It clearly depends on the initial situation and the outcome of the specific process taking place, i.e., $\boldsymbol{k}$ and $\boldsymbol{k}'$. In the form given in (6.5) it only depends on the probability of finding a particle with $\boldsymbol{k}$ for the out-scattering process or with $\boldsymbol{k}'$ for the in-scattering process. In principle the equation could be more complicated because the probability for the state of the scattering outcome enters in addition. This results in a nonlinear scattering term of the form

$$\sum_{\boldsymbol{k}'}(W(\boldsymbol{k}'\rightarrow\boldsymbol{k})f(\boldsymbol{k}'(t),\boldsymbol{x}(t),t)(1-f(\boldsymbol{k}(t),\boldsymbol{x}(t),t))$$
$$-W(\boldsymbol{k}\rightarrow\boldsymbol{k}')f(\boldsymbol{k}(t),\boldsymbol{x}(t),t)(1-f(\boldsymbol{k}'(t),\boldsymbol{x}(t),t))] \quad (6.6)$$

where $(1-f(\boldsymbol{k}(t),\boldsymbol{x}(t),t))$ and $(1-f(\boldsymbol{k}'(t),\boldsymbol{x}(t),t))$ account for the probability of finding the state $\boldsymbol{k}$ and $\boldsymbol{k}'$ unoccupied. This is the case for quantum mechanical descriptions where the Pauli principle becomes active, i.e., for fermions.

6.1.3 The BTE for Phonons

In the case of phonons we know that in equilibrium they follow a Bose–Einstein distribution

$$g_0 = \bar{n}(\omega_q) = \frac{1}{e^{\beta\hbar\omega_q} - 1} \tag{6.7}$$

which is the average number of phonons as given in (5.28) with the frequency $\omega_q = \omega(\boldsymbol{q})$, where $\boldsymbol{q}$ is the wave vector. From the average number of phonons we define a distribution function $g(\boldsymbol{q}, \boldsymbol{x}, t)$, i.e we allow for spatial variations of the phonon number density. The only term that gives a driving force for phonons will be a gradient with respect to the spatial coordinates. The BTE for phonons thus is given by

$$\frac{\partial}{\partial t}g(\boldsymbol{q}, \boldsymbol{x}, t) + \dot{\boldsymbol{x}}\nabla_x g(\boldsymbol{q}, \boldsymbol{x}, t) = C(\boldsymbol{q}, \boldsymbol{x}, t) \tag{6.8}$$

We know that the group velocity $\boldsymbol{v}_q = \dot{\boldsymbol{x}} = \nabla_q\omega(\boldsymbol{q})$ is the gradient of the frequency with respect to the wave-vector $\boldsymbol{q}$. To go further we have to take into account spatial variations of the phonon frequency $\delta\omega(\boldsymbol{q}, \boldsymbol{x}, t)$, which then results in an equation of the form

$$\begin{aligned} &C(\boldsymbol{q}, \boldsymbol{x}, t) \\ &= \left(\frac{\partial}{\partial t} + (\boldsymbol{v}_q + \nabla_q\delta\omega(\boldsymbol{q}, \boldsymbol{x}, t))\nabla_x + \nabla_x\delta\omega(\boldsymbol{q}, \boldsymbol{x}, t)\nabla_q\right)g(\boldsymbol{q}, \boldsymbol{x}, t) \end{aligned} \tag{6.9}$$

The reason for the appearance of $\delta\omega(\boldsymbol{q}, \boldsymbol{x}, t)$ is a perturbation of the system by a momentum dependent potential, e.g., long wavelength elastic distortions. Therefore, an anharmonic theory of phonons must be performed [6.6] and a phonon-phonon interaction included (see Section 7.1).

6.1.4 Balance Equations for Distribution Function Moments

In (6.1) the interpretation of the function $f(\boldsymbol{k}(t), \boldsymbol{x}(t), t)$ as a density became clear by the integration over velocities and positions. If we only integrate over the wave vector space we discard any information about the velocity of the respective particles and retain only information about the density in positional space

Density

$$n(\boldsymbol{x}, t) = \frac{1}{4\pi^3} \int_{\Omega_k} f(\boldsymbol{k}, \boldsymbol{x}, t) d^3\boldsymbol{k} \tag{6.10}$$

This is equivalent to the zeroth order moment of $f(\boldsymbol{k}, \boldsymbol{x}, t)$ with respect to the wave vector and we call $n(\boldsymbol{x}, t)$ the spatial density of the system. The first-order moment

Current Density

$$\boldsymbol{j}_n(\boldsymbol{x}, t) = \frac{1}{4\pi^3} \int_{\Omega_k} \frac{1}{\hbar}(\nabla_k E(\boldsymbol{k})) f(\boldsymbol{k}, \boldsymbol{x}, t) d^3\boldsymbol{k} = \boldsymbol{v}_a n(\boldsymbol{x}, t) \tag{6.11}$$

represents the particle current density, where we know that $\nabla_k E(\boldsymbol{k})/\hbar = \boldsymbol{v}$ is the group velocity of a particle with a specific $\mathbf{k}$. $\boldsymbol{v}_a$ is the average velocity of a particle. This velocity times the particle density results in the current density as given in (6.11). The energy density

Energy Density

$$u(\boldsymbol{x}, t) = \frac{1}{4\pi^3} \int_{\Omega_k} E(\boldsymbol{k}) f(\boldsymbol{v}, \boldsymbol{x}, t) d^3 v \tag{6.12}$$

is a second order moment if the energy given in the harmonic approximation $E = \hbar^2 k^2/(2m^*)$. The same approximation yields a third order moment

Energy Current Density

$$\boldsymbol{j}_u(\boldsymbol{x}, t) = \frac{1}{4\pi^3} \int_{\Omega_k} E(\boldsymbol{k}) \boldsymbol{v} f(\boldsymbol{k}, \boldsymbol{x}, t) d^3\boldsymbol{k} \tag{6.13}$$

that is called the energy current density. In general we have the r-th order moment m_r given by

$$m_r(\boldsymbol{x}, t) = \frac{a_r}{4\pi^3} \int_{\Omega_k} \boldsymbol{v}^r f(\boldsymbol{k}, \boldsymbol{x}, t) d^3\boldsymbol{k} \tag{6.14}$$

A special second-order moment not contained in (6.12) is the momentum current density, a tensorial quantity given by

Momentum Current Density Tensor

$$\Pi(x, t) = \frac{1}{4\pi^3}\int_{\Omega_k} v \otimes v f(k, x, t) d^3 k \tag{6.15}$$

where the tensor product denotes $(v \otimes v)_{ij} = v_i v_j$. The sum of diagonal elements (6.15) is proportional to the energy density in the harmonic approximation. The infinite sequence of moments represents the distribution of the system as well as $f(k, x, t)$ does. So far there is no advantage in looking at the moments of the distribution function. The only advantage arises when we are satisfied with the knowledge of a few of momenta, e.g., the particle density.

To derive equations of motions for the momenta given in (6.10)-(6.13) we integrate the BTE multiplied with powers of velocity v^r with respect to the wave vector–space. For this purpose we assume that Newton's law holds and the force acting on each particle fully determines the time derivative of the wave vector, i.e., we may set $\dot{k} = F(x, v)/\hbar$. Note that we leave the possibility of the force depending on the particle velocity open. This will enable us to include interactions with magnetic fields. For now we assume the force $F(x)$ to depend only on spatial coordinates, and thus for $r = 0$ we obtain

$$C_n(x) = \int_{\Omega_k}\left[\frac{\partial}{\partial t} f(k, x, t) + \frac{F(x)}{\hbar}\nabla_k f(k, x, t) + v\nabla_x f(k, x, t)\right] \mathrm{d}^3 k \tag{6.16}$$

The first term in (6.16) gives the time derivative of the carrier density $\dot{n}(x, t)$, the second term vanishes due to the Gauss theorem and the fact that $f(k(t), x(t), t)$ vanishes exponentially as $k \to \infty$

$$\int_{\Omega_k} F(x)\nabla_v f(k, x, t)\mathrm{d}^3 v = \int_{\Gamma(\Omega_k)} F(x) f(k, x, t) d\Gamma(\Omega_k) = 0 \tag{6.17}$$

The third term gives, by exchanging integration and spatial gradient,

$$\int_{\Omega_k}[v\nabla_x f(\boldsymbol{k},\boldsymbol{x},t)]\mathrm{d}^3\boldsymbol{k} = \nabla_x\int_{\Omega_k} v f(\boldsymbol{k},\boldsymbol{x},t)\mathrm{d}^3\boldsymbol{k} = \nabla_x \boldsymbol{j}_n(\boldsymbol{x},t) \tag{6.18}$$

the gradient of the particle current density. The r.h.s. of (6.16) results in a source term for particles which vanishes for systems that conserve the particle number. In this case the resulting equation is the well-known continuity equation. The general case we call the particle density balance equation

Density Balance

$$\frac{\partial}{\partial t}n(\boldsymbol{x},t) + \nabla_x \boldsymbol{j}_n(\boldsymbol{x},t) = C_n(\boldsymbol{x}). \tag{6.19}$$

If we integrate (6.19) over a finite region R in real space, the resulting equation states that the change in number of particles in R is given by the flux through its surface $\Gamma(R)$. Multiplying the BTE with v and integrating we obtain

$$C_{j_n}(\boldsymbol{x}) = \int_{\Omega_k} v\left[\frac{\partial}{\partial t}f(\boldsymbol{k},\boldsymbol{x},t) + \frac{\boldsymbol{F}(\boldsymbol{x})}{\hbar}\nabla_k f(\boldsymbol{k},\boldsymbol{x},t) + v\nabla_x f(\boldsymbol{k},\boldsymbol{x},t)\right]\mathrm{d}^3\boldsymbol{k} \tag{6.20}$$

The first term gives the time derivative of particle current density. The second term is an accelerating term that we obtain by applying the Gauss theorem. The third term results in the divergence of the momentum current density (6.15). Thus we obtain

Momentum balance

$$\frac{\partial}{\partial t}\boldsymbol{j}_n(\boldsymbol{x},t) - \frac{\boldsymbol{F}}{m}n(\boldsymbol{x},t) + \nabla_x \boldsymbol{\Pi}(\boldsymbol{x},t) = C_{j_n}(\boldsymbol{x}). \tag{6.21}$$

This corresponds to a continuity equation for the momentum and thus the r.h.s of (6.21) must be interpreted as a source of momentum or a momentum relaxation term. We multiply the BTE by $E(\boldsymbol{k})$. Integrating and applying the same arguments to the single terms as above, we obtain the energy balance equation

Energy Balance

$$\frac{\partial}{\partial t}u(\boldsymbol{x},t) - \boldsymbol{F}\boldsymbol{j}_n(\boldsymbol{x},t) + \nabla_x \boldsymbol{j}_u(\boldsymbol{x},t) = C_u(\boldsymbol{x}) \tag{6.22}$$

The single terms are easily interpreted: first there is the change in energy density in time, second we have an energy source term, which, with the electric field as the force term $\boldsymbol{F} = -q\boldsymbol{E}$, is the Joule heating of electrons, and third we have the divergence of the energy current density. The term on the r.h.s. we interpret as an energy relaxation term.

From the balance equations we see that the momentum procedure as applied to the BTE produces an infinite series of coupled equations: the continuity equation (6.19) which contains the particle current density that has its equation of motion given by the momentum balance (6.21) and contains the momentum current density tensor as given in (6.15). One part of the momentum current density tensor is the energy density, which has it equation of motion given by the energy balance (6.22), which in turn contains the energy current density, i.e., a moment of third order, which has its own equation of motion coupled to higher moments, etc.

This is an infinite hierarchy of equations that overall correspond to the BTE. Here the same arguments hold as given for the infinite series of moments representing the distribution function. Let us assume that for our purpose additional information about higher r-th moments does only contribute to marginal changes in the description of our system. This means that we may break the hierarchy at a certain point and discard or approximate all higher order moments. Therefore, some knowledge about approximating or modeling the current densities is needed.

6.1.5 Relaxation Time Approximation

Up to know we did not specify the scattering term of the BTE in more detail than given by (6.5) or (6.6). The scattering or transition probabilities $W(\boldsymbol{k} \rightarrow \boldsymbol{k}')$ must be calculated by taking into account the microscopic interaction process between particles. This is where again quantum mechanics enters. In this section we will do another approximation step which is in good agreement with the above–derived moment equations. We assume the distribution function to be an equilibrium distribution at least locally in space. In addition we assume that the scattering term may be written as

$$C(f(\boldsymbol{k}, \boldsymbol{x}, t)) = -\frac{f(\boldsymbol{k}, \boldsymbol{x}, t) - f_0(\boldsymbol{k}, \boldsymbol{x}, t)}{\tau} = \tau^{-1}\delta f(\boldsymbol{k}, \boldsymbol{x}, t), \quad (6.23)$$

i.e., a rate τ^{-1} multiplied by the deviation $\delta f(\boldsymbol{k}, \boldsymbol{x}, t)$ of the actual distribution function $f(\boldsymbol{k}, \boldsymbol{x}, t)$ from its equilibrium value $f_0(\boldsymbol{k}, \boldsymbol{x}, t)$, see Figure 6.3.

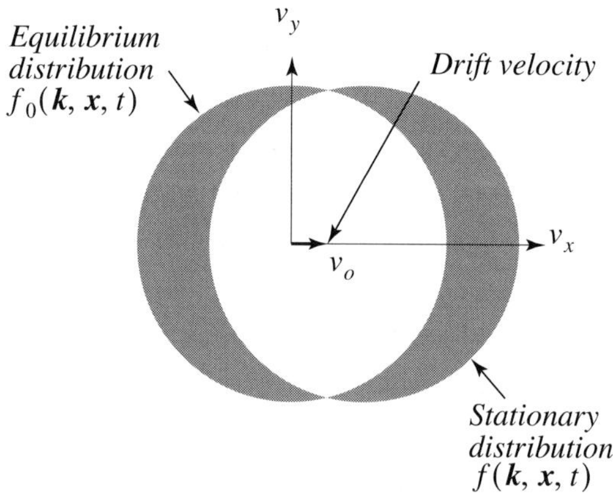

Figure 6.3. Schematic diagram of equilibrium and stationary deformed distribution function as a result of the relaxation time approximation for the scattering term in the BTE.

From the form (6.23) of the scattering term we immediately derive the following statements: without any external force term the system assumes a spatially homogeneous equilibrium state provided that $f_0(\boldsymbol{k}, t)$ does not vary in space. This is called thermodynamic equilibrium. Once an external field is applied the distribution deforms until relaxation term and streaming motion term balance. Schematically this situation is shown in Figure 6.3. The hatched region represents the difference $\delta f(\boldsymbol{k}, \boldsymbol{x}, t)$. This will be the only part of the distribution function that results in finite fluxes of moments of any order as calculated above.

Note that the relaxation time has been assumed constant, which may result in a more restricted approximation than intended. Keeping the relaxation time $\tau(\boldsymbol{k})$ as a function of $\boldsymbol{k}$ may even allow for anisotropic effects and different relaxation laws for the moments of a specific order. This is why in the balance equations we might distinguish between particle relaxation, momentum relaxation and energy relaxation terms. This in turn makes a difference for the dynamics of momentum and energy and may allow for further distinguishing timescales of the respective moments.

Bloch Oscillations

Before proceeding with local descriptions we look at the phenomenon called Bloch oscillations. Therefore, we drop our assumption of parabolic band and look at a the entire conduction band periodic in momentum space that may be described by $E = (\sin\hbar k)^2/(2m) = (\sin p)^2/(2m)$. Suppose an electron starts at $t = 0$ with $\boldsymbol{p} = \hbar\boldsymbol{k} = 0$ at the conduction band edge at the space point $\boldsymbol{x}_0$. Owing to an applied electric field $\boldsymbol{E}$ it accelerates and gains momentum according to Newton's law $\dot{\boldsymbol{p}} = \boldsymbol{F} = -q\boldsymbol{E}$. For small momenta $E = \boldsymbol{p}^2/(2m)$ and thus the velocity of the electron is given by $\boldsymbol{v} = \partial E/\partial\boldsymbol{p} = \boldsymbol{p}/m$. For the given dispersion relation after half of the Brillouin zone the velocity of the particle starts to decrease with increasing momentum until at the zone edge the velocity is zero. Moreover further acceleration results in a negative velocity, i.e., the particle turns back to its starting point $\boldsymbol{x}_0$ where the whole process repeats. This is a striking consequence of the conduction band picture of electrons since a constant electric field applied results in an alternating current. It is contrary to our experience of electronic behavior under normal circumstances and thus there must rather be something wrong with the picture. Indeed, a very important thing is missing in the description above: the interaction of the electron with phonons. At room temperature the huge number of phonons provide many scattering events so that the above scenario is not observable. Even at very low temperatures where only a few phonons exist the emission of phonons impedes this effect. As soon as an electron gains enough energy it will bounce back in energy emitting a phonon and restart to accelerate until it newly loses energy to the phonon system.

Since optical phonons have energies of about 50 meV and the conduction band has a width of the order of 1 eV and more there, will be almost no electrons able to overcome this barrier and turn back. Therefore, in a bulk semiconductor crystal Bloch oscillation are not observed.

On the other hand, the above example tells us that all material properties of electrons needed to describe fluxes of moments like the energy current density or the particle current density are in this special case due to phonon interaction, i.e., momentum transfer between the phonon system and the electronic system. Since there is a dissipation channel opened for the electronic system, these fluxes are termed irreversible. The term in the BTE responsible for this process is the collision term, which we have approximated in this section by a relaxation time. Therefore, all constitutive laws for the current densities will be connected to this relaxation time.

6.2 Local Equilibrium Description

As already pointed out in the previous section, we want to analyze what kind of constitutive equations follow from the deviation from thermal equilibrium if we apply a relaxation time approximation. Therefore, we introduce the picture of local equilibrium. This means that at least locally in position space a thermodynamic equilibrium distribution $f_0(\boldsymbol{k}, \boldsymbol{x}, t)$ exists to which $f(\boldsymbol{k}, \boldsymbol{x}, t)$ relaxes. Let us assume that the equilibrium distribution is given by a Fermi distribution

$$f_0(\boldsymbol{k}, \boldsymbol{x}, t) = \frac{1}{\exp\left(\frac{E(\boldsymbol{k}) - \mu(\boldsymbol{x}, t)}{kT(\boldsymbol{x}, t)}\right) + 1} \tag{6.24}$$

We see that in (6.24) the chemical potential μ and the temperature T appear time and space dependent.

6.2.1 Irreversible Fluxes and Thermodynamic Forces

The distribution function that locally deviates from its equilibrium value is given by $f(\boldsymbol{k}, \boldsymbol{x}, t) = f_0(\boldsymbol{k}, \boldsymbol{x}, t) + \delta f(\boldsymbol{k}, \boldsymbol{x}, t)$. Once we have $\delta f(\boldsymbol{k}, \boldsymbol{x}, t)$ it is possible to calculate all the current densities introduced in 6.1.4 as moments of $\delta f(\boldsymbol{k}, \boldsymbol{x}, t)$. We insert the expansion of the distribution function into the BTE (6.2) with the scattering term given by a relaxation time approximation (6.23) and obtain

$$\frac{\partial}{\partial t} f(\boldsymbol{k}, \boldsymbol{x}, t) + \frac{\boldsymbol{F}}{\hbar} \nabla_k f(\boldsymbol{k}, \boldsymbol{x}, t) + \boldsymbol{v} \nabla_x f(\boldsymbol{k}, \boldsymbol{x}, t) = -\frac{\delta f(\boldsymbol{k}, \boldsymbol{x}, t)}{\tau}. \quad (6.25)$$

The deviation is assumed to be small and thus to a first approximation its contribution to the streaming motion term is neglected. We obtain

$$\frac{\partial}{\partial t} f_0(\boldsymbol{k}, \boldsymbol{x}, t) + \frac{\boldsymbol{F}}{\hbar} \nabla_k f_0(\boldsymbol{k}, \boldsymbol{x}, t) + \boldsymbol{v} \nabla_x f_0(\boldsymbol{k}, \boldsymbol{x}, t) = -\frac{\delta f(\boldsymbol{k}, \boldsymbol{x}, t)}{\tau} \quad (6.26)$$

Inserting (6.24) in (6.26) we can immediately solve for $\delta f(\boldsymbol{k}, \boldsymbol{x}, t)$. To this end we calculate the effect of the streaming motion operator on the Fermi distribution as given by (6.24), where the gradient in k-space gives

$$\nabla_k f_0(\boldsymbol{k}, \boldsymbol{x}, t) = \frac{\partial f_0(E(\boldsymbol{k}), \boldsymbol{x}, t)}{\partial E} \nabla_k E(\boldsymbol{k}). \quad (6.27)$$

The gradient in real space is

$$\nabla_x f_0(\boldsymbol{k}, \boldsymbol{x}, t) = \frac{\partial f_0(E(\boldsymbol{k}), \boldsymbol{x}, t)}{\partial E} k_B T \left[-\frac{E - \mu}{k_B T^2} \frac{\partial T}{\partial \boldsymbol{x}} - \frac{1}{k_B T} \frac{\partial \mu}{\partial \boldsymbol{x}} \right] \quad (6.28)$$

The partial derivative with respect to time gives

$$\frac{\partial f_0(E(\boldsymbol{k}), \boldsymbol{x}, t)}{\partial t} = \frac{\partial f_0(E(\boldsymbol{k}), \boldsymbol{x}, t)}{\partial E} \left[-\frac{E - \mu}{T} \frac{\partial T}{\partial t} - \frac{\partial \mu}{\partial t} \right] \quad (6.29)$$

Solving (6.26) for δf we obtain

$$\delta f(\boldsymbol{k}, \boldsymbol{x}, t) = -\tau\frac{\partial f_0(E(\boldsymbol{k}), \boldsymbol{x}, t)}{\partial E}\left[-\frac{E-\mu}{T}\frac{\partial T}{\partial t} - \frac{\partial \mu}{\partial t} + \boldsymbol{v}\left(\boldsymbol{F} - \frac{\partial \mu}{\partial \boldsymbol{x}} - \frac{E-\mu}{T}\frac{\partial T}{\partial \boldsymbol{x}}\right)\right] \tag{6.30}$$

We restrict the situation to a stationary analysis, where the time derivatives in (6.30) vanish. Let us assume that the external force $\boldsymbol{F}$ is given by an electric field $\boldsymbol{E}$, which itself can be expressed as the gradient of an electrostatic potential (see (4.34)). The force acting on an electron is then $\boldsymbol{F} = -q\boldsymbol{E} = q\nabla\psi$. Note that the minus sign is due to its negative charge. This allows us to combine the chemical potential μ and the electrostatic potential ψ into one term forming the electrochemical potential

Electrochemical Potential

$$\eta = \mu - q\psi \tag{6.31}$$

Then the current densities become

Current Density

$$\boldsymbol{j}_n(x, t) = -\tau\int_{V_k}\boldsymbol{v}\frac{\partial f_0}{\partial E}\boldsymbol{v}\left(-\frac{\partial \eta}{\partial \boldsymbol{x}} - \frac{E-\mu}{T}\frac{\partial T}{\partial \boldsymbol{x}}\right)d^3\boldsymbol{k} \tag{6.32}$$

and

Energy Current Density

$$\boldsymbol{j}_u(x, t) = -\tau\int_{V_k}\boldsymbol{v}E\frac{\partial f_0}{\partial E}\boldsymbol{v}\left(-\frac{\partial \eta}{\partial \boldsymbol{x}} - \frac{E-\mu}{T}\frac{\partial T}{\partial \boldsymbol{x}}\right)d^3\boldsymbol{k} \tag{6.33}$$

By a simple manipulation we ca see that the latter consists of two terms

$$\boldsymbol{j}_u(x, t) = \mu\boldsymbol{j}_n \underbrace{-\tau\int_{V_k}\boldsymbol{v}(E-\mu)\frac{\partial f_0}{\partial E}\boldsymbol{v}\left(-\frac{\partial \eta}{\partial \boldsymbol{x}} - \frac{E-\mu}{T}\frac{\partial T}{\partial \boldsymbol{x}}\right)d^3\boldsymbol{k}}_{\boldsymbol{j}_Q} \tag{6.34}$$

With (6.34) we see that the current density consists of two parts: one that is due to particle transport through spatial regions $\mu\boldsymbol{j}_n$ and a second that is due to heat transport $\boldsymbol{j}_Q$. Since the first part of the energy current is proportional to the particle current, in the following we focus on $\boldsymbol{j}_Q$ and $\boldsymbol{j}_n$. Both current densities suggest a general structure of the form

$$\begin{bmatrix} \boldsymbol{j}_n \\ \boldsymbol{j}_Q \end{bmatrix} = -\begin{bmatrix} \boldsymbol{N}_{11} & \boldsymbol{N}_{12} \\ \boldsymbol{N}_{21} & \boldsymbol{N}_{22} \end{bmatrix} \begin{bmatrix} \boldsymbol{X}_1 \\ \boldsymbol{X}_2 \end{bmatrix} \tag{6.35}$$

$\boldsymbol{X}_j$ represents the respective thermodynamic force, which in this special case is

$$\boldsymbol{X}_1 = \frac{\partial \eta}{\partial \boldsymbol{x}} \tag{6.36}$$

the gradient of the electrochemical potential and the thermal driving force

$$\boldsymbol{X}_2 = \frac{1}{T}\frac{\partial T}{\partial \boldsymbol{x}} \tag{6.37}$$

which is the temperature gradient per unit temperature. $\boldsymbol{N}_{ij}$, the so–called transport coefficients, connect fluxes and forces. In real anisotropic materials they are found to be tensors. The first coefficient to be calculated relates the particle current density to the applied electrochemical potential

$$\boldsymbol{N}_{11} = -\tau \int_V \frac{\partial f_0}{\partial E}(\boldsymbol{v} \otimes \boldsymbol{v}) d^3 v \tag{6.38}$$

The next two coefficients are symmetric and given by

$$\boldsymbol{N}_{12} = -\tau \int_V \frac{\partial f_0}{\partial E}(E - \mu)(\boldsymbol{v} \otimes \boldsymbol{v}) d^3 v = \boldsymbol{N}_{21} \tag{6.39}$$

They tell us, in principle, what part of the particle current flows due to the temperature gradient applied, or what part of the heat current results from an applied electrochemical potential gradient. In reality they are not directly observed, and we shall discuss this later. The last coefficient gives the heat current due to a thermal driving force

$$\boldsymbol{N}_{22} = -\tau \int_V \frac{\partial f_0}{\partial E}(E - \mu)^2(\boldsymbol{v} \otimes \boldsymbol{v}) d^3 v \tag{6.40}$$

We know from the discussion about Bloch oscillations that for real band structures the velocity vectors in (6.38) have to be replaced by $\boldsymbol{v} = \partial E / \partial \boldsymbol{p}$. This leads to more complicated integrals for the transport coefficients. In the same way a momentum–dependent relaxation time τ must be kept in the integrand. Thus we must be aware of the fact that we are in the regime of a harmonic approximation with constant relaxation time. Let us rewrite the two equations (6.35) as

$$\boldsymbol{j}_n = -\boldsymbol{N}_{11}\nabla\eta - \boldsymbol{N}_{12}\frac{\nabla T}{T} = -\frac{\boldsymbol{i}}{e} \tag{6.41a}$$

$$\boldsymbol{j}_Q = -\boldsymbol{N}_{21}\nabla\eta - \boldsymbol{N}_{22}\frac{\nabla T}{T} \tag{6.41b}$$

where $\boldsymbol{i}$ is the electron current density.

Transport Coefficients and their Significance

Two special cases help us to understand the significance of the transport coefficients:

1. In the absence of a thermal gradient and for a homogeneous material ($\nabla\mu = 0$) we can easily verify that $\boldsymbol{N}_{11}$ is proportional to the electrical conductivity σ, exactly

Electrical Conductivity

$$\sigma = q^2 \boldsymbol{N}_{11} \tag{6.42}$$

From (6.11) we know that the electrical current density for electrons may be written as $-q\boldsymbol{j}_n = -q\boldsymbol{v}_d n = \sigma\boldsymbol{E}$. Thus the electron drift velocity is given by $\boldsymbol{v}_d = -\mu_e \boldsymbol{E}$ and the constant of proportionality is called the electron mobility

Carrier Mobility

$$\mu_e = \frac{\sigma}{nq} \tag{6.43}$$

We are already well familiar with the coefficient σ and now understand the microscopic origin from which it comes. The absence of a thermal gradient may nevertheless leave us with an electrical current induced heat flow, i.e.

$$j_Q = N_{21}\frac{\partial\eta}{\partial x} = \frac{N_{21}}{N_{11}}j_n = -\frac{N_{21}}{qN_{11}}i = \Pi i \qquad (6.44)$$

The last factor of proportionality between the thermal current density j_Q and the electrical current density i is termed the Peltier coefficient

Peltier Coefficient

$$\Pi = -\frac{N_{21}}{qN_{11}} \qquad (6.45)$$

Since thermal currents are not directly observable but rather their gradients, we recall the energy balance equation (6.22) in the form

$$\frac{\partial}{\partial t}u(x, t) = F j_n(x, t) - \nabla_x(j_Q(x, t) + \mu j_n(x, t)) \qquad (6.46)$$

In terms of the electrical current density, using the definition of the electrochemical potential (6.31), we therefore have

$$\frac{\partial}{\partial t}u(x, t) = -i\nabla\Pi + \frac{1}{\sigma}i^2 \qquad (6.47)$$

The second term in (6.47) we identify as the Joule heating that a current produces. Note that it is quadratic in the current density and its coefficient is the inverse of the electrical conductivity as given by (6.42). After the discussion of Bloch oscillations it became clear that electrical conductivity is due to a coupling of electrons to a new dissipation channel, namely the phonon system. This coupling was introduced phenomenologically by means of the relaxation time approximation. The first term in (6.47) shows that no effect will be observed in a homogeneous material, since with $\Pi = \text{const}$, $\nabla\Pi = 0$. No prediction about Π is possible. Therefore, we look at the interface or two materials A and B with different Peltier coefficients Π_A and Π_B, and two different electrical conductivities σ_A and σ_B.

Suppose we have a constant current flowing from material A to B, as indicated in Figure 6.4.

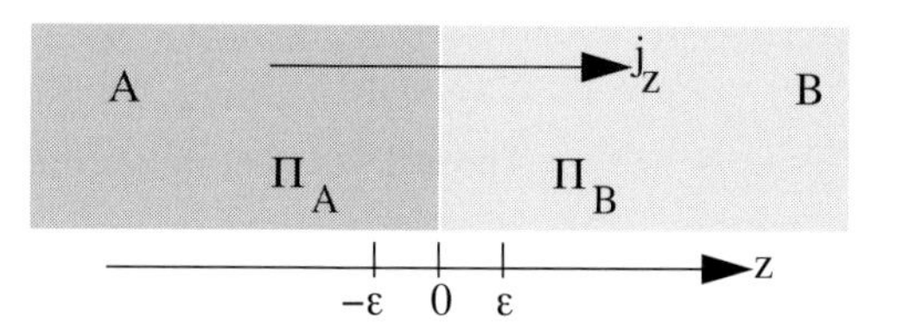

Figure 6.4. Contact between two different materials showing the Peltier effect.

Integrate (6.47) over small volume around the interface between A and B

$$\lim_{\varepsilon \to 0} \int_{-\varepsilon}^{\varepsilon} \frac{\partial}{\partial t} u(x, t) dz = \lim_{\varepsilon \to 0} \left(- \boldsymbol{i} \int_{-\varepsilon}^{\varepsilon} \nabla \Pi dz + \boldsymbol{i}^2 \int_{-\varepsilon}^{\varepsilon} \frac{1}{\sigma} dz \right)$$
$$= -i_z(\Pi_B - \Pi_A) \tag{6.48}$$

Peltier Effect

Integration and taking the limit in (6.48) are equivalent to asking for the amount of heat produced in a very small region around the interface. We see that in the limit $\varepsilon \to 0$ the second integral vanishes, i.e., the Joule heating is a volume effect, whereas the first integral gives a finite contribution in this limit if $\Pi_B \neq \Pi_A$, i.e., it is a surface or interface effect. This means that we have a local heating or cooling at the interface between material A and B, which is called the Peltier effect.

2. If there is no particle current present, i.e., $\boldsymbol{j}_n = 0$, then, according to (6.41a), the electrochemical potential gradient is completely determined by the temperature gradient

$$\nabla \eta = -\frac{\boldsymbol{N}_{12}}{T \boldsymbol{N}_{11}} \nabla T \tag{6.49}$$

Their constant of proportionality divided by the unit charge

Absolute Thermo-electric Power

$$\varepsilon = -\frac{N_{12}}{qTN_{11}} \tag{6.50}$$

is called the absolute thermoelectric power of a material. If we insert this into (6.41b) we obtain

$$j_Q = \frac{N_{11}N_{22} - N_{12}N_{21}}{N_{11}T}\nabla T \tag{6.51}$$

where the coefficient of the temperature gradient is called the thermal conductivity

Thermal Conductivity

$$\kappa = \frac{N_{11}N_{22} - N_{12}N_{21}}{N_{11}T} \tag{6.52}$$

Let us go back to the case where there are two different materials A and B in a ring like structure as shown in Figure 6.5. The ring is open

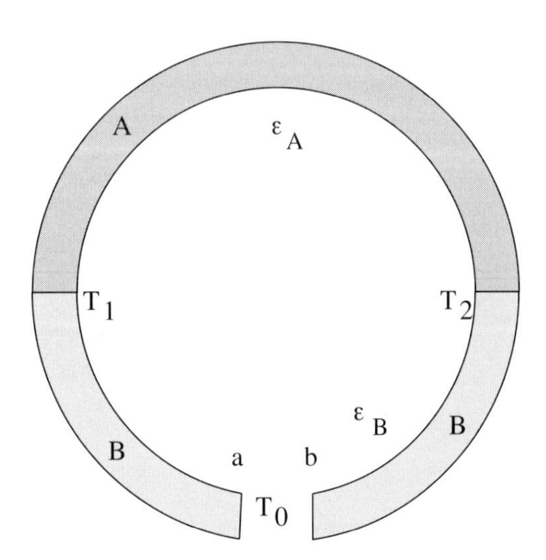

Figure 6.5. Contact between two different materials showing Seebeck effect.

anywhere in the material B, where there is a temperature T_0. The interfaces between the two different materials are kept at temperatures T_1 and T_2. We integrate (6.49) along a path s from a to b

$$\int_a^b \nabla\eta ds = \int_a^b q\varepsilon\nabla T ds \tag{6.53}$$

This yields

$$\frac{\eta_b - \eta_a}{q} = \int_{T_0}^{T_1} \varepsilon dT + \int_{T_1}^{T_2} \varepsilon dT + \int_{T_2}^{T_0} \varepsilon dT = -(\varepsilon_A - \varepsilon_B)(T_1 - T_2) \tag{6.54}$$

Since a and b are located in the same material B, $\mu_a = \mu_b$ holds and (6.54) gives us the so–called absolute differential thermovoltage

Absolute Differential Thermo-voltage

$$\delta\psi = -(\varepsilon_A - \varepsilon_B)\delta T \tag{6.55}$$

which is the inverse of the Peltier effect, i.e., an electrostatic potential difference builds up in a system of two different materials where no current is flowing due to a temperature gradient.

6.2.2 Formal Transport Theory

Our analysis here will be strictly macroscopic, and for this we invoke equilibrium thermodynamics. Assuming local equilibrium we might still say that the temperature and the electrochemical potential are "good" quantities, at least locally. We allow them to vary in space. In particular, we start with the fundamental energy relation (B 7.2.3) which we write using density extensive variables and for multiple component particle systems with densities n_i and their respective intensive chemical potentials μ_i

$$du = Tds + \sum_i \mu_i dn_i \tag{6.56a}$$

$$ds = \frac{1}{T}du - \sum_i \frac{\mu_i}{T} dn_i \tag{6.56b}$$

When this expression is taken per unit time, we obtain the rate equations

$$\dot{s} = \frac{1}{T}\dot{u} - \sum_i \frac{\mu_i}{T}\dot{n}_i \tag{6.57a}$$

$$\boldsymbol{j}_s = \frac{1}{T}\boldsymbol{j}_u - \frac{\mu_n}{T}\boldsymbol{j}_n \tag{6.57b}$$

where in (6.57b) we are restricted to a single type of particle. It employs the current densities for entropy, energy, and particles, and is also obtainable from the entropic fundamental relation (B 7.2.7). Since energy density and particle density are both conserved, we have that $\partial u / \partial t = -\nabla \bullet \boldsymbol{j}_u$, and $\partial n / \partial t = -\nabla \bullet \boldsymbol{j}_n$. Entropy density is not conserved, so that

$$\dot{s} = \frac{\partial s}{\partial t} + \nabla \bullet \boldsymbol{j}_s \tag{6.58}$$

From (B 7.2.7) we obtain that

$$\frac{\partial s}{\partial t} = \sum_{k=0}^{2} F_k \frac{\partial x_k}{\partial t} = \left(\frac{1}{T}\right)\frac{\partial u}{\partial t} - \left(\frac{\mu_n}{T}\right)\frac{\partial n}{\partial t} = -\left(\frac{1}{T}\right)\nabla \bullet \boldsymbol{j}_u + \left(\frac{\mu_n}{T}\right)\nabla \bullet \boldsymbol{j}_n \tag{6.59}$$

with the entropic intensive parameters $F_0 = 1/T$, and $F_1 = -\mu_n/T$. The divergence of (6.57b) gives

$$\begin{aligned} \nabla \bullet \boldsymbol{j}_s &= \nabla \bullet \left(\frac{1}{T}\boldsymbol{j}_u\right) - \nabla \bullet \left(\frac{\mu_n}{T}\boldsymbol{j}_n\right) \\ &= \left(\nabla \frac{1}{T}\right) \bullet \boldsymbol{j}_u + \left(\frac{1}{T}\right)\nabla \bullet \boldsymbol{j}_u - \left(\nabla \frac{\mu_n}{T}\right) \bullet \boldsymbol{j}_n - \left(\frac{\mu_n}{T}\right)\nabla \bullet \boldsymbol{j}_n \end{aligned} \tag{6.60}$$

Inserting (6.59) and (6.60) into (6.58) results in

Entropy Production Defines the Affinities and Fluxes

$$\dot{s} = \left(\nabla \frac{1}{T}\right) \bullet \boldsymbol{j}_u - \left(\nabla \frac{\mu_n}{T}\right) \bullet \boldsymbol{j}_n \tag{6.61}$$

At this point we allow the inclusion of external forces as well. In particular, we have to do so for the charged particles, which are also subject to the Lorenz force

$$f_L = q_n(\boldsymbol{E} + \boldsymbol{v}_n \times \boldsymbol{B}) \quad (6.62)$$

that is caused by an externally applied electromagnetic field. This means that we have to modify the particle affinities to yield

$$\dot{s} = \left(\nabla\frac{1}{T}\right) \bullet \boldsymbol{j}_u + \left(\frac{\boldsymbol{f}_L}{T} - \nabla\frac{\mu_n}{T}\right) \bullet \boldsymbol{j}_n \quad (6.63)$$

However, since by definition the particle current densities and the particle velocity vectors are parallel, the cross products drop out of the result and we obtain

$$\dot{s} = \left(\nabla\frac{1}{T}\right) \bullet \boldsymbol{j}_u + \left(\frac{q_n\boldsymbol{E}}{T} - \nabla\frac{\mu_n}{T}\right) \bullet \boldsymbol{j}_n\text{, or} \quad (6.64a)$$

$$\dot{s} = \left(\nabla\frac{1}{T}\right) \bullet \boldsymbol{j}_u + \left(-\frac{q_n\nabla\psi}{T} - \nabla\frac{\mu_n}{T}\right) \bullet \boldsymbol{j}_n \quad (6.64b)$$

if we make use of the relation between the electric field and the electrostatic potential ψ. We see that entropy production by the electrons or holes is naturally "driven" by two forces, their concentration gradient (or chemical potential gradient) represented by $\nabla\mu_n$, and the externally applied field (or electrical potential gradient) represented by $\nabla\psi$. We can now simplify matters by introducing the electrochemical potential η_n for the charge carriers to obtain

$$\dot{s} = \left(\nabla\frac{1}{T}\right) \bullet \boldsymbol{j}_u + \left(-\nabla\frac{\eta_n}{T}\right) \bullet \boldsymbol{j}_n \quad (6.64c)$$

Noting that

$$\nabla(1/T) = -\nabla T/T^2 \quad (6.65)$$

and

$$\nabla(\eta_n/T) = \nabla\eta_n/T - \eta_n\nabla T/T^2 \quad (6.66)$$

we obtain

$$\dot{s} = \frac{1}{T}\left\{-\left(\frac{\nabla T}{T}\right) \bullet (j_u - \eta_n \bullet j_n) + (-\nabla \eta_n) \bullet j_n\right\} \tag{6.66a}$$

Using $dQ = TdS$ in the fundamental energy relation (B 7.2.4) we obtain $j_q = j_u - \mu_n \cdot j_n$, which, when inserted into (6.66a), gives

$$\dot{s} = \frac{1}{T}\left\{-\left(\frac{\nabla T}{T}\right) \bullet j_q + (-\nabla \eta_n) \bullet j_n\right\} \tag{6.66b}$$

and which defines the affinities associated with the fluxes (see Box 7.2) of this system. Equation (6.66a) also tells us that the heat flux is comprised of energy transport as well as particle energy transport, and that the affinities needed for the following steps are the gradients of temperature and particle electrochemical potentials.

Assumptions Needed to Define the Transport Coefficients

We now summarize the assumptions underlying the previous analysis for transport coefficients

- The system has no "memory" and is purely resistive. The fluxes depend only on the local values of the intensive parameters and on the affinities, and do so "instantaneously".
- Each flux is zero for zero affinity. If we add the assumption of linear dependence, we can truncate a series expansion of the fluxes w.r.t. the affinities after the first nonzero term, i.e.,

$$j_k = \sum_{l=1}^{\mathrm{nc}} L_{kl} \bullet \mathrm{F}_l \tag{6.67}$$

for nc transport entities, and assuming that the transport coefficients L_{kl} are of tensor nature.

The remarkable Onsager theory proves that, in the presence of a magnetic field $L_{kl}(B) = L_{kl}^T(-B)$, and hence that, in the absence of a magnetic field $L_{kl} = L_{kl}^T$. The approach in the above section was to write down formal expression for the transport components, and then to associate the terms from the thermodynamic expressions that we have obtained before. Many solid-state silicon devices have been invented that exploit these effects. We discuss some of these in Section 7.3.6.

6.2.3 The Hall Effect

Taking a magnetic field into account in the carrier drift force term $\boldsymbol{F} = -q(\boldsymbol{E} + \boldsymbol{v} \times \boldsymbol{B})$, one part of $\boldsymbol{F}$ is given by the electric field and the other by the Lorenz force. This Lorenz force causes the electrons to move to the one side of the conductor, while the electrons move to the opposite side (see Figure 6.6). This results in a potential difference perpendicular to the current direction. Therefore, the equipotential lines are shifted by an angle Θ. Placing two contacts on the boundary of the conductor as shown in Figure 6.6, the potential difference can be measured. The so–

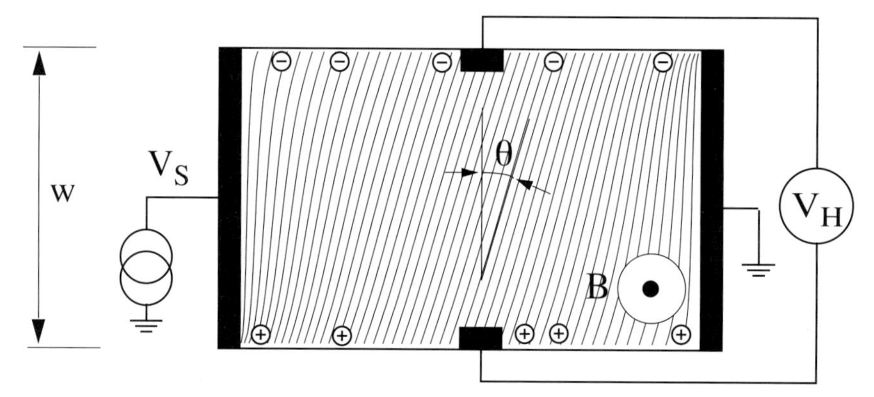

Figure 6.6. Schematic diagram of a Hall plate showing the electric potential distribution due to an out–of–plane magnetic field.

called Hall voltage is directly proportional to the component of the applied magnetic field perpendicular to the Hall plate.

To describe this effect we first insert the expression for $\boldsymbol{F}$ containing the Lorenz force directly into (6.30) and see that the effect of the magnetic field drops out. Thus for the inclusion of magnetic field dependence in (6.30) higher order terms must be considered. Neglecting the whole streaming motion term of δf was therefore a too restrictive approximation. To this end, we keep the keep the gradient with respect to **k** from the streaming motion part

$$\delta f = -\tau\frac{\partial f_0}{\partial E}\left(\boldsymbol{v}\nabla\eta + \frac{q}{\hbar}(\boldsymbol{v}\times\boldsymbol{B})\nabla_k\delta f\right) \tag{6.68}$$

where we assumed the isothermal case, hence the temperature gradient vanishes. Equation (6.68) can be solved iteratively by replacing $\nabla_k\delta f$ by a series expansion in ascending powers of $\boldsymbol{B}$. Then we have to the lowest order

$$\delta f = -\tau\frac{\partial f_0}{\partial E}\boldsymbol{v}\frac{\nabla\eta + (\boldsymbol{s}\times\nabla\eta) + \boldsymbol{s}(\boldsymbol{s}\nabla\eta)}{1+s^2} \tag{6.69}$$

with $\boldsymbol{s} = q\tau(m^*)^{-1}\boldsymbol{B}$, where we used $\boldsymbol{v} = \hbar\boldsymbol{k}/m^*$. For the case of an anisotropic material, $(m^*)^{-1}$ is the inverse of the mass tensor $\boldsymbol{M}$

$$\boldsymbol{M} = \begin{bmatrix} m_1 & 0 & 0 \\ 0 & m_2 & 0 \\ 0 & 0 & m_3 \end{bmatrix} \tag{6.70}$$

where we assumed the effective mass tensor to be diagonal. This means, furthermore, that in the iterative solution (6.69) all tensorial terms containing the effective mass have to be maintained. In this case we obtain

$$\delta f = -\tau\frac{\partial f_0}{\partial E}\frac{\nabla\eta + q\tau(\boldsymbol{B}\times\boldsymbol{M}^{-1}\nabla\eta) + \frac{q^2\tau^2}{|M|}(\boldsymbol{B}\nabla\eta)\boldsymbol{MB}}{1 + \frac{q^2\tau^2}{|M|}\boldsymbol{BMB}} \tag{6.71}$$

We use (6.71) to calculate the electric current density $\boldsymbol{i} = -q\boldsymbol{j}_n$ for electrons from the form given in (6.11) and assume that the material is homogeneous, i.e., $\nabla\mu = 0$. In this way we obtain three contributions

$$\boldsymbol{i} = q^2\boldsymbol{K}_1\boldsymbol{E} + q^3\boldsymbol{K}_2\boldsymbol{B} \times (\boldsymbol{M}^{-1}\boldsymbol{E}) + \frac{q^4}{|M|}\boldsymbol{K}_3(\boldsymbol{EB})\boldsymbol{MB} \tag{6.72}$$

where the components of the tensors $\boldsymbol{K}_i$ are given by [6.2]

$$\mathbf{K}_i = -\frac{1}{4\pi^3}\int_{V_k}\frac{\tau^i}{1 + \frac{q^2\tau^2}{|M|}\boldsymbol{BMB}}\frac{\partial f_0}{\partial E}\boldsymbol{v} \otimes \boldsymbol{v}d^3\boldsymbol{k} \tag{6.73}$$

The rank two tensors $\boldsymbol{K}_i$ are called the generalized transport coefficients. They reflect the fully anisotropic character of constitutive equations in a crystal. To simplify the discussion, we now write (6.72) by defining a generalized constitutive equation for the current density

$$\boldsymbol{i} = \sigma^*\boldsymbol{E} \tag{6.74}$$

where the effective conductivity σ^* is defined by

$$\sigma^* = q^2\boldsymbol{K}_1\boldsymbol{I} + q^3\boldsymbol{K}_2\boldsymbol{A}_2 + \frac{q^4}{|M|}\boldsymbol{K}_3\boldsymbol{A}_3 \tag{6.75}$$

and where $\boldsymbol{I}$ is the identity matrix. The matrices $\boldsymbol{A}_2$ and $\boldsymbol{A}_3$ are given by

$$\boldsymbol{A}_2 = \begin{bmatrix} 0 & -B_3m_2 & B_2m_3 \\ -B_3m_1 & 0 & -B_1m_3 \\ -B_2m_1 & B_1m_2 & 0 \end{bmatrix} \tag{6.76a}$$

$$\boldsymbol{A}_3 = \begin{bmatrix} B_1^2m_1 & B_1B_2m_1 & B_1B_3m_1 \\ B_2B_1m_2 & B_2^2m_2 & B_2B_3m_2 \\ B_3B_1m_3 & B_3B_2m_3 & B_3^2m_3 \end{bmatrix} \tag{6.76b}$$

Note that these equations are given in coordinates of the crystal directions. Thus, for the effective masses we may chose the two different masses of Silicon according to the crystal direction. The relaxation time is taken to be constant in (6.73). We may roughly estimate its value $\tau = 5 \times 10^{-14}\mathrm{s}$: this corresponds to an average electron mobility of about $0.145\ \mathrm{m^2/Vs}$, as is the case for low–doped silicon.

6.3 From Global Balance to Local Non-Equilibrium

The route from global to local equilibrium and further through different length scales is best explained in the schematic diagram of Figure 6.7.

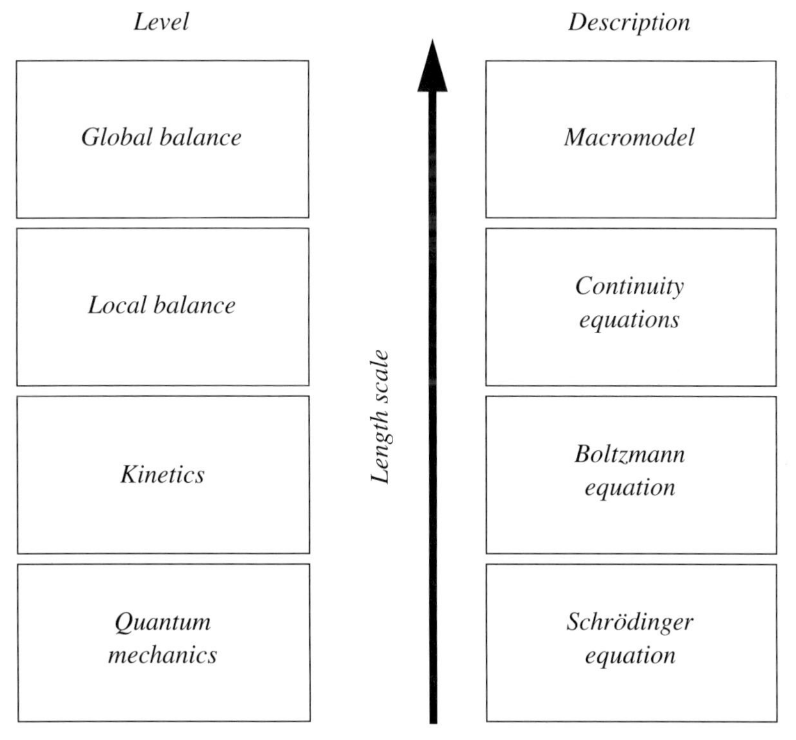

Figure 6.7. This diagram compares the transport theory levels with their respective equations. The ordering of the levels follow the applicable length scale.

We have encountered several of the levels and descriptions up to now. The only job left is to describe the global balance picture.

6.3.1 Global Balance Equation Systems

In a global balance equation system we consider only the contacts and integrate over their cross-sections to obtain currents instead of current densities. At this level we have lost the distributed nature of transport phenomena. Global balance simulation programs treat networks of devices that are connected through one–dimensional idealized wires. These wires do not dissipate energy and all functionality is concentrated in the single devices. The resulting equation of motion is no longer a partial differential equation but rather a system of ordinary differential equations. This is why we call such systems concentrated parameter systems or lumped models.

6.3.2 Local Balance: The Hydrodynamic Equations

Recall the balance equations for the moments (6.19), 6.21) and 6.21), and rewrite them in a slightly different manner, i.e., put $\boldsymbol{j}_n = n\boldsymbol{v}_a$, $u = nw$, and $\boldsymbol{j}_u = nw\boldsymbol{v}_a$. Here we introduced w as the average energy of the electrons and used the average velocity $\boldsymbol{v}_a$ as given in (6.11). The collision terms on the r.h.s. of the balance equations are interpreted as rates of changes due to collisions for the carrier density, the current density and the energy, respectively. Therefore, we write

$$C_n = \left(\frac{\partial n}{\partial t}\right)_C \tag{6.77a}$$

$$C_{j_n} = C_p = \left(\frac{\partial n\boldsymbol{v}_d}{\partial t}\right)_C = \boldsymbol{v}_d\left(\frac{\partial n}{\partial t}\right)_C + n\left(\frac{\partial \boldsymbol{v}_d}{\partial t}\right)_C \tag{6.77b}$$

$$C_u = C_w = \left(\frac{\partial nw}{\partial t}\right)_C = w\left(\frac{\partial n}{\partial t}\right)_C + n\left(\frac{\partial w}{\partial t}\right)_C \tag{6.77c}$$

Thus we have for the density balance equation

$$\frac{\partial n}{\partial t} + \nabla_x(n\boldsymbol{v}_a) = \left(\frac{\partial n}{\partial t}\right)_C \tag{6.78}$$

We use the force term $\boldsymbol{F} = -q\boldsymbol{E}$ and performing the time derivative of $\boldsymbol{j}_n = n\boldsymbol{v}_a$ taking (6.78) into account the momentum balance equation will read

$$\frac{\partial n\boldsymbol{v}_a}{\partial t} - \boldsymbol{v}_a\nabla_x(n\boldsymbol{v}_a) + \frac{q}{m}\boldsymbol{E}n + \nabla_x\boldsymbol{\Pi} = n\left(\frac{\partial \boldsymbol{v}_a}{\partial t}\right)_C \tag{6.79}$$

We write $-\boldsymbol{v}_a\nabla_x(n\boldsymbol{v}_a) = -\nabla_x(n\boldsymbol{v}_a \otimes \boldsymbol{v}_a) + n\boldsymbol{v}_a(\nabla_x \otimes \boldsymbol{v}_a)$ and then obtain

$$n\frac{\partial \boldsymbol{v}_a}{\partial t} + n\boldsymbol{v}_a(\nabla_x \otimes \boldsymbol{v}_a) + \frac{q}{m}\boldsymbol{E}n + \frac{1}{m}\nabla_x nk_B\boldsymbol{T} = n\left(\frac{\partial \boldsymbol{v}_a}{\partial t}\right)_C \tag{6.80}$$

where we have replaced the momentum current tensor $\boldsymbol{\Pi}$ by the temperature tensor $\boldsymbol{T}$. This is done by performing the integral in (6.15) over the differences between the microscopic group velocities and the average velocities

$$nk_B\boldsymbol{T} = \frac{m}{4\pi^3}\int_{\Omega_k}(\boldsymbol{v} - \boldsymbol{v}_a) \otimes (\boldsymbol{v} - \boldsymbol{v}_a)f(\boldsymbol{k}, \boldsymbol{x}, t)d^3\boldsymbol{k} \tag{6.81}$$

We interpret the r.h.s. of (6.81) as this part of the average energy that remains when we subtract the average kinetic energy in harmonic approximation $E_{kin} = 1/2m^*\boldsymbol{v}_a^2$ from the total average energy. This is a term that shows random motion with vanishing average velocity, namely the temperature. This means that we can write the average energy consistent of two parts as $w = m/2v_a^2 + 1/2\mathrm{Tr}(k_B\boldsymbol{T})$, where the trace operator on a tensor means $\mathrm{Tr}(k_B\boldsymbol{T}) = k_B\sum_i T_{ii}$. With the same arguments we obtain

$$n\frac{\partial w}{\partial t} + \nabla_x\boldsymbol{Q} + n\boldsymbol{v}_a(\nabla_x w) + q\boldsymbol{E}n\boldsymbol{v}_a + \frac{1}{n}\nabla_x nk_B\boldsymbol{T} = n\left(\frac{\partial w}{\partial t}\right)_C \tag{6.82}$$

where we set the heat flow $\boldsymbol{Q}$ as

$$Q = \frac{1}{4\pi^3}\frac{m}{2}\int_{\Omega_k}(\boldsymbol{v}-\boldsymbol{v}_a)^2(\boldsymbol{v}-\boldsymbol{v}_a)f(\boldsymbol{k},\boldsymbol{x},t)\mathrm{d}^3\boldsymbol{k} \tag{6.83}$$

Equations (6.78), (6.80), and (6.82) are referred to as the hydrodynamic model for the motion of electrons in a semiconductor crystal. Together with the same set of equations that can be derived for the defect electron or hole density p they form the basis for bipolar charge carrier transport in semiconductors. However, note that (6.78), (6.80) and (6.82) are a finite set of equations. It contains more unknowns than equations, i.e., it is underdetermined. Therefore, simplifying assumptions are needed to give a closed form. For specific forms of simplifying assumptions regarding the temperature tensor $\boldsymbol{T}$ we refer the reader to the literature [6.4, 6.5].

6.3.3 Solving the Drift-Diffusion Equations

There is only one important equation missing to complete the set of hydrodynamic equations, the Poisson equation (4.35). This equation relates the moving charge carriers to the formation of the internal electrostatic potential in a crystal and is written as

$$\nabla\bullet(\varepsilon\nabla\psi) = q(n-p+N_A-N_D) = -\rho \tag{6.84}$$

where N_A and N_D stand for the ionized acceptor and donor concentration respectively. So far there is nothing special about these equations and one could think that it is straight forward to solve them via a simple finite difference scheme. This approach is far from having any chance to result in a solution of realistic problems in semiconductor transport.

The subset of the hydrodynamic equations that contains only the density balance for electrons and holes together with (6.84) is referred to as the drift-diffusion model. Let us focus on the stationary problem ($\dot{n} = 0$, $\dot{p} = 0$), that results in the following three partial differential equations

Drift–Diffusion Equations

$$\nabla \cdot \boldsymbol{D} = \rho \tag{6.85}$$

where $\boldsymbol{D}$ is the displacement vector with the constitutive equation

$$\boldsymbol{D} = -\varepsilon_r \varepsilon_0 \nabla \psi \tag{6.86}$$

where the dielectric constants ε_r and ε_0 are defined in Chapter 4.

$$\nabla \cdot \boldsymbol{j}_n = qR \tag{6.87a}$$

$$\nabla \cdot \boldsymbol{j}_p = -qR \tag{6.87b}$$

The net recombination rate R describes a destruction of holes and electrons in pairs, i.e., the events per unit time an electrons goes from the conduction band into the valence band. Therefore, the same number occurs in both equations (6.87a) and (6.87b). The respective constitutive equations for the currents are

$$\boldsymbol{j}_n = -q\mu_n n \nabla \psi + qD_n \nabla n \tag{6.88a}$$

$$\boldsymbol{j}_p = -q\mu_p p \nabla \psi - qD_p \nabla p \tag{6.88b}$$

where μ_n and μ_p are the electron and hole mobility, respectively, and D_n and D_p denote the respective diffusion constants. The Einstein relations relate these two quantities by

$$D_n = \frac{k_B T_0}{q} \mu_n \tag{6.89a}$$

and

$$D_p = \frac{k_B T_0}{q} \mu_p \tag{6.89b}$$

In (6.89a) and (6.89b) we encounter the temperature T_0. This means that our model discards any information about the temperature distribution in positional space, i.e., we have assumed a global uniform temperature of electrons and holes that coincides with the lattice temperature.

In order to solve equations (6.85)–(6.87b) we use the box integration method. The idea is to discretize the simulation domain in terms of boxes surrounding each discretization node this way that the whole simulation domain is completely covered by the box subdomains. These subdomains do not overlap. A schematic view is given in Figure 6.8.

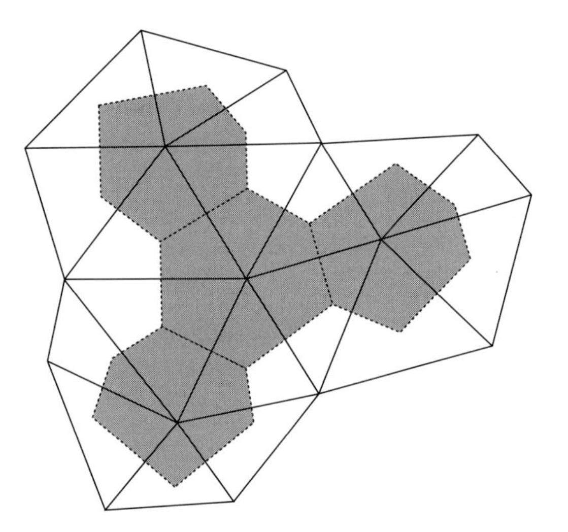

Figure 6.8. Schematic view of a two–dimensional triangular grid with the boxes (shaded) defined by the normal bisectors the sides emanating from a given node.

We rewrite equations (6.85)–(6.87b) as follows using Gauss' theorem

$$\int_S \boldsymbol{D}\boldsymbol{s}dS = \int_V \rho dV \tag{6.90a}$$

$$\int_S \boldsymbol{j}_n\boldsymbol{s}dS = \int_V qRdV \tag{6.90b}$$

$$\int_S \boldsymbol{j}_p\boldsymbol{s}dS = -\int_V qRdV \tag{6.90c}$$

which is more suitable to explain the idea behind the box integration method. V and S denote the volume and the boundary of a box and $\boldsymbol{s}$ is the normal vector on the boundary of the box, shown in Figure 6.9.

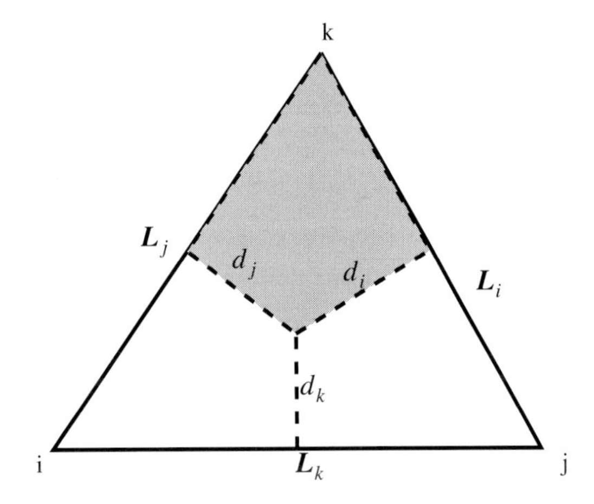

Figure 6.9. Two–dimensional triangular element with nodal indices i, j, k. The three vectors $\boldsymbol{L}_i$, $\boldsymbol{L}_j$ and $\boldsymbol{L}_k$ represent the edges connecting nodes j to k, k to i and i to j, respectively. The cross sections for the fluxes are d_i, d_i and d_i.

The discrete form of (6.90a) is found by assuming that ψ varies linearly along the edges connecting two nodes and thus is a linear function on the two dimensional triangular element, e.g., defined by the nodes i, j and k in Figure 6.9. Therefore, the corresponding flux $\boldsymbol{D}$ is constant. Let us integrate on the part of the box associated with note k that lies in the element defined by i, j and k. This gives us two contributions from the surface integral form the flux displacement vector

$$\int_{S_k} \boldsymbol{D} s dS = \boldsymbol{D}_i \boldsymbol{l}_i d_i + \boldsymbol{D}_j \boldsymbol{l}_j d_j \tag{6.91}$$

where S_k is the surface of the part of the box around node k lying in the triangle (i, j, k). Here we introduced the unit vector in the respective directions $\boldsymbol{l}_i = \boldsymbol{L}_i / L_i$ and $\boldsymbol{l}_j = \boldsymbol{L}_j / L_j$, that are associated with the box surface elements d_i and d_j. In terms of the electrostatic potential (6.91) reads

$$\int_{S_k} \boldsymbol{D} s dS = \frac{k_B T \varepsilon_0 \varepsilon_r}{q} \left[\frac{d_i}{L_i}(u_k - u_j) + \frac{d_j}{L_j}(u_k - u_i) \right] = \rho_k \frac{d_i L_i + d_j L_j}{2} \tag{6.92}$$

where we scaled the electrostatic potential in terms of the thermal voltage according to

$$u = \frac{q\psi}{k_B T} \tag{6.93}$$

The r.h.s. of (6.92) is a product of the nodal value ρ_k of the charge density and the area of the portion of the box surrounding node k that lies in the element. The discretization of (6.90b) and (6.90c) has to be performed with care since we have already made assumptions on the potential. Introducing (6.89a) and (6.93) into (6.88a) yields the electron current density

$$\boldsymbol{j}_n = qD_n(\nabla n - n\nabla u) \tag{6.94}$$

$\boldsymbol{j}_n$ may be projected on an edge of the triangular element, say $\boldsymbol{L}_k$, to give

$$j_{nk} = qD_n\left(\frac{\partial n}{\partial l_k} - n\frac{du}{dl_k}\right) = \boldsymbol{j}_n \boldsymbol{l}_k \tag{6.95}$$

The projection j_{nk} is assumed to be constant along the edge $\boldsymbol{L}_k$. This constrains the element size to a length scale where the assumption is fulfilled. On the other hand, the linear potential function on the edge

$$u(l_k) = \frac{u_j - u_i}{L_k} l_k + u_i = a_k l_k + u_i \tag{6.96}$$

leads to a differential equation for n along $\boldsymbol{L}_k$

$$J_{nk} = \frac{dn}{dl_k} - \alpha_k n = \frac{j_{nk}}{qD_n} \tag{6.97}$$

Integrating (6.97) from node i to node j yields $J_{nk} = [B(u_{ji})n_j - B(u_{ij})n_i]/L_k$. n_i and n_j are the values of n at nodes i and j, respectively. $B(u_{ji}) = x/(\exp(u_{ji}) - 1)$ is the Bernoulli function for the argument $u_{ji} = u_j - u_i$. Thus the flux of $\boldsymbol{j}_n$ related to node k enters (6.90b) to give

$$qD_{ni}\frac{d_i}{L_i}[B(u_{jk})n_i - B(u_{kj})n_k] + qD_{nj}\frac{d_j}{L_j}[B(u_{ik})n_i - B(u_{ki})n_k] \\ = qR_k\frac{d_iL_i + d_jL_j}{2} \tag{6.98}$$

Here we used D_{ni} and D_{nj} as the values of the diffusion coefficient along the edges i and j. R_k is the value of the recombination rate at node k. For inhomogeneous materials these values have to be determined by appropriate averaged values. The same equation as (6.98) holds for the defect electron with the restriction that all n have to be replaced by p and the indices at u have to be reversed.

6.3.4 Kinetic Theory and Methods for Solving the BTE

The term *kinetic theory* means that a description of the system is given where no assumptions about the distribution function are made. There is no parametrization by a few parameters, e.g., the temperature. Moreover, the non-equilibrium situation could even prohibit a definition of parameters like the temperature. This implies that a full solution of the Boltzmann transport equation must be obtained. There are several methods to solve this task that require more or less computational effort and lead to more or less accurate results. In the following we shall discuss four of them.

Iterative Method

The iterative method makes use of a transformation of the BTE into an integral equation. Write (6.2) in the form

$$\frac{df_k}{dt} = \frac{\partial f_k}{\partial t} - q\boldsymbol{E}\nabla_k f_k + \frac{\hbar\boldsymbol{k}}{m}\nabla_x f_k = -\sum_{k'}(W_{kk'}f_k - W_{k'k}f_{k'}) \\ = \left(-\Gamma_k f_k + \sum_{k'} W_{k'k}f_{k'}\right) \tag{6.99}$$

where we used

$$\dot{\boldsymbol{k}} = \frac{-q\boldsymbol{E}}{\hbar} \text{ and } \dot{\boldsymbol{x}} = \frac{\hbar\boldsymbol{k}}{m} \tag{6.100}$$

With (6.100) we can immediately solve the homogeneous equation (6.99), to give

$$f_h(\boldsymbol{k}, \boldsymbol{x}, t) = f(\boldsymbol{k}(t), \boldsymbol{x}(t), 0)\exp\left(-\int_0^t \Gamma(k(t'))dt'\right) \tag{6.101}$$

This equation tells us that with $\Gamma = 0$ all the phase space density $f(\boldsymbol{k}(0), \boldsymbol{x}(0), 0)$ would move according to (6.100) to the new position at $\boldsymbol{k}(t)$ and $\boldsymbol{x}(t)$ to give $f(\boldsymbol{k}(t), \boldsymbol{x}(t), 0)$. The fact that $\Gamma \neq 0$ lowers the arriving density at time t. This is because while moving some of the density is scattered out from the deterministic path. There is a particular solution of the differential equation found that reads

$$f_p(\boldsymbol{k}, \boldsymbol{x}, t) = \int_0^t \exp\left(-\int_\tau^t \Gamma(k(t'))dt'\right) \sum_{k'} W_{k'k}(f(\boldsymbol{k}'(\tau), \boldsymbol{x}(\tau), \tau))d\tau \tag{6.102}$$

The total solution given by $f = f_h + f_p$ is obtained by iteratively inserting f into (6.101) and (6.102). This procedure is very interesting for stationary problems in spatially homogeneous materials.

Direct Integration

The direct integration requires a discretization of the partial differential equation, the BTE. Remember, however, that this requires a discretization in a six-dimensional phase space. The problem then grows as N^6, where N is the number of discretization points for each of the phase space directions. Therefore, this method is not suitable for problems that require the full dimensionality because no sufficient accuracy may be achieved.

Monte Carlo Simulation

This is a stochastic method that works by creating stochastic free flight between the collisions (see Box 6.1). Since all the scattering probabilities are known, one can generate a random series of collision–free deterministic motions of a particle, e.g., according to (6.100). At the and of each deterministic motion stands a scattering process, which in turn is also generated stochastically. The distribution function f is generated either

Box 6.1. The Monte Carlo method.

Principle. The conduction of charge carriers through a crystal is well approximated by a classical free flight of the carrier under the influence of its own momentum and the applied electromagnetic fields, interspersed by scattering events which change the momentum and energy of the carrier. As we have seen, the scattering has many causes, including the phonons, the impurity ions and the other carriers.

The basis of the Monte Carlo method is to simulate carrier transport by generating random scattering events (hence the name Monte Carlo or MC) and numerically integrating the equations of motion of the carrier once the new momentum vector is known. Because in principle the MC needs to consider each carrier, and must collect enough data to be statistically relevant, it is very time consuming to calculate.

Integration. During free flight over a time interval t the carriers behave classically, hence we can write the classical Newton relationship

$$\left.\begin{aligned} \boldsymbol{x}(t) &= \boldsymbol{x}(0) + t\dot{\boldsymbol{x}} \\ \boldsymbol{p}(t) &= \boldsymbol{p}(0) + t\boldsymbol{F} \end{aligned}\right\} \quad \text{(B 6.1.1)}$$

for the position $\boldsymbol{x}(t)$ and momentum $\boldsymbol{p}(t)$. The the force acting on a carrier is caused by the electric field $\boldsymbol{E}$ acting on the carrier

$$\boldsymbol{F}(t) = \partial\boldsymbol{p}(t)/\partial t = q_c\boldsymbol{E} \quad \text{(B 6.1.2)}$$

Just before the next collision, the position and momentum have the following values

$$\begin{aligned} \boldsymbol{p}(t) &= \boldsymbol{p}(0) + q_c\boldsymbol{E} \\ \boldsymbol{x}(t) &= \boldsymbol{x}(0) + \Delta E/\boldsymbol{F} \end{aligned} \quad \text{(B 6.1.3)}$$

for an energy increase of ΔE during the interval. Typical 1D trajectories are shown in Figure B6.1.1.

Event generation. The scattering events are the key to the MC method, and are generated by a series of pseudo random numbers: the onset of scattering; the choice of scattering mechanism; the post-scattering momentum value; the post-scattering momentum direction. The scattering events typically included are: ionized impurity scattering, inter-valley absorption and emission and electron–phonon scattering. The terms also take into consideration that the scattering rate is energy dependent and that photons can be absorbed and generated. In this way a high degree of realism is achieved.

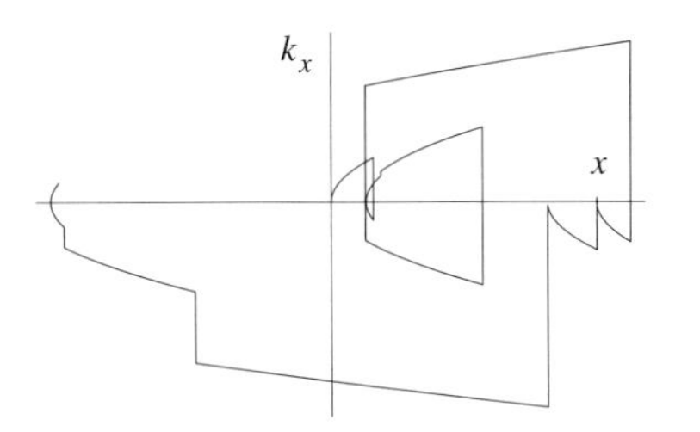

Figure B6.1.1. Typical 1D position and momentum trajectories for the Monte Carlo method.

Self-consistency. After initializing the simulation domain with carriers with position and momentum, the algorithm achieves a balance between classical electrostatics and the transport of carriers by the following repeated steps for each carrier: generate a lifetime τ; compute the position $\boldsymbol{x}(\tau)^-$ and momentum $\boldsymbol{p}(\tau)^-$ at the end of the free-flight step; select the next type of scattering event; compute the post–scattering position $\boldsymbol{x}(\tau)^+$ and momentum $\boldsymbol{p}(\tau)^+$; decide if the trajectory has reached a boundary or contact, in which case it should be reflected or discontinued. If a contact is reached, inject a new carrier at the complementary contact; At regular intervals, recompute the electric field with the Poisson equation.

by time average or by an ensemble average, depending on whether the problem is ergodic or not. This method is very convenient since it follows the microscopic motion of a particle. Also, depending on how accurate the result must be, it is more or less time consuming. Modern Monte Carlo techniques are refined with respect to the questions asked. For high accuracy needed in only specific parts of the phase space there are already highly optimized algorithms and even commercially available computer programs.

Spherical Harmonics Expansion

The idea behind this method is to expand the distribution function in terms of spherical harmonic functions using the coordinate reference system (k, θ, φ) in k-space.

$$f(\boldsymbol{k}, \boldsymbol{x}) = Y_0 f_0(k, \boldsymbol{x}) + \sum_n Y_1^n f_1^n(k, \boldsymbol{x}) + \sum_m Y_2^m f_2^m(k, \boldsymbol{x}) + \dots \quad (6.103)$$

The information about the angular dependencies is now put into the spherical harmonics Y_l^m with $l = 0, 1, 2, \dots$ given by

$$Y_l^m = \begin{cases} P_l^m(\cos\theta)\cos(m\varphi), \ m = 0, 1, 2, \dots \\ P_l^{|m|}(\cos\theta)\sin(m\varphi), \ m = -0, -1, -2, \dots \end{cases} \quad (6.104)$$

These are orthogonal functions in the (θ, φ) space. Therefore, inserting (6.103) into the BTE and multiplying by a given Y_o^k gives after integration an equation for $f_o^k(k, \boldsymbol{x})$. In this way we obtain a whole hierarchy of equations without the assumption of a local equilibrium like for the moment equations. The disadvantage is that it is still a four-dimensional problem and thus is best suited for two-dimensional problems in positional space. An additional drawback is that not the whole hierarchy can be solved for and higher order objects have to be neglected at some point. Nevertheless, this method gives good results if a resolution in k-space is needed.

6.4 Summary for Chapter 6

The chapter started with the Boltzmann transport equation. It describes the way in which a particle density moves through configuration space along a phase space trajectory. The BTE is a balance between the streaming motion of particles along their phase space trajectories and the scattering forces that modify this path. We saw that, depending on the requirements of our model, the one or the other transport description has to be applied. Modern electronic device simulators for electron transport in semiconductor combine the different levels of description in one application. This so called mixed–mode simulation applies the detailed description of local non–equilibrium where it is necessary, manages a local equilibrium description where it is adequate, and includes boundary conditions from a circuit connected to the device defined on a global balance level.

6.5 References for Chapter 6

6.1 C. Cercigniani, *The Boltzmann Equation and Its Application*, Springer (1988)

6.2 P.S.Kireev, *Semiconductor Physics*, MIR Publishers Moscow (1978)

6.3 D. K. Ferry, *Semiconductors*, Macmillan Publishing Company (1991)

6.4 S. Selberherr, *Analysis and Simulation of Semiconductor Devices*, Springer (1984)

6.5 M. Rudan and F. Odeh, *Multidimensional Discretization Scheme for the Hydrodynamic Model of Semiconductor Devices*, COMPEL, Vol. 5, No. 3, 149-183 (1986)

6.6 W. Jones, N. H. March, *Theoretical Solid State Physics*, Vols. I and II, John Wiley & Sons (1973)

Chapter 7 Interacting Subsystems

The interactions among the phonon, electron, and photon pseudo particles and with the classical fields of strain, electromagnetics and thermal gradients are the cause of a range of effects that are exploited in many amazing semiconductor devices we know today. In this chapter we consider those solid-state effects that are used in the well-known devices of microelectronics, such as the diode and the transistor, and we also take a look at the solid-state effects that are exploited in the devices of micro-technology, and ever increasingly, the devices of nano-technology. In fact, some effects may hamper devices from one application area, and and hence be termed parasitic, whereas in another field they become of primary interest, providing a way to measure an external quantity or provide a source of actuation. Mechanical stress or the temperature are two typical examples of effects that are either parasitic or useful, depending on how and where they arise. In the pressure sensor, or the atomic microscope beam, the level of stress is measured in a resistor to indicate the level of deflection of a membrane or the beam—here stress is useful. However, when a magnetic field sensor is packaged carelessly, it shows an offset in the measured magnetic field sensor that is more indicative of packaging stress levels and the temperature—in this case the stress is parasitic.

The chapter shows how interactions have successfully lead to innovative semiconductor devices, and how the preceding analysis can help both to understand the phenomena, as well as to extract design rules that help create better device and system designs. Ultimately, in our view, engineering theory is justified when it leads to better design methods. Of course we cannot deal with all possibilities, and as a result we have made a selection based on our own interests. We hope, however, that one concept will remain more strongly than others: that it is the *interaction* or *coupling* of natural "systems" that lead to *useful* (or annoyingly *parasitic*) effects.

Chapter Goal The goal of this chapter is to explore the interactions that arises between some of the "pure" effects of the preceding chapters.

Chapter Roadmap The road map for this chapter, which is the longest of the book, is illustrated in Figure 7.3. In each major section is inspired by a block in the

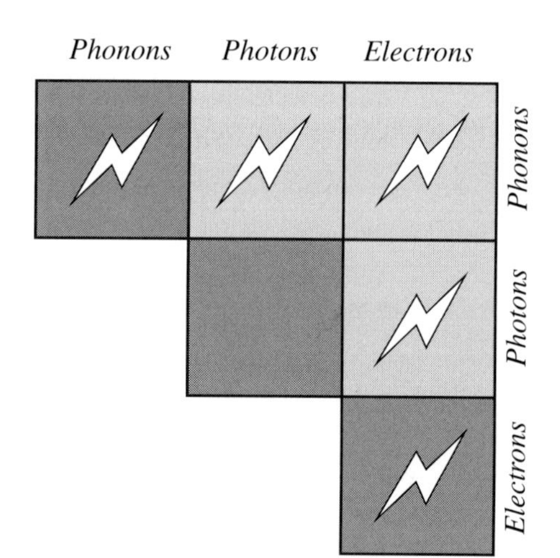

Figure 7.1. The basic subsystems that can interact to produce special effects: Phonon-phonon, phonon-photon, etc. The boxes marked by a flash are treated in this chapter. Also see [7.9].

diagram. The dark blocks indicate interactions between similar particles, i.e., phonons with phonons. The lighter blocks indicate interactions between different particle families. We do not restrict ourselves only to quantum phenomena, but also include effects that are comfortably described by a more classical model. At the end, we look at effects that arise due to material inhomogeneities, i.e., when we depart from the "infinite", perfect silicon crystal.

7.1 Phonon-Phonon

The harmonic potential is in fact a truncated series expansion model (see (2.13)) of the true crystal inter-ion potential, and as such is not capable of explaining all lattice phenomena satisfactorily. If we include further terms of the series, we say that the potential is anharmonic.

An-harmonicity

This has a host of consequences: the phonon frequencies become shifted from their harmonic frequencies; anharmonic phonons interact with each other; a phonon in a particular state has a finite lifetime before it disappears, and this fact is due to an anharmonic phonon-phonon interaction. We will briefly discuss two important further consequences in the sections to follow: the lattice thermal conductivity and the expansion of the lattice with temperature, both of which cannot be explained with the harmonic potential. But first we take a closer look at the concepts of interaction and lifetime.

7.1.1 Phonon Lifetimes

Phonon Lifetimes

The theory required for an adequate description of phonon lifetimes is fairly involved, going beyond the level and goals of this text. Instead, the interested reader is referred to Madelung [7.10]. Here we provide only a brief description of the phenomena. Each of the third and fourth terms of the series expansion of the interatomic potential can be written in a manner that expresses it as a three, respectively a four phonon interaction. For the three phonon interaction, four different actions are possible from this description:

- Absorption of a phonon, with two photons created;
- Absorption of two phonons, with one phonon created;
- Absorption of three phonons (energy balance violation);
- Creation of three phonons (energy balance violation);

The lifetime τ is then defined as the inverse of the probability that a particular phonon state j with wavevector $\boldsymbol{k}_j$ will, through one of the interactions above, disappear. The probability is formed by summing over all possible interactions to destroy a phonon given all possible distributions among the phonon states, minus all possible interactions to create a phonon given all possible distributions among the phonon states. In analogy to other "particles", given the existence of a phonon lifetime τ_{p}, the phonons are now also endowed with a mean free path l_p.

7.1.2 Heat Transport

Thermal Conductivity

Because the phonons transport energy through the crystal, and are generated by raising the temperature of a crystal, we expect that the macroscopic laws of heat conduction will be predicted by the microscopic lattice phonon model, at least for electrically insulating materials in which heat transport is not dominated by freely moving energetic electrons. Macroscopically, heat conduction in a solid is observed to be governed by the Fourier law

$$\boldsymbol{q} = -\kappa \cdot \nabla T \tag{7.1}$$

which relates the heat flux density $\boldsymbol{q}$ in Wm^{-2} to the negative temperature gradient $-\nabla T$ in Km^{-1} through the material's thermal conductivity (tensor) κ in $Wm^{-1}K^{-1}$.

The harmonic model for the lattice predicts that once a lattice wave is set into motion, it continues forever, i.e., there is no resistance to energy transport. We know from observations that this is not true. Already the fact that a crystal is limited in extent introduces reflections and disturbances to the perfectly homogeneous lattice model. However, all "particles" travelling through a real lattice also experience scattering centers, i.e., phonons interacting with each other, and this is what ultimately causes finite thermal resistances. Since a full derivation of the heat conductivity goes well beyond the level of this text (again, the interested reader should consult Madelung [7.10]), we instead look at an alternative formulation [7.2].

Our model assumes that we have a small, one-dimensional temperature gradient along the negative x-axis, as illustrated in Figure 7.2. For the

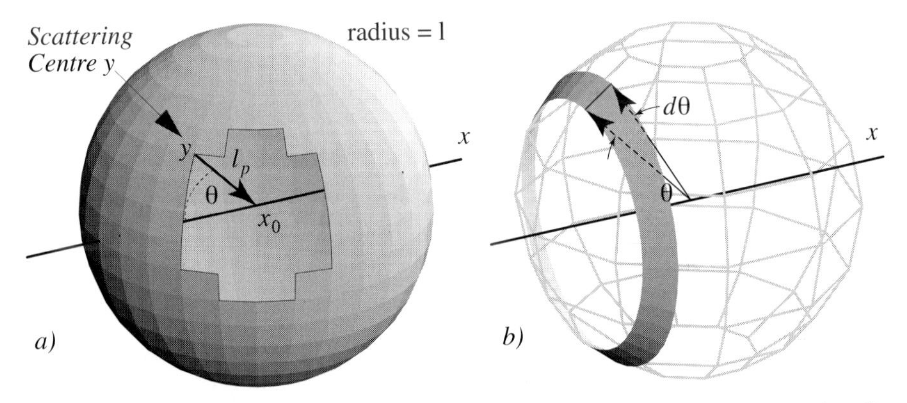

Figure 7.2. (a) Phonons arrive at point x_0 from a sphere of scattering centers with radius equal to the mean free path length. (b) Nomenclature for the solid angle integration.

point x_0 along the axis, our approach will be to estimate the 1D phonon energy flux q_x that arrives due to the scattered phonons. For a phonon mean free path of l_p, we may assume that all phonons arriving at x_0 come from the surface of a sphere of radius l_p. These phonons will have on average the velocity $\bar{v}_p$. If we call θ the angle that a point of phonon origin (a scattering centre) on the sphere, say y, makes at x_0 with respect to the negative x-axis, then that phonon has a velocity component along the x-axis of $\bar{v}\cos\theta$. Like the temperature, the internal energy is also assumed to vary only along the x-axis. This means that the phonons arrive at x_0 with an internal energy density of $u(x_0 - l\cos\theta)$. The average energy current in the x-direction is now the solid angle integral of the the product of velocity with energy density $\bar{v}_p\cos\theta\ u(x_0 - l\cos\theta)$. To obtain the differential solid angle as a function of θ we take the ratio of the differential surface area to the total spherical area, i.e.,

$$\frac{(2\pi r)(r\sin\theta)d\theta}{4\pi r^2} = \frac{1}{2}\sin\theta d\theta \tag{7.2}$$

Thus we have to evaluate

$$\begin{aligned} q_x &= \int_0^{\pi} \bar{v}_p\cos\theta\ u(x_0 - l\cos\theta)\frac{1}{2}\sin\theta d\theta \\ &= \frac{\bar{v}_p}{2}\int_{-1}^{1} \mu u(x_0 - l\mu)d\mu \end{aligned} \tag{7.3}$$

Note the change of variables $\mu = \cos\theta$ and the new integration limits. We next expand the energy density u about the position x_0. We drop the quadratic and higher terms to obtain $u = u|_{x_0} + \partial u/\partial x|_{x_0}\mu l$, and insert this into (7.3) and evaluate

$$q_x = \frac{\bar{v}}{2}\int_{-1}^{1}\left(u|_{x_0}\mu + \left.\frac{\partial u}{\partial x}\right|_{x_0}\mu^2 l\right)d\mu = 0 + \left.\frac{\bar{v}_p l}{3}\frac{\partial u}{\partial x}\right|_{x_0} \tag{7.4}$$

Remembering that $\partial u/\partial x = (\partial u/\partial T)(\partial T/\partial x) = c_v\partial T/\partial x$, we can relate the derived to the phenomenological equation

$$q = -\kappa\frac{\partial T}{\partial x} = \frac{\bar{v}_p l}{3}c_v\frac{\partial T}{\partial x} \text{ and hence} \tag{7.5a}$$

$$\kappa = \frac{1}{3}\bar{v}_p l c_v = \frac{1}{3}\bar{v}_p^2\tau_p c_v \tag{7.5b}$$

where τ_p is the phonon lifetime or relaxation time, which in turn is the the inverse of the phonon scattering rate.

Clearly, (7.5a) predicts that κ is temperature dependent, with three potential sources. We do not expect the phonon velocity $\bar{v}_p$ to show a strong temperature dependence for the solid state, since its value should be proportional to the root of the material density. In Section 2.4.2 we have seen that the specific heat at high temperatures tends to a constant value c_v. We are thus left with the phonon lifetime as only temperature-dependent influence at high temperatures. No simple theoretical explanation of these scattering processes is possible, and experimental evidence suggest that $\kappa \propto T^{-\beta}$ for $1 \leq \beta \leq 2$. This makes sense, since we expect that as the number of generated phonons increase due to the raised temperature, so does the number of inter-phonon scattering events, and hence the thermal conductivity should decrease with raised temperature.

7.2 Electron-Electron

For this section we assume that all electron-electron interaction of ground state electrons has already been considered in calculating the occupied ground states, i.e., the valence band structure. Therefore, we only consider the interaction of excited electrons. The density of excited electrons is assumed to be this low that a single electron description is justified. Remember that this single electron description allows the description of a completely occupied valence band missing one electron as a single defect electron or hole in the same manner as for a single electron in the conduction band.

In the light of the above simplifying assumptions we calculate the transition rates between two electronic states as they enter the Boltzmann transport equation (6.2) by means of the Fermi golden rule (3.82). For the electron-electron interaction this yields a nonlinear term in the Boltzmann transport equation. For this section we only want to describe the electronic system as non-interacting particles for which only an effective interaction due to density fluctuations is put into the streaming motion term of the Boltzmann transport equation.

7.2.1 The Coulomb Potential (Poisson Equation)

In 6.1.1 we wrote $\dot{\boldsymbol{k}} = \boldsymbol{F}/\hbar$ considering Newton's law. Here $\boldsymbol{F}$ was supposed to be an external force field. Electrons are charged particles and thus even at low densities will interact with each other through the electrostatic potential.

Let us assume that the force term is due to an externally applied electrostatic potential, which gives an electric field of the form $\boldsymbol{E} = -\nabla\Psi(\boldsymbol{x})$ (see (4.34)) and the force term gets $\boldsymbol{F} = e\nabla\Psi(\boldsymbol{x})$. For the streaming motion term of the Boltzmann transport equation (6.2) we therefore write

$$\frac{\partial}{\partial t}f(\boldsymbol{k}, \boldsymbol{x}, t) + \frac{e}{\hbar}\nabla\Psi(\boldsymbol{x})\nabla_{\boldsymbol{k}}f(\boldsymbol{k}, \boldsymbol{x}, t) + \frac{1}{\hbar}\nabla_{\boldsymbol{k}}E(\boldsymbol{k})\nabla_{\boldsymbol{x}}f(\boldsymbol{k}, \boldsymbol{x}, t) = 0 \quad (7.6)$$

in the absence of scattering, where we introduced a real band structure $E(\boldsymbol{k})$ through the group velocity $v_g = \nabla_{\boldsymbol{k}}E(\boldsymbol{k})/\hbar$. Let us further assume that we are near equilibrium and the distribution function $f(\boldsymbol{k}, \boldsymbol{x}, t)$ is given by a well known equilibrium distribution $f_0(E(\boldsymbol{k}))$. $\Psi(\boldsymbol{x}, t)$ is a weak perturbation that might also depend on time and we want to calculate $\delta f(\boldsymbol{k}, \boldsymbol{x}, t) = f(\boldsymbol{k}, \boldsymbol{x}, t) - f_0(E(\boldsymbol{k}))$ the variation in the distribution function due to this perturbational potential. For this purpose we write both $\Psi(\boldsymbol{x}, t)$ and $\delta f(\boldsymbol{k}, \boldsymbol{x}, t)$ as Fourier transforms in the spatial coordinate and Laplace transforms in time. This yields

$$\Psi(\boldsymbol{x}, t) = Re[\Psi(\boldsymbol{p}, z)e^{\boldsymbol{px} - zt}] \quad (7.7)$$

$$\delta f(\boldsymbol{k}, \boldsymbol{x}, t) = f(\boldsymbol{k}, \boldsymbol{x}, t) - f_0(E(\boldsymbol{k})) = Re[\delta f(\boldsymbol{k}, \boldsymbol{p}, z)e^{\boldsymbol{px} - zt}] \quad (7.8)$$

where $z = \omega + i\eta$ and η introduces an adiabatic switching on of the perturbation in time. We insert (7.7) and (7.8) in (7.6) and neglect the second order term $\delta f(\boldsymbol{k}, \boldsymbol{p}, z)\Psi(\boldsymbol{p}, z)$ arising due to the gradient term $\nabla_{\boldsymbol{k}}f(\boldsymbol{k}, \boldsymbol{x}, t)$ in (7.6). This yields

$$-iz\delta f(\boldsymbol{k},\boldsymbol{p},z) + i\boldsymbol{p}\frac{e}{\hbar}\Psi(\boldsymbol{p},z)\nabla_k f_0(E(\boldsymbol{k})) + i\boldsymbol{p}\boldsymbol{v}_k\delta f(\boldsymbol{k},\boldsymbol{p},z) = 0 \quad (7.9)$$

which we may solve for $\delta f(\boldsymbol{k},\boldsymbol{p},z)$ to give

$$\delta f(\boldsymbol{k},\boldsymbol{p},z) = \frac{e\boldsymbol{p}\boldsymbol{v}_k\frac{\partial f_0}{\partial E(\boldsymbol{k})}\Psi(\boldsymbol{p},z)}{\omega + i\eta - \boldsymbol{p}\boldsymbol{v}_k} \quad (7.10)$$

(7.10) must be interpreted as the variation or response of the distribution function due to an perturbation. From (6.10) we calculate the density response function by integration of (7.10) over k-space and thus obtain

$$\delta n(\boldsymbol{p},\omega) = \frac{1}{4\pi^3}\int_{\Omega_k}\frac{e\boldsymbol{p}\boldsymbol{v}_k\frac{\partial f_0}{\partial E(\boldsymbol{k})}\Psi(\boldsymbol{p},\omega)}{\omega + i\eta - \boldsymbol{p}\boldsymbol{v}_k}d^3\boldsymbol{k} = \chi_n^{(0)}(\boldsymbol{p},\omega)\Psi(\boldsymbol{p},\omega) \quad (7.11)$$

$\chi_n^{(0)}(\boldsymbol{p},\omega)$ is the zero order density susceptibility which gives us the spatial variations in carrier density due to lowest order approximation. This function is essential for calculating the dielectric properties of the semiconductor material.

7.2.2 The Dielectric Function

We now turn to the initial point where electrons as charged particles are interacting. This interaction appears to create an internal potential $\Psi^{int}(\boldsymbol{x},t)$ in addition to the external potential $\Psi^{ext}(\boldsymbol{x},t)$ so that we have to calculate with an effective potential $\Psi^{eff}(\boldsymbol{x},t)$ of the form

Effective Potential

$$\Psi^{eff}(\boldsymbol{x},t) = \Psi^{ext}(\boldsymbol{x},t) + \Psi^{int}(\boldsymbol{x},t) \quad (7.12)$$

The additional internal potential is due to charge density variations and according to (4.37) may be written as

$$\Psi^{int}(\boldsymbol{x},t) = -\frac{e}{4\pi\varepsilon_0}\int_{\Omega_r}\frac{\delta n(\mathbf{x}')}{|\boldsymbol{x}-\boldsymbol{x}'|}d^3\boldsymbol{x}' \quad (7.13)$$

A Fourier transform yields

$$\Psi^{eff}(\boldsymbol{p}, \omega) = \Psi^{ext}(\boldsymbol{p}, \omega) - \frac{e}{p^2}\delta n(\boldsymbol{p}, \omega) \tag{7.14}$$

Using (7.11) we have

$$\left[1 + \frac{e}{p^2}\chi_n^{(0)}(\boldsymbol{p}, \omega)\right]\Psi^{eff}(\boldsymbol{p}, \omega) = \Psi^{ext}(\boldsymbol{p}, \omega) \tag{7.15}$$

which gives us the relation between the external potential and the effective potential and which we call the dynamic dielectric function

$$\varepsilon(\boldsymbol{p}, \omega) = \frac{\Psi^{ext}(\boldsymbol{p}, \omega)}{\Psi^{eff}(\boldsymbol{p}, \omega)} = 1 + \frac{e}{p^2}\chi_n^{(0)}(\boldsymbol{p}, \omega) \tag{7.16}$$

Random Phase approximation

in *random phase approximation* which means that only the internal Coulomb potential in addition to an external potential is taken into account and all other possible interactions are neglected.

7.2.3 Screening

The term in the definition of the dynamic dielectric function (7.16) that arises from the density susceptibility acts as a screening for the external potential. Let us, for this purpose, analyze the static dielectric function $\varepsilon(\boldsymbol{p}, 0)$, which, taking into account (7.11), reads

$$\varepsilon(\boldsymbol{p}, 0) = 1 - \frac{e^2}{p^2 4\pi^3}\int_{\Omega_k}\frac{\partial f_0}{\partial E(\boldsymbol{k})}d^3\boldsymbol{k} = 1 + \frac{\lambda^2}{p^2} \tag{7.17}$$

We defined the screening parameter λ

$$\lambda^2 = -\frac{e^2}{4\pi^3}\int_{\Omega_k}\frac{\partial f_0}{\partial E(\boldsymbol{k})}d^3\boldsymbol{k} \tag{7.18}$$

Screening Length

which is a wavelength the so called screening wavelength. Its inverse may interpreted as the screening length. This can be easily seen for the example of a static external Coulomb potential $\Psi^{ext}(\boldsymbol{x}) = e(4\pi\varepsilon_o|\boldsymbol{x}|)^{-1}$. Its Fourier transform is given by

$$\Psi^{ext}(\mathbf{p}) = \frac{e}{p^2} \tag{7.19}$$

To obtain the effective potential we have to divide this external potential by the static dielectric function, to give

$$\Psi^{eff}(\boldsymbol{p}) = \frac{e}{p^2\left(1 + \frac{\lambda^2}{p^2}\right)} = \frac{e}{p^2 + \lambda^2} \tag{7.20}$$

Transforming back to real space we obtain

$$\Psi^{eff}(\boldsymbol{x}) = \frac{e}{4\pi\varepsilon_o|\boldsymbol{x}|}e^{-\lambda|\boldsymbol{x}|} \tag{7.21}$$

In this form we see the meaning of λ as the screening wavelength. The exponential function in (7.21) weakens the effect of the Coulomb potential at long distances. While the Coulomb potential in vacuum goes like $|\boldsymbol{x}|^{-1}$ in the spatial variables and is said to be a long ranging potential, in a medium where moving charge densities are present this behavior is covered in lowest order by the density response function of the charges that enters the dielectric function. Thus every source for Coulomb potentials in a semiconductor will finally suffer this screening effect.

7.2.4 Plasma Oscillations and Plasmons

For the screening effects we looked at the static properties of the dielectric function. Let us now assume that the spatial variations present are of very long wavelength and we are interested in the frequency behavior of the system. Therefore, we set $\eta = 0$ in (7.11), since the adiabatic switching on of the perturbation is assumed to be finished. What influence does the frequency behavior of the perturbation have on the system? We expand the denominator of (7.11) for small $\boldsymbol{p} \rightarrow 0$, to give

$$\frac{1}{\omega - \boldsymbol{p}\boldsymbol{v}_k} = \frac{1}{\omega}\left[1 + \frac{1}{\omega}\boldsymbol{p}\boldsymbol{v}_k + \dots\right] \tag{7.22}$$

Thus we have for the dielectric function

$$\varepsilon(\boldsymbol{p}, \omega) \approx 1 + \frac{e^2}{p^2 4\pi^3}\int_{\Omega_k} \frac{\boldsymbol{p}\boldsymbol{v}_k}{\omega}\left[1 + \frac{1}{\omega}\boldsymbol{p}\boldsymbol{v}_k + \dots\right]\frac{\partial f_0}{\partial E(\boldsymbol{k})}d^3\boldsymbol{k} \tag{7.23}$$

in which the first term in the power series vanishes due to $E(\boldsymbol{k}) = E(-\boldsymbol{k})$. Now let us assume parabolic bands with $E(\boldsymbol{k}) = \hbar^2 k^2/(2m^*)$ which gives us a second order term of the form

$$(\boldsymbol{p}\boldsymbol{v}_k)(\boldsymbol{p}\boldsymbol{v}_k) = \left(\sum_i \frac{\hbar}{m}p_i k_i\right)\left(\sum_j \frac{\hbar}{m}p_j k_j\right) = \frac{\hbar^2}{3m^2}p^2 k^2 \tag{7.24}$$

which, inserted into (7.23), results in a dielectric function in the long wavelength limit $\boldsymbol{p} \to 0$

$$\varepsilon(0, \omega) = 1 + \frac{e^2\hbar^2}{3m^2\omega^2\pi^2}\int_{\Omega_k} k^4 \frac{\partial f_0}{\partial E(\boldsymbol{k})}dk \tag{7.25}$$

This we write another way to read

$$\varepsilon(\omega) = 1 - \frac{\omega_{pl}^2}{\omega^2} \tag{7.26}$$

where we defined the plasma frequency by

$$\omega_{pl}^2 = \frac{e^2\hbar^2}{3m^2\pi^2}\int_{\Omega_k} k^4\left(-\frac{\partial f_0}{\partial E(\boldsymbol{k})}\right)dk \tag{7.27}$$

The dielectric function vanishes at the plasma frequency and thus the effective potential shows a singularity at the plasma frequency since $\Psi^{eff}(\omega) = \Psi^{ext}(\omega)/\varepsilon(\omega)$. Let us calculate the plasma frequency for a degenerate Fermi gas, for which $f_0 = \Theta(E_F - E_k)$ is the distribution function and we have $-\partial f_0/\partial E(\boldsymbol{k}) = \delta(E_k - E_F)$ to give a plasma frequency

$$\omega_{pl}^2 = \frac{e^2\hbar^2}{3m^2\pi^2}\int_{\Omega_k} k^4\delta\left(\frac{\hbar^2k^2}{(2m^*)}-E_F\right)dk = \frac{e^2\hbar^2}{3m^2\pi^2}k_F^4\frac{m}{\hbar^2k_F}$$
$$= \frac{e^2k_F^3}{3m\pi^2} = \frac{ne^2}{m} \tag{7.28}$$

where we used the electron density in a degenerate system defined as $n = k_F^3/3\pi^2$. We conclude that the occurrence of plasma oscillations is due to the long range Coulomb interaction. Typical frequencies are found for metals at energies $\hbar\omega_{pl} \approx 10eV$ which, compared to the thermal energy k_BT at room temperature, is several orders of magnitude larger. Hence plasma oscillations will not be contained in the thermal density fluctuations at room temperature.

7.3 Electron-Phonon

Electron-phonon scattering is the major effect in electronic transport at high temperatures, i.e., in the range of 300 K. There is a dissipation channel arising that forms the basis for many different transport coefficients of the electronic system in a semiconductor. The dissipated energy will change the thermal properties of the lattice system.

The description of electron-phonon interaction on a quantum mechanical level therefore must include both the electron wavefunction and the lattice wavefunction. The transition of scattering rate between initial and final states in this interaction will be governed by the *Fermi golden rule* (3.82). Suppose that initially the electron occupies a state with wavevector **k**. In Figure 7.3 the two possible processes are schematically drawn: absorption of a phonon with wavevector **q** and emission of such a phonon. In both cases the electron wavevector changes by the modulus of q provided momentum conservation holds. The transition rate reads

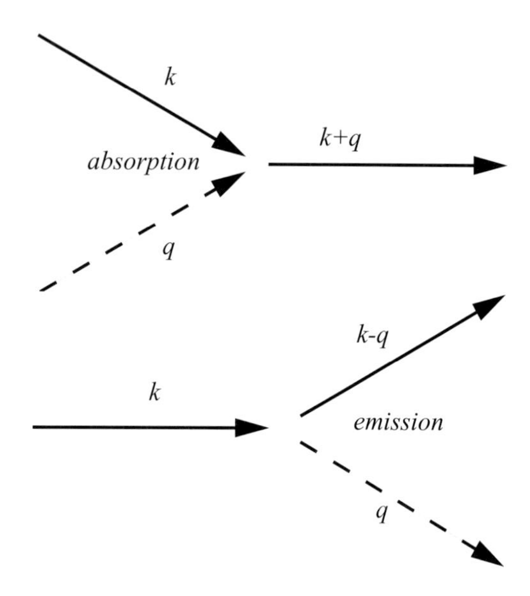

Figure 7.3. Schematic representation of phonon absorption and phonon emission processes.

$$W(\boldsymbol{k}, \boldsymbol{k}';\boldsymbol{q}) = \frac{2\pi}{\hbar}|V(\boldsymbol{k}, \boldsymbol{k}';\boldsymbol{q})|^2\delta(E_k - E_{k'} \pm \hbar\omega_q) \tag{7.29}$$

where the transition matrix element is given by

$$V(\boldsymbol{k}, \boldsymbol{k}';\boldsymbol{q}) = \langle \boldsymbol{k}', n_{\boldsymbol{q}} \pm 1|\hat{V}_{k,k',q}|\boldsymbol{k}, n_{\boldsymbol{q}}\rangle \tag{7.30}$$

$|\boldsymbol{k}, n_{\boldsymbol{q}}\rangle$ indicates the Dirac notation of a product of two wavefunctions one from the electronic system and one from the lattice vibrational system. We must further specify the interaction potential operator $\hat{V}_{k,k',q}$. This operator characterizes the nature of the different electron-phonon interactions given in the following sections.

7.3.1 Acoustic Phonons and Deformation Potential Scattering

The description of a periodic potential structure as given in Section 3.3 is appropriate for situations where the ions forming the potential are not subject to displacement. From Sections 2.3 and 2.4 we know that this does not correspond to the real situation. There is a permanent motion of the ions that form the crystal lattice. Thus we analyze the periodic potential in which electrons moves $V(\boldsymbol{x}) = \sum V(\boldsymbol{x} - \boldsymbol{R}_i)$ with respect to small displacements of the ions, where $\boldsymbol{x}$ is the position of the electron. $\boldsymbol{R}_i = \boldsymbol{R}_i^0 + \boldsymbol{s}_i(t)$ is the time dependent position of the ion as described in Sections 2.3 and 2.4 and the $\boldsymbol{s}_i(t)$ are longitudinal vibrations of the crystal lattice. The periodic potential structure results in a certain energy E_c of the conduction band edge. In an effective mass description the variation of the conduction band energy is to lowest order given by the change δE_c in the conduction band edge energy. Longitudinal acoustic phonons are compression waves $s(\mathbf{r}, t)$ for the set of ions in the solid with a relative volume change given by $\Delta(\mathbf{r}, t) = \nabla s(\mathbf{r}, t) = \delta V / V$. This leads to a change in the lattice constant and therefore in the valence band edge energy. Thus we write

$$\delta E_c = V \frac{\partial E_c}{\partial V} \nabla s(\mathbf{r}, t) = \Xi_1 \nabla s(\mathbf{r}, t) \tag{7.31}$$

where Ξ_1 is called the deformation potential which takes a characteristic material dependent value. Let us write the longitudinal vibrations in terms of normal coordinates and look at one Fourier component for the wavevector $\mathbf{q}$

$$s_q(\mathbf{r}, t) = \left(\frac{\hbar}{2MN\omega_q}\right)^{1/2} (b_q e^{iqr} + b_q^+ e^{-iqr}) \boldsymbol{e}_q e^{-i\omega_q t} \tag{7.32}$$

where N is the total number of atoms in the lattice, M is the total mass and $\boldsymbol{e}_q$ is the polarization of the lattice vibration. Thus (7.31) reads

$$\delta E_c = V_{def}^{e-ph}(\boldsymbol{r}, t) = \frac{\Xi_1}{\sqrt{MN}} \sum_q \left(\frac{\hbar}{2\omega_q}\right)^{1/2} i\boldsymbol{q}\boldsymbol{e}_q (b_q(t) + b_{-q}^+(t)) e^{qr} \tag{7.33}$$

Let us use the fermion and boson number representations to write down the interaction operator, where the fermions are expanded in terms of Bloch waves (3.92)

$$\hat{V}_{k,k',q} = i\boldsymbol{q}\boldsymbol{e}_q\sqrt{\frac{\hbar\Xi_1^2}{2MN\omega_q}}[b_q + b_{-q}^+]c_{k'}^+c_k \times \int_{Lattice} e^{i(\boldsymbol{k}+\boldsymbol{q}-\boldsymbol{k}')\boldsymbol{r}} u^*(\boldsymbol{k}', \boldsymbol{r})u(\boldsymbol{k}, \boldsymbol{r})dV \tag{7.34}$$

The integration in (7.34) has to be performed over the whole crystal lattice. Since the $u(\boldsymbol{k}, \boldsymbol{r})$ are defined in a single lattice cell this integral may be split into a integration over a such a cell summed over all lattice cell sites to give

$$\int_{Lattice} e^{i(\boldsymbol{k}+\boldsymbol{q}-\boldsymbol{k}')\boldsymbol{r}} u^*(\boldsymbol{k}', \boldsymbol{r})u(\boldsymbol{k}, \boldsymbol{r})dV = \sum_{\text{sites}} \int_{cell} e^{i(\boldsymbol{k}+\boldsymbol{q}-\boldsymbol{k}')\boldsymbol{r}} u^*(\boldsymbol{k}', \boldsymbol{r})u(\boldsymbol{k}, \boldsymbol{r})dV \tag{7.35}$$

With the substitution $\boldsymbol{r} = \boldsymbol{R}_i + \boldsymbol{r}'$ we obtain

$$\delta(\boldsymbol{k}+\boldsymbol{q}-\boldsymbol{k}')F_{\boldsymbol{k},\boldsymbol{k}'} = \sum_i e^{i(\boldsymbol{k}+\boldsymbol{q}-\boldsymbol{k}')\boldsymbol{R}_i} \times \int_{\text{cell}} e^{i(\boldsymbol{k}+\boldsymbol{q}-\boldsymbol{k}')\boldsymbol{r}'} u^*(\boldsymbol{k}', \boldsymbol{r}')u(\boldsymbol{k}, \boldsymbol{r}')dV' \tag{7.36}$$

where we set

$$F_{\boldsymbol{k},\boldsymbol{k}'} = \int_{\text{cell}} u^*(\boldsymbol{k}', \boldsymbol{r})u(\boldsymbol{k}, \boldsymbol{r})dV \tag{7.37}$$

Thus we have for the interaction operator

$$\hat{V}_{k,q} = i\boldsymbol{q}\boldsymbol{e}_q(\hbar\Xi_1^2/(2MN\omega_q))^{1/2}F_{\boldsymbol{k}+\boldsymbol{q},\boldsymbol{k}}[b_q + b_{-q}^+]c_{\boldsymbol{k}+\boldsymbol{q}}^+c_{\boldsymbol{k}} \tag{7.38}$$

We observe two terms in (7.38) and interpret them according to the creation and destruction operator definitions previously made in Chapter 3: $b_q c_{\boldsymbol{k}+\boldsymbol{q}}^+ c_{\boldsymbol{k}}$ destroys an electron with wavevector $\boldsymbol{k}$ creates one $\boldsymbol{k}+\boldsymbol{q}$ and destroys a phonon with wavevector $\boldsymbol{q}$, while $b^{+}_{-q}c^{+}_{k+q}c_k$ destroys an electron with wavevector $\boldsymbol{k}$, creates one with $\boldsymbol{k}-\boldsymbol{q}$ and creates a phonon with wavevector $-\boldsymbol{q}$. This is why we call the first term phonon absorption and the second one phonon emission. Thus the square of the matrix element in (7.29) yields according to (7.30) for the emission of a phonon

$$|V(\boldsymbol{k},\boldsymbol{k}';\boldsymbol{q})|^2 = \frac{\hbar\Xi_1^2 q^2}{2MN\omega_{\boldsymbol{q}}}(n_q+1) \tag{7.39}$$

For the absorption matrix element in (7.39) n_q+1 has to be replaced by n_q. The n_q follow a Bose statistics as described in Section 5.2.2 and thus we may replace n_q by

$$n_q = \frac{1}{\exp(\hbar\omega_q/k_BT)-1} \tag{7.40}$$

which in the high temperature limit yields $n_q = k_BT/\hbar\omega_q$. This means that the acoustic phonon scattering rate increases linearly with the temperature.

7.3.2 Optical Phonon Scattering

This interaction mechanism occurs in two different ways which both are due to the optical lattice vibrations in which one sub-lattice moves relative to a second one.

In silicon this motion creates a shift of the potential field of each ion. The bond charges follow this motion and slightly accumulate where the two ions approach increasing the negative charge, while a positive remains where the ions have moved apart. This is a macroscopic electric field due to the relative displacement field $\boldsymbol{w}_q$ of the sub-lattices and thus results in an interaction potential

Non-Polar Materials

$$\delta E = D\boldsymbol{w}_q \tag{7.41}$$

Again the macroscopic deformation field parameter D depends on the respective material properties. We see that the interaction itself is not q-dependent as in the case of acoustic phonons. Since the optical phonons are rather energetic the high temperature limit lies far above 300K and is not to be applied as in the acoustic case. The relative displacement field Fourier component with wavevector **q** is given similar to (7.32) by

$$\boldsymbol{w}_q = \sqrt{\frac{\hbar D^2}{2MN\omega_q}}[b_q e^{iqr} + b_q^+ e^{-iqr}]\boldsymbol{e}_q e^{-i\omega t} \tag{7.42}$$

Note the difference in the pre-factor, where $\overline{M}^{-1} = M_+^{-1} + M_-^{-1}$. M_+^{-1} is the mass of the sub-lattice with displacement $\boldsymbol{s}_+$ in positive direction and M_-^{-1} the respective masses of the sub-lattice with displacement $\boldsymbol{s}_-$ in negative direction. We may think of the total displacement as $\boldsymbol{w}_q = \boldsymbol{s}_+ - \boldsymbol{s}_-$. Thus the respective matrix element has the same structure as (7.38), it reads for the optical phonon scattering

$$V(\boldsymbol{k}, \boldsymbol{k}';\boldsymbol{q}) = \sqrt{\frac{\hbar D^2}{2MN\omega_q}} \times \\ [\sqrt{n_q}\delta(E_{k'} - E_{k+q} + \hbar\omega_q) + \sqrt{n_q + 1}\delta(E_{k'} - E_{k+q} - \hbar\omega_q)] \tag{7.43}$$

Polar Materials

In polar crystals like GaAs a very effective interaction mechanism is present due to the opposite polarization of the two sub-lattices. The displacement of positive and negative ions $s_\pm$ with the effective charges $\pm e^*$ leads to a local dipole moment $\boldsymbol{p} = e^*(\boldsymbol{s}_+ - \boldsymbol{s}_-)$ from which we derive a macroscopic polarization of the form

$$\boldsymbol{P} = \left[\frac{N\overline{M}\omega_{opt}^2}{4\pi V}\left(\frac{1}{\varepsilon_\infty} - \frac{1}{\varepsilon_0}\right)\right]^{1/2} \boldsymbol{w}_q \tag{7.44}$$

where $\boldsymbol{w}_q$ is the relative displacement of positive and negative charges as in (7.42) and $\overline{M}^{-1} = M_+^{-1} + M_-^{-1}$ is given by their respective masses. ω_{opt} is the optical phonon frequency. The interaction energy in this case is given by the product of the induced electric field $\boldsymbol{E}$ and the polarization $\boldsymbol{P}$

$$\delta E = -\varepsilon_{\infty} \boldsymbol{E} \boldsymbol{P} \tag{7.45}$$

where ε_{∞} is the optical or high frequency dielectric constant. For the induced electric field we have to take into account that the respective electrostatic potential is due to screening effects because of the movable charges present and thus will be of the form as given in (7.20) or (7.21) and therefore (7.45) reads

$$\delta E = \left[\frac{N\bar{M}\omega_{opt}^2}{4\pi V}\left(\frac{1}{\varepsilon_{\infty}} - \frac{1}{\varepsilon_0}\right)\right]^{1/2} \frac{q^2}{q^2 + \lambda^2}[b_q e^{i\boldsymbol{qr}} + b_q^+ e^{-i\boldsymbol{qr}}]\boldsymbol{e}_q e^{-i\omega t} \tag{7.46}$$

7.3.3 Piezoelectricity

Simple Piezoelectric Model

Consider the 1D latticc with a basis of Section 2.4.1 (also see Figure 2.20), but now endow the two atoms each with equal but opposing charges. The kinetic co-energy is added up from the contributions of the individual ions

$$T^* = \frac{1}{2}\sum_{i\alpha}^{n,\,2} m_{\alpha}\dot{u}_{i\alpha}^2 \tag{7.47}$$

The index α counts over the ions in an elementary basis cell, and the index i counts over the lattice cells. Similarly, the harmonic bond potential energy is dependent on the stretching of the inter-ion bonds

$$U = \frac{1}{4}\sum_{i\alpha,\,\beta j}^{n,\,2,\,n,\,2} \frac{E_{\alpha i\beta j}}{a}(u_{i\alpha} - u_{j\beta})^2 \tag{7.48}$$

to which we now add the ionic energy due to the nearest-neighbor charge interaction

$$U_q = \frac{1}{4\pi\varepsilon_0}\frac{q_{\alpha}}{r_{\alpha}} \tag{7.49}$$

Actually, the charge interactions are long-range. Since the crystal is neutral, we assume that the remaining crystal screens off the rest of the charges, and the local disturbance is only "felt" by the direct neighbors of the moving charge. Clearly, this is a simplistic model used only to show the features of the piezoelectric interaction.

The restriction to next-neighbor interactions and identical interatomic force constants yields the following equations of motion when the expressions (7.47) and (7.48) are inserted into the Lagrange equations

$$m\ddot{u}_{im} = -\frac{E}{a}(2u_{im} - u_{iM} - u_{(i-1)M}) \tag{7.50a}$$

$$M\ddot{u}_{iM} = -\frac{E}{a}(2u_{iM} - u_{(i+1)m} - u_{im}) \tag{7.50b}$$

Reversible Thermo-dynamic Derivation

For a more general model, we first consider a general thermodynamic phenomenological model of piezoelectricity (see e.g. [7.5] or [7.12]). The internal energy of the crystal, U, is defined as (for the general idea, see Box 7.1 and Box 7.2)

Box 7.1. The thermodynamic postulates [7.11].

The Four *Callen* Postulates. Thermodynamics is founded upon a set of "laws". Modern theorists have reformulated these laws as a set of postulates. We closely follow [7.11]:

I. *There exist particular states (called equilibrium states) that, macroscopically, are characterized completely by the specification of the internal energy* U*, and a set of extensive parameters* $X_1, X_2, \ldots, X_r$.

II. *There exists a function (called the entropy* S*) of the extensive parameters, defined for all equilibrium states, and having the following property: The values assumed by the extensive parameters in the absence of a constraint are those that maximize the entropy over the manifold of constrained equilibrium states.*

III. *The entropy of a composite system is additive over the constituent subsystems (whence the entropy of each subsystem is a homogeneous first-order function of the extensive parameters). The entropy is continuous and differentiable and is a monotonically increasing function of the energy.*

IV. *The entropy of any system vanishes in the state for which* $T \equiv (\partial U/\partial S)_{X_1, X_2, \ldots} = 0$.

$$dU = \sigma{:}d\varepsilon + \boldsymbol{E}\cdot d\boldsymbol{D} + \boldsymbol{H}\cdot d\boldsymbol{B} + TdS \tag{7.51}$$

Box 7.2. Basic thermodynamic relations [7.11].

Extensive and Intensive Parameters. The extensive parameters are denoted by

$$\{S, V, N_1, N_2, \ldots, N_r\} \equiv \{X_0, X_1, X_2, \ldots, X_t\} \quad \text{(B 7.2.1)}$$

for the volume V and the N_k the mole numbers of the constituent chemical species. The intensive parameters are denoted by

$$\{T, P, \eta_1, \eta_2, \ldots, \eta_r\} \equiv \{P_0, P_1, P_2, \ldots, P_t\} \quad \text{(B 7.2.2)}$$

Here, T is the temperature, P is the pressure, and the η_k are the electrochemical potentials of the constituent species.

The Fundamental Equations. The fundamental energy equation of a simple system is

$$U = U(X_0, X_1, X_2, \ldots, X_t) \quad \text{(B 7.2.3)}$$

In its differential form, the fundamental thermodynamic equation for the energy is

$$dU = TdS + \sum_{k=1}^{t} P_k dX_k = \sum_{k=0}^{t} P_k dX_k \quad \text{(B 7.2.4)}$$

The intensive parameters P_k are defined by

$$P_k = \partial U / \partial X_k \quad \text{(B 7.2.5)}$$

The fundamental entropic relation is

$$S = S(X_0, X_1, X_2, \ldots, X_t) \quad \text{(B 7.2.6)}$$

In its differential form, the fundamental thermodynamic equation for the entropy is

$$dS = \frac{1}{T} dU + \sum_{k=1}^{t} F_k dX_k = \sum_{k=0}^{t} F_k dX_k \quad \text{(B 7.2.7)}$$

The intensive parameters F_k are defined by

$$F_k = \partial S / \partial X_k \quad \text{(B 7.2.8)}$$

The Euler Relation. $U = \sum_{k=0}^{t} P_k X_k$ (B 7.2.9)

The Gibbs-Duheim Relation.

$$\sum_{k=0}^{t} X_k dP_k = 0 \quad \text{(B 7.2.10)}$$

The Thermodynamic Potentials. By the Legendre transformation we can replace any extensive parameter in (B 7.2.3) by an intensive parameter, yielding a new fundamental relation that is closer to laboratory conditions. The most familiar of these "potentials" are the *Helmholtz free energy* (replacing S with T)

$$F = U - TS , \quad \text{(B 7.2.11)}$$

the *Enthalpy* (replacing V with P)

$$H = U + PV , \quad \text{(B 7.2.12)}$$

the *Gibbs free energy* (combining the above)

$$G = U - TS + PV , \quad \text{(B 7.2.13)}$$

and the *Grand canonical potential* (replacing S with T, and N with $\bar{\mu}$)

$$K = U - TS - \eta N . \quad \text{(B 7.2.14)}$$

The Maxwell Relations. We have that

$$\partial X_j / \partial P_k = \partial X_k / \partial P_j \quad \text{(B 7.2.15)}$$

$$\partial X_j / \partial X_k = -\partial P_k / \partial P_j \quad \text{(B 7.2.16)}$$

$$\partial P_j / \partial X_k = \partial P_k / \partial X_j \quad \text{(B 7.2.17)}$$

These define the material properties in convenient forms. Also called reciprocal relations.

Affinities and Fluxes. In a discrete system, an extensive parameter *flux* is defined by

$$J_k = dX_k / dt \quad \text{(B 7.2.18)}$$

Taking the time derivative of the entropic fundamental relation we obtain

$$dS/dt = \sum_k \frac{dS}{dX_k} \frac{dX_k}{dt} = \sum_k \mathfrak{F}_k J_k \quad \text{(B 7.2.19)}$$

where we have defined the extensive parameter's associated affinity $\mathfrak{F}_k = dS/dX_k$. For continuous systems we have that

$$\dot{s} = \sum_k \nabla F_k \bullet \mathbf{j}_k \quad \text{(B 7.2.20)}$$

In both instances the rate of entropy production is the sum of the products of affinities with their respective fluxes.

This fundamental relation includes the effects of mechanical deformation, electric and magnetic fields, as well as thermal phenomena. In (7.51) the intensive variables: the stress tensor σ, the electric field vector $\boldsymbol{E}$, the magnetic field vector $\boldsymbol{H}$ and the scalar temperature T, are independent, and the extensive variables: the strain tensor ε, the dielectric displacement vector $\boldsymbol{D}$, the magnetic induction vector $\boldsymbol{B}$ and the scalar entropy S, are dependent. By this property we can write that

$$d\varepsilon = \left.\frac{\partial\varepsilon}{\partial\sigma}\right|_{\boldsymbol{E}\boldsymbol{H}T} d\sigma + \left.\frac{\partial\varepsilon}{\partial\boldsymbol{E}}\right|_{\sigma\boldsymbol{H}T} d\boldsymbol{E} + \left.\frac{\partial\varepsilon}{\partial\boldsymbol{H}}\right|_{\boldsymbol{E}\sigma T} d\boldsymbol{H} + \left.\frac{\partial\varepsilon}{\partial T}\right|_{\boldsymbol{E}\boldsymbol{H}\sigma} dT \quad (7.52a)$$

$$d\boldsymbol{D} = \left.\frac{\partial\boldsymbol{D}}{\partial\sigma}\right|_{\boldsymbol{E}\boldsymbol{H}T} d\sigma + \left.\frac{\partial\boldsymbol{D}}{\partial\boldsymbol{E}}\right|_{\sigma\boldsymbol{H}T} d\boldsymbol{E} + \left.\frac{\partial\boldsymbol{D}}{\partial\boldsymbol{H}}\right|_{\boldsymbol{E}\sigma T} d\boldsymbol{H} + \left.\frac{\partial\boldsymbol{D}}{\partial T}\right|_{\boldsymbol{E}\boldsymbol{H}\sigma} dT \quad (7.52b)$$

$$d\boldsymbol{B} = \left.\frac{\partial\boldsymbol{B}}{\partial\sigma}\right|_{\boldsymbol{E}\boldsymbol{H}T} d\sigma + \left.\frac{\partial\boldsymbol{B}}{\partial\boldsymbol{E}}\right|_{\sigma\boldsymbol{H}T} d\boldsymbol{E} + \left.\frac{\partial\boldsymbol{B}}{\partial\boldsymbol{H}}\right|_{\boldsymbol{E}\sigma T} d\boldsymbol{H} + \left.\frac{\partial\boldsymbol{B}}{\partial T}\right|_{\boldsymbol{E}\boldsymbol{H}\sigma} dT \quad (7.52c)$$

$$dS = \left.\frac{\partial S}{\partial\sigma}\right|_{\boldsymbol{E}\boldsymbol{H}T} d\sigma + \left.\frac{\partial S}{\partial\boldsymbol{E}}\right|_{\sigma\boldsymbol{H}T} d\boldsymbol{E} + \left.\frac{\partial S}{\partial\boldsymbol{H}}\right|_{\boldsymbol{E}\sigma T} d\boldsymbol{H} + \left.\frac{\partial S}{\partial T}\right|_{\boldsymbol{E}\boldsymbol{H}\sigma} dT \quad (7.52d)$$

With this starting point, we use a Legendre transformation (see Box 7.2) to convert the fundamental relation from an internal energy formalism to a Gibbs free energy formalism. In doing so, we switch the roles of pressure and volume, of entropy and temperature, of electrical displacement and electric field, and of magnetic induction and magnetic field, to obtain the full and differential form of the Gibbs free energy

$$G = U - \sigma{:}\varepsilon - \boldsymbol{E}\cdot\boldsymbol{D} - \boldsymbol{H}\cdot\boldsymbol{B} - TS \quad (7.53a)$$

$$dG = -\varepsilon{:}d\sigma - \boldsymbol{D}\cdot d\boldsymbol{E} - \boldsymbol{B}\cdot d\boldsymbol{H} - SdT \quad (7.53b)$$

But, by the first order homogeneous property of fundamental relations (i.e., (B 7.2.4)), we can also write that

$$dG = \left.\frac{\partial G}{\partial\sigma}\right|_{\boldsymbol{E}\boldsymbol{H}T} {:}d\sigma + \left.\frac{\partial G}{\partial\boldsymbol{E}}\right|_{\sigma\boldsymbol{H}T} \cdot d\boldsymbol{E} + \left.\frac{\partial G}{\partial\boldsymbol{H}}\right|_{\boldsymbol{E}\sigma T} \cdot d\boldsymbol{H} + \left.\frac{\partial G}{\partial T}\right|_{\boldsymbol{E}\boldsymbol{H}\sigma} dT \quad (7.54)$$

so that the following four associations may be made

$$\varepsilon = -\frac{\partial G}{\partial \sigma}\bigg|_{EHT} \qquad (7.55a)$$

$$\boldsymbol{D} = -(\nabla_E G)_{\sigma HT} \qquad (7.55b)$$

$$\boldsymbol{B} = -(\nabla_H G)_{E\sigma T} \qquad (7.55c)$$

$$S = -\frac{\partial G}{\partial T}\bigg|_{EH\sigma} \qquad (7.55d)$$

The subscripts indicate which parameters are to be kept constant during differentiation. The Maxwell relation (B 7.2.15) enables us to obtain constitutive equations from (7.55a)–(7.55d), and also to determine the symmetries inherent in the system of constitutive equations

$$(\nabla_E \varepsilon)_{\sigma HT} = (\nabla_\sigma \boldsymbol{D})_{EHT} = (d)_{HT} \qquad (7.56a)$$

$$(\nabla_E \boldsymbol{B})_{\sigma HT} = (\nabla_H \boldsymbol{D})_{E\sigma T} = (n)_{\sigma T} \qquad (7.56b)$$

defining the piezoelectric coefficients d and the refractive index n, and so on. The relations rely on the fact that second derivatives of first order relations are not dependent on the order of differentiation. We obtain sixteen such constitutive relations, summarized by the following four equations

$$\varepsilon = s_{EHT}{:}\sigma + d_{HT} \cdot \boldsymbol{E} + d_{ET} \cdot \boldsymbol{H} + \alpha_{EH}\Delta T \qquad (7.57a)$$

$$\boldsymbol{D} = d_{HT}{:}\sigma + \kappa_{\sigma HT} \cdot \boldsymbol{E} + n_{\sigma T} \cdot \boldsymbol{H} + p_{\sigma H}\Delta T \qquad (7.57b)$$

$$\boldsymbol{B} = d_{ET}{:}\sigma + n_{\sigma T} \cdot \boldsymbol{E} + \mu_{\sigma ET} \cdot \boldsymbol{H} + i_{\sigma E}\Delta T \qquad (7.57c)$$

$$\Delta S = \alpha_{EH}{:}\sigma + p_{\sigma H} \cdot \boldsymbol{E} + i_{\sigma E} \cdot \boldsymbol{H} + C_{\sigma EH}\Delta T \qquad (7.57d)$$

The symbols have the following meaning:

- s_{EHT} is the rank four elastic compliance tensor (the inverse of the stiffness tensor derived in Chapter 2) measured at isothermal conditions and at constant electromagnetic field.
- d_{HT} and d_{ET} are rank three piezoelectric tensors. It is unusual to have both in the same formulation. Typically we can apply a quasi-

static approximation (see Section 4.1.2), in which case one of the two terms drop from the formalism.

- α_{EH} is the rank two tensor of thermal expansion coefficients (see Section 2.4.2)
- $\kappa_{\sigma HT}$ is the rank two tensor of dielectric permittivities.
- $p_{\sigma H}$ is the vector of pyroelectric coefficients.
- $\mu_{\sigma ET}$ is the rank two tensor of magnetic permeabilities.
- $C_{\sigma EH}$ is the scalar heat capacity (see Section 2.4.2).

We see that reversible thermodynamics provides us with the *form* of the possible functional relationships. For the *content* of the tensors we have to turn to detailed theories, as has been done in the text for a number of the coefficients above. For example, if we include only the symmetry properties of the crystalline materials under consideration, we can already greatly reduce the number of possible nonzero entries in the above tensor coefficients. For a detailed example considering the elastic stiffness tensor, see Section 2.3.

When analyzing piezoelectric transducers it is usual to greatly reduce the above constructive relationship. In particular, since the velocity of sound is so much smaller than the velocity of light, regardless of the medium, we may assume that as far as the mechanical deformation field is concerned the electromagnetic field changes almost instantaneously. Furthermore, unless parasitically dominant or required as an effect, we assume that the piezoelectric phenomena is operated under isothermal conditions. Note here that these assumptions are merely a matter of computational convenience. In fact, nowadays many computer programs exist that implement the full theory and so allow the designer a considerable amount of more detailed investigative possibilities. Sticking to our simplifications, we obtain

$$\varepsilon = s_{EHT}:\sigma + d_{HT} \cdot \boldsymbol{E} \tag{7.58a}$$

$$\boldsymbol{D} = d_{HT}:\sigma + k_{\sigma HT} \cdot \boldsymbol{E} \tag{7.58b}$$

From (7.58a) we see that the electric field and the stress cause a strain ε to appear in the material. Alternatively, from (7.58b), the stress and the electric field cause a dielectric displacement $\boldsymbol{D}$. For further analysis we now multiply (7.58a) by the mechanical stiffness tensor $E_{EHT} = s_{EHT}^{-1}$ to obtain

$$E_{EHT}:\varepsilon = s_{EHT}^{-1}:s_{EHT}:\sigma + E_{EHT}:d_{HT} \cdot \boldsymbol{E} \text{ or} \tag{7.59a}$$

$$\sigma = E_{EHT}:\varepsilon - e_{HT} \cdot \boldsymbol{E} \tag{7.59b}$$

For (7.58b) we use the fact that $\sigma = E_{EHT}:\varepsilon_m$

$$\boldsymbol{D} = d_{HT}:E_{EHT}:\varepsilon_m + \kappa_{\sigma HT} \cdot \boldsymbol{E} = e_{HT}^T:\varepsilon_m + \kappa_{\sigma HT} \cdot \boldsymbol{E} \text{ or} \tag{7.59c}$$

$$\boldsymbol{E} = \kappa_{\sigma HT}^{-1} \cdot \boldsymbol{D} - \kappa_{\sigma HT}^{-1} \cdot e_{HT}^T:\varepsilon_m \tag{7.59d}$$

We now have the constitutive equations in the most convenient format, and can insert them into the mechanical and the electromagnetic equations of motion, which we first restate

$$\rho \frac{\partial^2 \boldsymbol{u}}{\partial t^2} = \nabla \cdot \sigma - \boldsymbol{f} \tag{7.60a}$$

$$\nabla \times \nabla \times \boldsymbol{E} = -\mu\kappa \frac{\partial^2 \boldsymbol{E}}{\partial t^2} \tag{7.60b}$$

Taking the divergence of equation (7.59b)

$$\begin{aligned} \nabla \cdot \sigma &= \nabla \cdot (E_{EHT}:\varepsilon - e_{HT} \cdot \boldsymbol{E}) \\ &= (\nabla \cdot E_{EHT}):\varepsilon + E_{EHT}:(\nabla \cdot \varepsilon) - (\nabla \cdot e_{HT}) \cdot \boldsymbol{E} - e_{HT} \cdot (\nabla \cdot \boldsymbol{E}) \end{aligned} \tag{7.61}$$

If the material properties are piece-wise constant, the above equation simplifies greatly, because then the material property divergences are nonzero only across material interfaces and appear therefore as jump conditions.

7.3.4 Piezoelectric Transducers

Let us now consider an application of an integrated viscosity sensor based on a piezoelectric thin film [7.5]. The idea is to generate shear surface waves in a transducer that is in contact with a liquid. The shear waves are attenuated by the liquid as a function of the liquid's viscosity. The best sensor avoids generating out-of-plane displacements, for these are radiated as acoustic waves and cause huge energy losses. The piezoelectric thin film, such as PZT or ZnO, deposited on the chip's surface, is contacted with evaporated gold electrodes that are formed in an interdigitated pattern, see Figure 7.4. The exact placement of the electrodes

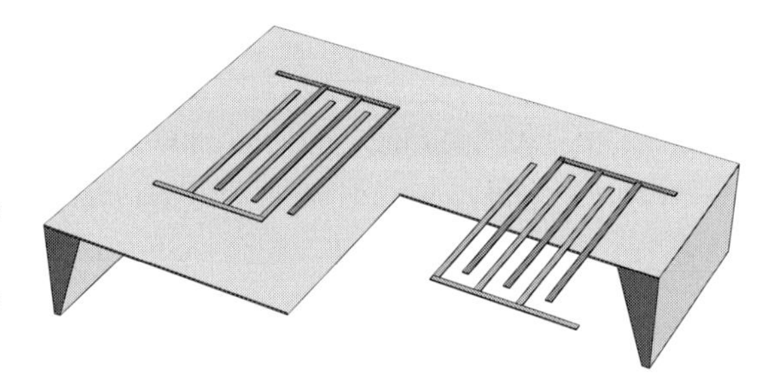

Figure 7.4. Geometric arrangement of an interdigitated electrode thin film piezoelectric transducer with one quarter of the membrane removed. Not to scale.

depends on the crystal orientation of the thin film, so that when a potential is applied to the electrodes, mainly shear surface waves are generated.

The wave velocity in the sensor layers determines the spacing of the electrodes, so that a set of them achieves a cumulative effect. Thus if d is the separation of the electrodes, then

$$d = \frac{\lambda}{2} \tag{7.62}$$

where λ is the elastic wavelength. Excitation of the electrodes by

$$v = v_o \exp[i\omega t] \tag{7.63}$$

produces a strain field with a spatial period of $2d$ and a frequency of $\pi c/d$ where c is the wave propagation velocity. At least part of this strain field is of the right shape to propagate as a Love surface acoustic wave, and clever design will maximize this ratio.

For our purposes it is sufficient to view the wave as generated by a row of alternating point sources, each source placed exactly between the electrode pairs [7.5]. At some point x along the surface we now add up the contributions of each electrode pair's point source

$$u = \sum_{i=1}^{N-1} (-1)^n A_n \exp\left[\frac{i\omega(x_n - x)}{c}\right] \exp[i\omega t] \tag{7.64}$$

where N is the number of electrode pairs, $(-1)^n$ expresses the fact that the pairs alternate in excitation direction, x_n is the position of the source, and A_0 is the amplitude of the excited wave. If we now denote the time delay for a pulse at pair n to arrive at x as

$$t_n = \frac{(x_o + nd)}{c} \tag{7.65}$$

then we can write that the total displacement at x is

$$u = A_n \exp[-i2\pi f t_o] \frac{1 - \exp\left[-i(N-1)\pi\left(\frac{f}{f_o} - 1\right)\right]}{1 - \exp\left[i\pi\left(\frac{f}{f_o} - 1\right)\right]} \tag{7.66}$$

where $f_o = c/2d$ is the natural frequency of the interdigitated electrodes, see Figure 7.5.

Usually such sensors are arranged as a pair of sending electrodes separated by a sensor space (the delay line) and an identical pair of receiving electrodes. The delay line is the space where the signal is allowed to attenuate against a fluid. The receiver pair works in the reverse fashion of the sender electrodes, i.e., the elastic wave induces a voltage in the elec-

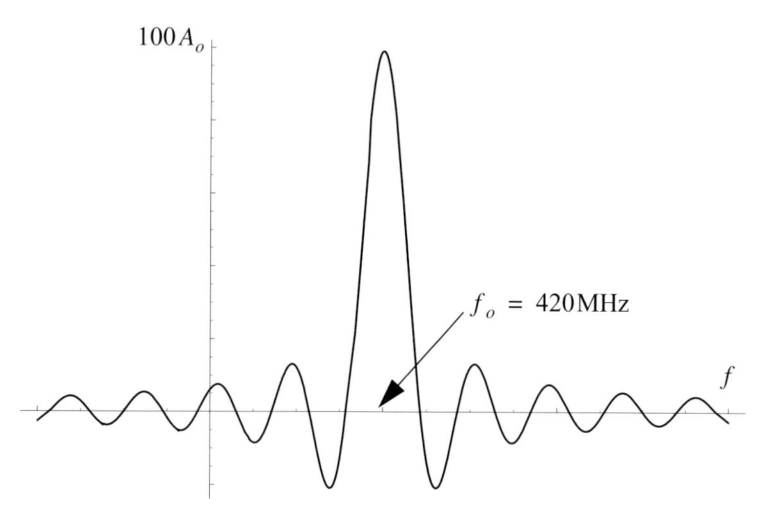

Figure 7.5. Relative frequency response of a 100 interdigitated pair $10\ \mu m$ *pitch transducer on a silicon substrate with* $c = 8400\ ms^{-1}$ *and* $f_o = 420\ MHz$*, evaluated at the position of the outermost electrode pair* x_o.

trodes. Usually a careful circuit setup is required so that the measurement is made in a time window so as to avoid spurious reflections in the result.

7.3.5 Stress Induced Sensor Effects: Piezoresistivity

Recall that the shape of the conduction band minima are used to obtain an expression for the effective electron mass tensor, and the valence band maxima for the hole effective mass tensor, and are based on a Taylor expansion of the energy about a point $\boldsymbol{k}_0$ in k-space

$$E(\boldsymbol{k}_0 + \kappa) \approx E(\boldsymbol{k}_0) + \frac{1}{2}\sum_{ij}\frac{\partial^2 E}{\partial k_i \partial k_j}\kappa_i\kappa_j = E(\boldsymbol{k}_0) + \frac{1}{2}\sum_{ij} m_{ij}{}^*\kappa_i\kappa_j \quad (7.67a)$$

$$m_{ij}{}^* = \frac{1}{2}\sum_{ij}\frac{\partial^2 E}{\partial k_i \partial k_j} \quad (7.67b)$$

The effective mass enters the expression for the carrier mobilities as

$$\mu_{ij} \propto \frac{1}{(m_{ij}*)^a} \tag{7.68}$$

where the exponent a depends on the scattering mechanisms that are considered. The point here is that if the band's curvature changes, as is illustrated in Figure 7.6, then so does the effective mass due to the

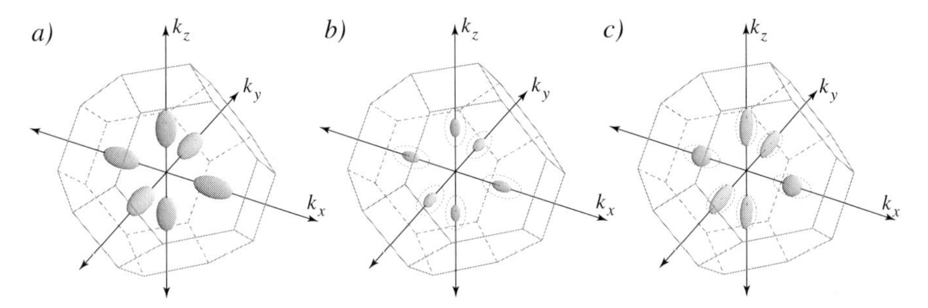

*Figure 7.6. The conduction band minima of silicon are ellipsoidal in shape and lie along the "k" axes. a) Reference state. b) Hydrostatic stress causes the minima to equally decrease in size. c) Uniaxial stress along the x-axis has no appreciable effect on the conduction band minima of the y and z axes. The figures are **not** to scale.*

change in curvature. Hence we also expect the mobility and the carrier conductivities to change. In addition, depending on the direction of the stress, the conductivity can change from an effectively isotropic value to an anisotropic tensor. This is the basis of the piezoresistive effect, and is exploited in piezoresistors, as shown for the integrated pressure sensor of Figure 7.7. Whereas piezoresistors are useful, there is also a parasitic effect. Electronic packaging can cause resistance changes within the chip due to package stress, and special measures must be taken to reduce this effect, including the addition of extra circuitry in the case of sensor devices. When the crystal deformation is linear, we can write the piezoresistive effect as a tensor equation

Figure 7.7. Schematic cross section of a silicon pressure sensor. The back of the wafer is anisotropically etched away to leave a thin membrane. On top the membrane contains diffused resistors arranged in a Wheatstone bridge circuit. When a gas or liquid pressure deforms the membrane, the resistance of the diffused resistors changes linearly with the stress level. Not to scale.

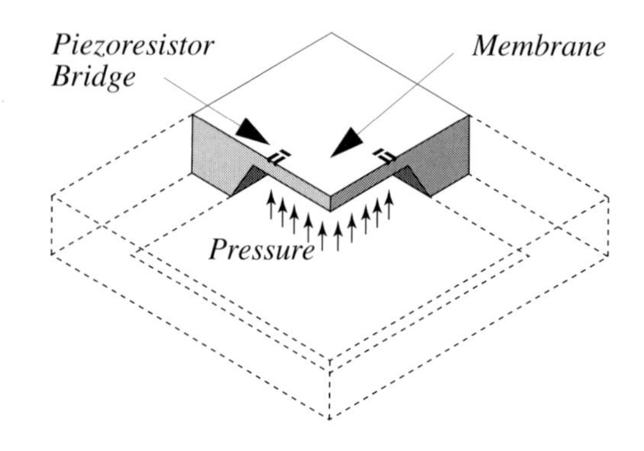

$$\left(\frac{\Delta\rho}{\rho}\right)_{ij} = -\left(\frac{\Delta\mu}{\mu}\right)_{ij} = \sum_{k=1}^{3}\sum_{l=1}^{3}\pi_{ijkl}\sigma_{kl} \tag{7.69}$$

where $\Delta\rho/\rho$ is the relative change in resistivity, π is a fourth-rank tensor of piezoresistive coefficients and σ is the mechanical stress. For silicon the piezoresistive coefficients can be reduced in number in exactly the same way as we did for the elastic coefficients, and for the same reasons of symmetry. Therefore, we are left with three unique values in the Voight (or engineering) notation. Values depend on the impurity doping level and should be measured (here $\rho_n = 11.7$ Ω cm and $\rho_p = 7.8$ Ω cm). The piezoresistance coefficients can be computed from scratch using the $\boldsymbol{k} \cdot \boldsymbol{p}$ method of Cordona et al [7.7]. The computation is rather involved, and experience shows that unless spin-orbit coupling is taken into consideration, the predicted values do not compare well with measurements [7.8].

7.3.6 Thermoelectric Effects

The crosstalk between charge and heat transport gives rise to a number of electro-thermal effects that can be used to locally heat or cool, or to measure thermal quantities in the conductor. As we shall see, the addition of a magnetic field produces an ever richer structure of possibilities.

Integrated Thermopiles: The Seebeck Effect

The integrated thermopile exploits the thermoelectric power of different materials to produce a sophisticated temperature sensor. It also relies on massive parallelism and careful accounting of heat losses. One particularly successful design [7.4] employs many alternating n-doped and p-doped connected polysilicon wires patterned on top of a chip surface, see Figure 7.8. The inter-metal contacts are alternately at T_{hot} and T_{cold}. If

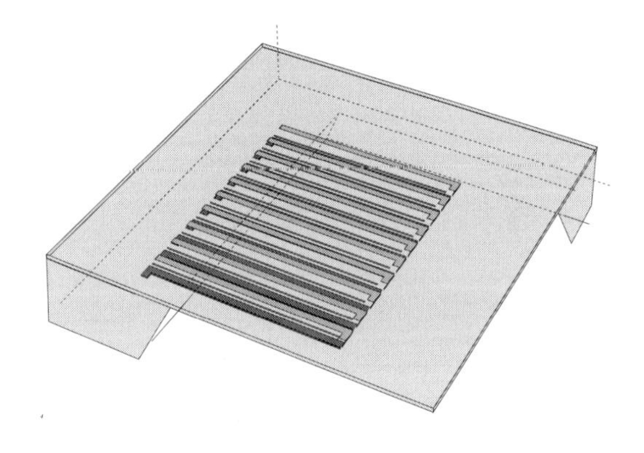

Figure 7.8. The geometric layout of CMOS integrated thermopiles on a thermally insulating dielectric membrane. One quarter of the device is shown. The cold contacts lie over the bulk silicon, the hot contacts on the membrane. Various techniques to generate heat on the membrane, i.e., by absorbing infrared radiation, air cooling, gas absorption, power dissipation, etc., make this a very successful device structure [7.4].

we mark the mid-way points on the p-doped wires consecutively as $\boldsymbol{x}_i$, where i counts over the individual p-doped wires, then we can use (6.54) to obtain

$$\begin{aligned}\frac{\eta_i - \eta_{i-1}}{q} &= \int_{\boldsymbol{x}_{i-1}}^{\boldsymbol{x}_i} (\varepsilon_n \cdot \nabla T - \varepsilon_p \cdot \nabla T) \cdot \mathrm{d}\boldsymbol{r} \\ &= (\varepsilon_n - \varepsilon_p) \cdot \int_{\boldsymbol{x}_{i-1}}^{\boldsymbol{x}_i} \nabla T \cdot \mathrm{d}\boldsymbol{r} = \varepsilon_{i,\,i-1} \Delta T_{i,\,i-1}\end{aligned} \tag{7.70}$$

Since the thermopile wires are uniformly doped, (7.70) immediately describes the incremental voltage that we can expect from each wire pair. The thermoelectric couple of the thermopile, $\varepsilon_{i,\,i-1}$ sets the design objective. For n-doped and p-doped polysilicon wires in commercial CMOS technology we can expect to achieve a maximum of $\varepsilon_{i,\,i-1} = 900 \ \mu V K^{-1}$, with typical values lying at half that value [Nathan]. For design purposes, this implies that we need to ensure that the hot and cold contacts are very well isolated from each other, which is usually achieved by positioning the cold contacts on the bulk of the chip, and the hot contacts on a thermally-isolated membrane. In addition, we need many thermocouples in the thermopile, and typical designs have 50 or more pairs. Note that the large number of metal lines connecting the thermally insulated hot and cold contacts is a misnomer, since metal is an excellent conductor of heat. This makes the design of real thermopiles in silicon technology very challenging, and requires the use of on-chip amplification of the small voltages that arise when small temperature differences have to be measured.

Seebeck Effect in Silicon

Formal transport theory predicts the existence of the Seebeck effect, and of course determines which mathematical form it should have. At the microscopic level, the cause of the Seebeck effect in Silicon begins with a non-local shift in the position of the Fermi level due to the temperature gradient [7.9]. In hotter regions, the Fermi level is shifted closer towards the middle or intrinsic position of the bandgap than in colder regions. At the same time, the charge carriers in the hot regions have a higher velocity than in the cold regions. In this way the hot regions deplete somewhat, and the cold regions gain in carrier concentration. There is also a net flow of phonons from hot to cold regions. The drag force exerted by the phonons on the electrons further enhance the concentration of carriers towards colder regions. The value of the Seebeck coefficient is computed from material parameters as

$$\varepsilon_e = -\frac{k_B}{q}\left(\ln\left(\frac{N_C}{n_e}\right) + \frac{5}{2} + s_e + \phi_e\right) \qquad (7.71a)$$

$$\varepsilon_h = \frac{k_B}{q}\left(\ln\left(\frac{N_C}{n_h}\right) + \frac{5}{2} + s_h + \phi_h\right) \qquad (7.71b)$$

where q is the fundamental charge, N_C is the conduction band effective density of states, N_V is the valence band effective density of states, n_e is the electron concentration, n_h is the hole concentration, s_e and s_h are corrections due to enhanced electron and hole carrier scattering, and ϕ_e and ϕ_h are corrections that take into account the phonon drag on the electrons and holes.

Peltier Heating and Cooling

At a material junction we observe a further thermoelectric effect that is reversible. Depending on the direction of current flowing through the junction, heat is either generated or absorbed according to

$$\Delta J_q = \Pi_{AB} J_e \tag{7.72}$$

The Peltier coefficient Π_{AB} between materials A and B defines both the sign and quantity of heat current transported per unit carrier current, and is obtained from (6.41b) for isothermal conditions

$$j_q = -(T\varepsilon_n\sigma_n)\nabla\frac{\eta_n}{q_n} - (T\varepsilon_p\sigma_p)\nabla\frac{\eta_p}{q_p} \tag{7.73a}$$

$$i_n = -(\sigma_n)\nabla\frac{\eta_n}{q_n} \tag{7.73b}$$

$$i_p = -(\sigma_p)\nabla\frac{\eta_p}{q_p} \tag{7.73c}$$

When we combine the equations in favour of the heat and carrier current densities, we obtain

$$j_q = (T\varepsilon_n)i_n + (T\varepsilon_p)i_p \tag{7.74}$$

In an uniformly doped region, (7.74) reduces to

$$j_q = (T\varepsilon)i = \Pi i \tag{7.75}$$

where $\Pi = T\varepsilon$ is the absolute thermoelectric power of the material. Usually we only deal with differences between two materials adjoining at a junction, where it is usual to define the relative thermoelectric power of the junction by

$$\Pi_{AB} = \Pi_B - \Pi_A = T(\varepsilon_B - \varepsilon_A) \tag{7.76}$$

This effect can be used to pump heat from one region to another, for example, to cool electronic circuits or even to refrigerate a space.

Bridgeman Effect

Heat is absorbed or liberated in a homogeneous conductor when a current passes across an interface where crystal orientation changes, or when a current distribution is strongly nonuniform. This effect is sometimes also called the internal Peltier effect.

Thomson Effect

Heat is absorbed or liberated in a homogeneous conductor, subject to a temperature gradient, that has a current flowing through it. The effect is primarily caused by the temperature-dependence of the Seebeck coefficient. In this case, the temperature gradient of the Peltier coefficient $\partial(\varepsilon T)/\partial T = \varepsilon + T\partial\varepsilon/\partial T$ leads to an additional term that we denote the Thomson coefficient after its discoverer

$$\tau_T = T\frac{\partial\varepsilon}{\partial T} \tag{7.77}$$

The direction of heat flow is defined by the direction of the current flow $\boldsymbol{i}$; its value is determined by

$$j_q = \int_{T_0}^{T_1} \tau_T \boldsymbol{i} dT \tag{7.78}$$

Other Similar Effects

When a thermoelectrically active substance is subject to a penetrating magnetic induction, a whole range of further "classical" effects are observed:

- *Nernst* effect. When heat flows across lines of magnetic force, we observe a voltage perpendicular to both the heat current and the magnetic field.

- *Righi-Leduc* effect. When heat flows across lines of magnetic force, we observe another heat current perpendicular to both the original heat current and the magnetic field.
- *Hall* effect, discussed in 6.2.3.
- *Ettinghauser-Nernst* effect. When an electric current flows across the lines of force of a magnetic field, we observe a voltage (called the Hall voltage) which is perpendicular to both the electric current and the magnetic field. In addition, we observe a temperature gradient in the opposite direction to the Hall voltage.

As mentioned before, in formal transport theory, the Onsager coefficients are anti-symmetric w.r.t. the presence of the field, or

$$\boldsymbol{L}_{kl}(\boldsymbol{B}) = \boldsymbol{L}_{kl}^{T}(-\boldsymbol{B}) \tag{7.79}$$

For vector quantities, this anti-symmetry expresses itself through the cross product and through the rotation operator, and is again the reason for the recurring "perpendicular" fields.

7.4 Electron-Photon

There exists an optically induced transition between electronic states. The energy of absorbed photons is transferred to the electronic system, such that the electron goes from the state of lower energy to the one with higher energy. The inverse process exists as well, delivering the energy from the electronic system to the photon system. These two processes we call absorption and emission. Therefore, the basics of electron-photon interactions, as they enter the time dependent perturbation given in 3.2.7, are important for the description of arising phenomena.

As in the preceding sections we ask how the interaction operator looks like. Suppose there is an external harmonic electric wave $\boldsymbol{E}(\boldsymbol{r}, t) = \boldsymbol{E}_0 \cos(\boldsymbol{k}\boldsymbol{r} - \omega t)$ that acts on an electron. Assume that the electrons had two possible states ψ_1 and ψ_2 where they may be found in, e.g., the valence or conduction band edge. As we know from electrostatics and from the polar optical phonon interaction, the energy is given by the product of the incident electric field $\boldsymbol{E}$ and the dipole moment $\boldsymbol{p}$ of the electronic system. In classical terms the dipole moment is given as charge times distance $\boldsymbol{p} = e\boldsymbol{r}$. Assume that the electric field varies little in space with respect to the extension of the electronic states. Since the charge of the electron has a density distribution according to the wavefunction $\Psi(\boldsymbol{r}, t)$, which for the unperturbed problem is a superposition $\Psi(\boldsymbol{r}, t) = \sum_j c_j(t)\exp(-i\omega_j t)\psi_j(\boldsymbol{r})$. The matrix elements read

Dipole approximation

$$V_{mn} = \int \psi_m^*(\boldsymbol{r}) e\boldsymbol{r}\boldsymbol{E}(\boldsymbol{R}, t)\psi_n(\boldsymbol{r})dV \tag{7.80}$$

where **R** is the position of the center of mass of the dipole. We may interpret the integral in (7.80) as a dipole moment that couples to an external field. In most cases the wavefunctions are of this kind that $V_{mm} = 0$, i.e., there is no static dipole moment present in the system. The only possibility to have a non vanishing matrix element is for $m \neq n$. We return to a harmonic wave and assume the electronic system to be rather localized, e.g., atomic electron states. In addition we deal with wavelengths of the electric field that are much larger than the extension of the atom. So that the dipole approximation holds. Then we have

Transition Dipole Matrix Element

$$V_{12} = V_{21}^* = \boldsymbol{\mu}_{12}(\exp(i\omega t) + \exp(-i\omega t))\boldsymbol{E}(\boldsymbol{R})/2 \tag{7.81}$$

with $\boldsymbol{\mu}_{12} = \int \psi_1^*(\boldsymbol{r}) e\boldsymbol{r}\psi_2(\boldsymbol{r})dV$, which usually are called the transition dipole matrix elements.

7.4.1 Intra- and Interband Effects

There are three basic processes involved in the interaction: interband transition, intra-band transitions, and free carrier absorption (see Figure 7.9).

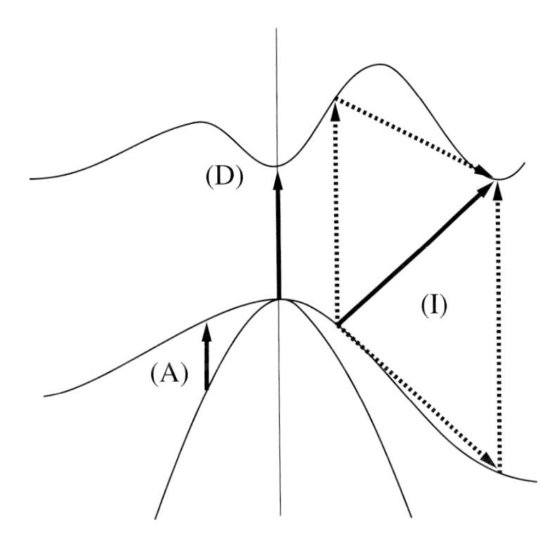

Figure 7.9. Direct (D) and indirect (I) interband transitions. Intra-band transitions (A).

Interband transitions show two basic different characters depending on whether we look at direct or indirect band gap semiconductors. The common feature is that an electron changes from the valence band edge to the conduction band edge or vice versa. For direct semiconductors these band edges lie at the same wavevector. This is indicated with the vertical line at (D) in Figure 7.9. For indirect semiconductors the band edges lie at different wavevectors. The band gap energy in semiconductors is of the order of 1 eV. Thus the photon energy must have the same value and therefore we obtain $\hbar\omega = 1\mathrm{eV}$. With the frequency wavevector relation for photons $\omega = ck$ we immediately may calculate the transferred momentum as the one of the absorbed photon to give $k = 1\mathrm{eV}/\hbar c$. This corresponds to a light field wavevector of about $3 \cdot 10^6 m^{-1}$, which is roughly 3-4 orders of magnitude smaller than the first brillouin zone edge. Since we have to respect momentum and energy conservation and the transferred momentum is rather small, the processes are vertical in the band structure. Transitions at the band edge in indirect semiconductors therefore are not possible without the help of a third party system adding the momentum missing in the balance. It is the phonon system to allow for these processes indicated as (I) in Figure 7.9. In total there are two processes involved in sequence and thus we must go beyond first order perturbation theory. The consequence is that the two transition probabilities for the processes multiply to give the total transition probability and thus we may immediately identify indirect interband transitions as less probable.

Intra-band transitions are indicated as (A) in Figure 7.9, e.g., between the heavy hole and light hole valence band. In this case the electrons change from one subset to another and remain in the same band. We therefore count these processes for the same category as the interband transitions. A very important process is the so called free carrier absorption. The electrons remain in their subset of either the valence or conduction band. These processes have to be distinguished from transport effects happening inside a subset of a band.

Emission of photons from electron in the conduction band has been observed even from within industrially applied devices. Due to submicron gate lengths in modern Field Effect Transistors the electric field under certain operating conditions can exceed $10^5 V/cm$. Some of the carriers accelerated by such field strengths may gain up to 1 eV in energy. A small fraction of these carriers has been shown to be able to give up this kinetic energy by emission of a photon. A pretty illustration of this light emission is found in [7.18]. A ring oscillator consisting of 47 inverters, as it is usually found in microprocessor clocks, has been observed by means of a photomultiplier. A time integrated image of the ring oscillator shows infrared and far infrared activity spatially resolved in the single drain regions of the inverter devices. This observation technique is assumed to become even a standard diagnostic tool for monitoring the functionality of modern integrated circuit devices [7.18].

7.4.2 Semiconductor Lasers

We start our discussion on the laser effect in semiconductors with a short explanation of the principle effect by means of a two level atom model.

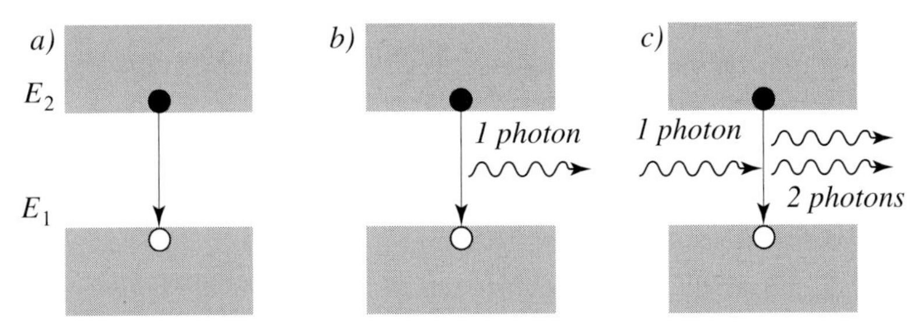

Figure 7.10. Electron-hole recombination. a) Non-radiative (without photon emission) recombination. The energy is dissipated into other degrees of freedom, e.g., phonons. b) Spontaneous emission of a photon. c) Emission is stimulated by an incoming photon that is in phase with the stimulation.

The energy levels of the respective states are E_1 and E_2, which means that the electromagnetic wave at $\omega_0 = (E_2 - E_1)/\hbar$ couples resonantly to the electronic system. We insert this into the equation of motion for the superposition coefficients as given from (3.74) in first order perturbation theory to give

$$\begin{aligned} \dot{c}_1(t) &= \frac{-i}{\hbar} V_{12} c_2(t) \exp(-i\omega_0 t) \\ \dot{c}_2(t) &= \frac{-i}{\hbar} V_{21} c_1(t) \exp(i\omega_0 t) \end{aligned} \tag{7.82}$$

Rotating Wave approximation

Remember that V_{mn} has a harmonic time dependence through the electric field vector with frequency ω, near the resonance frequency ω_0. Let us assume that $c_1(t)$ and $c_2(t)$ change very slowly on time scales where $\exp(\mp i(\omega + \omega_0)t)$ is changing. Integrating (7.82) over time will yield a vanishing contribution of the exponential term since it will perform many oscillations in the integration period that cancel each other. Therefore, we are led to neglect the terms with the summed frequencies $(\omega + \omega_0)$ in the exponents compared to those with the frequency difference $(\omega - \omega_0)$. This so called rotating wave approximation yields for the transition dipole matrix element the new form $V_{12} = \mu \exp(-i\omega t) E_0/2$, where μ is the projection of the dipole matrix element $\boldsymbol{\mu}_{12}$ along the electric field vector $\boldsymbol{E}$. Insert this into (7.82) and drive the electric field in perfect resonance, the solution is given by

$$\begin{aligned} c_1(t) &= \cos(\Omega t) \\ c_2(t) &= -i\exp(-i\varphi)\sin(\Omega t) \end{aligned} \tag{7.83}$$

Rabi Frequency

with $\Omega = E_0\mu/(2\hbar)$. Ω is called the Rabi frequency. It is best interpreted by looking at the square modulus of the coefficients: $|c_1(t)|^2 = (\cos(\Omega t))^2$ and $|c_2(t)|^2 = (\sin(\Omega t))^2$. These square moduli are the probabilities for the electron to occupy either quantum state 1 or 2. Thus the absorption process is going on in a time window from $t = 0$ to $t = \pi/(2\Omega)$. For $t > \pi/(2\Omega)$ the probability for the electron to occupy quantum state 2 decreases. This can be interpreted as an emission process. In order to have a well defined final state the external field excitation pulse has to be of finite duration. It is interesting to observe that the transition dipole matrix element oscillates with the frequency ω_0 and is modulated by harmonic function with twice the Rabi frequency.

In the case where there are multiple electrons in the system distributed onto the two energy levels the square moduli of the coefficients denote the number of electrons to be found in the respective level $|c_1(t)|^2 = N_1$, $|c_2(t)|^2 = N_2$, with the total number of electrons given by $N = N_1 + N_2$. Let us calculate the dipole moment of the electron charge distribution given by the wavefunction

$$\begin{aligned} \boldsymbol{p} &= \int \psi^*(\boldsymbol{r}, t)(-e)\boldsymbol{r}\psi(\boldsymbol{r}, t)dV \\ &= -(c_1{}^*c_2 e^{-i\omega_0 t}\boldsymbol{\mu}_{12} + c_1 c_2{}^* e^{i\omega_0 t}\boldsymbol{\mu}_{21}) \end{aligned} \tag{7.84}$$

With the definition $\alpha(t) = c_1{}^*c_2 e^{-i\omega_0 t}$, we have for the dipole moment $-\boldsymbol{p} = \alpha(t)\boldsymbol{\mu}_{12} + \alpha^*(t)\boldsymbol{\mu}_{21}$ and an equation of motion for $\alpha(t)$ to read

$$\dot{\alpha} = -(i\omega_0 + \gamma)\alpha + \frac{i}{\hbar}V_{21}D \tag{7.85}$$

where $D = N_2 - N_1$ is called the inversion and we have added a phenomenological damping term with damping constant γ the inverse of the so called dipole decay time. For the inversion we obtain an equation of motion of the following form

$$\dot{D} = \frac{2}{i\hbar}(\alpha^*(t)V_{21} - \alpha(t)V_{12}) - \frac{D - D_0}{T} \tag{7.86}$$

where we also have introduced a phenomenological decay time T and an equilibrium inversion D_0 which is given by the temperature distribution of the carriers.

We formally integrate (7.85) and make use of the rotating wave approximation to give

$$\alpha(t) = \frac{i}{\hbar}\int_{-\infty}^{t} \exp(-(i\omega_0 + \gamma)(t - t'))V_{21}(t')D(t')dt' \tag{7.87}$$

We now assume that the dipole decay time $1/\gamma$ is much smaller than times in which the inversion D or the electric field amplitude E_0 change. We therefore put these quantities outside the integration and obtain

$$\alpha(t) = -i\frac{\mu E_0}{2\hbar}\frac{1}{\gamma + i(\omega_0 - \omega)}D \tag{7.88}$$

Insert this in (7.86) and we have

$$\dot{D} = -\frac{D - D_0}{T} - \frac{\gamma^2}{\gamma^2 + (\omega_0 - \omega)^2}\frac{I}{I_s}D \tag{7.89}$$

where $I = c\varepsilon_0|E_0|^2$ is the light field intensity and $I_s = c\varepsilon_0|\hbar/\mu|^2 T^{-1}$ is the saturation intensity.

We have to find an equation for the laser field intensity, i.e., the $|E_0|^2$ that is generated by the inversion. This can be achieved by inserting the polarization generated from the inversion density into the amplitude equation of the propagating electromagnetic wave inside the laser cavity. A derivation is given in [7.20], we will explain the result

$$\dot{I} = -\frac{\omega}{Q}\left[1-\frac{\chi_0 Q\gamma^2}{\gamma^2+(\omega_0-\omega)^2}\frac{D}{D_0}\right]I \tag{7.90}$$

In (7.90) we see that light intensity is lost through the quality factor of the cavity. The factor Q/ω is the energy stored divided by the energy lost per second. So the first term gives a negative contribution to the time derivative of the intensity. The second term contains a product of inversion density and light intensity and is a nonlinear driving term for the light field intensity. So we end up with a set of two coupled nonlinear equations (7.89) and (7.90). They describe how the laser field intensity and the inversion density evolve in time. In a small signal analysis around the steady state $\dot{I} = 0$ and $\dot{D} = 0$ we see that the system exhibits relaxation oscillations which are typical for the switching on and off of such a device. An external control parameter is the steady state inversion density D_0. This density can be externally set by the current injection of carriers into the device and eventually controls the laser intensity.

For the laser to work, the laser field has to be kept in the active region, i.e., a cavity electromagnetic wave mode must be provided by specifying special boundary conditions for the electromagnetic field. The simplest way to do this is to place mirrors at the ends of the active region. This ensures that a high density of photons of the selected cavity wave mode is present where the active electronic system is situated. In terms of our simple two level picture, Figure 7.10, this means that there is a strong stimulated emission pumping almost all transition energies into the one electromagnetic wave mode, i.e., the stimulated photons have the same phase as the present electromagnetic wave and therefore this yields an amplification of the light field intensity I. This is what LASER stands for: Light Amplification by Stimulated Emission of Radiation. The rest of the recombination processes not coupling to the laser mode are losses that decrease the inversion density without pumping the laser mode. The loss process in the intensity rate equation is due to the finite reflectivity of the mirrors that make up the boundary conditions. In addition, this loss term allows the laser mode intensity to leave the cavity and to be used as a strong coherent light source. Omitting the boundary conditions would clearly lead to all emission processes going into spontaneous emission out of phase with each other. This in turn is how a light emitting diode would operate.

Let us now think of the carriers as electron densities in the conduction band n_{ec} and in the valence band n_{ev}. Where the electron density in the valence band is 1 minus the hole density n_h. Therefore, we may write

$$D = n_{ec} - n_{ev} = n_e - (1 - n_h) = n_e + n_h - 1 \tag{7.91}$$

In a semiconductor these densities depend on the wavevector **k** of the electron or hole. From Chapter 5 we know that these electrons and holes follow the respective distribution functions $f_e(\boldsymbol{k}, \boldsymbol{x}, t)$ and $f_h(\boldsymbol{k}, \boldsymbol{x}, t)$. Both distribution functions do have an equation of motion namely the Boltzmann transport equation. In this context it becomes clear that also the dipole moment becomes dependent of the wave vectors of electrons and holes. It will have an equation of motion too, with a relaxation term, and transport term and a coupling to the inversion. It is not the place to explain these processes in detail here, since the classical interpretation is only half of the truth and for an accurate description a full quantum mechanical treatment is needed. To explain the phenomena some simple considerations about p-n diodes will be enough as they are given in Section 7.6.4. The holes are transported from the one side to the active region, while the electrons arrive from the opposite side. Since both charge carriers have different transport properties all the requirements for carrier transport apply. A whole lot of phenomena may arise due to these differences which, at first glance, may be neglected. Nevertheless in special applications, e.g., in high frequency modulation of lasers they will have to be accounted for.

7.5 Phonon-Photon

The coupling of lattice deformations to light will be accentuated in the following by two effects showing the static and dynamic behavior of the interaction.

7.5.1 Elasto-Optic Effect

In a homogeneous material the coupling of an external electric field to the material manifests itself through the dielectric constant $\boldsymbol{\varepsilon}$ as indicated in Section 4.1. The dielectric constant is a tensorial quantity with components ε_{ij}, and is therefore related to the crystal symmetry. Note that the dielectric tensor is symmetric. Distortion in the material will change the dielectric tensor and therefore affect the properties of light travelling through a crystal.

The index of refraction for an isotropic material, as it is applied in Section 4.3.1, is simply the square root of the dielectric constant $n = \sqrt{\varepsilon}$. This gives rise to a change of the speed of light in material. As we know that the dielectric property of a material is a tensorial quantity and that this tensor is symmetric, there must be a speed of light related to the polarization of the electromagnetic wave. We remember from linear algebra that any symmetric tensor may be transformed in his proper coordinate system, where the bases are give by its eigenvectors and its eigenvalues are called the dielectric constants of the principle axes. So depending on the symmetry of the crystal there may be up to three different light propagation velocities. Therefore, the simple consideration with the square root above in general holds for any principle direction involving the respective dielectric constants of the axes.

Any change in the dielectric properties of the material will effect the components ε_{ij}. These changes may result from strains internal to the material caused by an external load. Let us denote the strain, as given in Chapter 2, by S_{ij} in order not to confuse with the dielectric tensor. Remember that this is also a symmetric tensorial quantity. The relation for the change of the dielectric tensor $\Delta\varepsilon_{ij}$ due to strain can be found in the literature [7.19]

$$\Delta\varepsilon_{il} = -\varepsilon_{ij} P_{jkmn} \varepsilon_{kl} S_{mn} \tag{7.92}$$

Acousto-Optic Tensor

The fourth order tensor P_{jkmn} that mediates between strain and dielectric tensor changes is called the acousto-optic tensor. Its values are specific for the respective material under investigation.

We restrict the following discussion to cubic crystal lattices. In such symmetries the dielectric tensor has only diagonal elements of the same magnitude and thus behaves like in an isotropic material, where we write $\varepsilon_{ij} = \varepsilon_r \delta_{ij}$, where we have $\delta_{ij} = 0$ for $i \neq j$ and $\delta_{ij} = 1$ for $i = j$ as usual. Inserting this into (7.92) we obtain a much simpler equation

$$\Delta\varepsilon_{il} = -\varepsilon_r^2 P_{ilmn} S_{mn} \tag{7.93}$$

For a cubic material like GaAs the acousto-optic tensor has only three non vanishing components out of 81. Those are [7.19] $P_{1111} = -0.165$, $P_{2222} = -0.140$, and $P_{2323} = -0.072$. Note that in this way the above isotropic behavior of the dielectric constant vanishes and we have to deal instead with a tensor. This also implies that the refractive index is no longer simply the square root of the dielectric constant but rather gets dependant on how the electromagnetic is polarized and what its propagation direction is. We will not go into detail how to calculate the changes in the refractive index. A description can be found in [7.12]. We notice that there is an effect that couples elastostatic properties to optics and thus gives rise to various applications for monitoring these properties or modulating light propagation through a crystal.

7.5.2 Light Propagation in Crystals: Phonon-Polaritons

Let us describe the electromagnetic field by the Maxwell equations and the polarization by an equation of motion for undamped oscillators with eigenfrequency ω_0. These polarization oscillations couple to the electric field via the susceptibility χ, so we have

$$\nabla \times \boldsymbol{H} = \varepsilon_0 \dot{\boldsymbol{E}} + \boldsymbol{P} \tag{7.94a}$$

$$\nabla \times \boldsymbol{E} = -\mu_0 \dot{\boldsymbol{H}} \tag{7.94b}$$

$$\ddot{\boldsymbol{P}} + \omega_0^2 \boldsymbol{P} = \chi \varepsilon_0 \boldsymbol{E} \tag{7.94c}$$

Assume that there are transverse plane waves for the fields that propagate in the z-direction. Suppose that the polarization vector and the electric field vector point to the x-direction than there is a magnetic field vector pointing in the y-direction. Thus we may do the ansatz for any of the fields E_x, H_y, and P_x of the form

$$F = F_0 \exp(i(kz - \omega t)) \tag{7.95}$$

If we insert this into the equations of motion (7.94a) this yields a characteristic equation

$$\omega^4 - \omega^2(\omega_0^2 + \chi + c^2k^2) + \omega_0^2 c^2 k^2 = 0 \tag{7.96}$$

Polariton Dispersion Relation

Solving (7.96) for ω we obtain the dispersion relation for a light field propagating in a polarizable medium. This polarizability may be due to optical phonons in an ionic crystal. The quantized electromagnetic-polarization wave is called polariton and thus (7.96) is known as the polariton dispersion relation. It consists of two branches divided by a frequency region where no real solution of (7.96) can be found for arbitrary values of k. This stop band ranges from $\omega = \omega_0$ to $\omega = \omega_0 + \chi$. For high values of the wavevector k the upper branch behaves like a light field with linear dispersion with the light velocity c/n_o, while for low k it is dispersion free as for an optical phonon. The lower branch shows inverse behavior, it saturates for high k and behaves like a light field dispersion for low k with a light velocity lower than c/n_o (see Figure 7.11).

This gives us a further insight in how light propagates in a polarizable medium. As we already saw from the discussion in Section 7.4 there is a

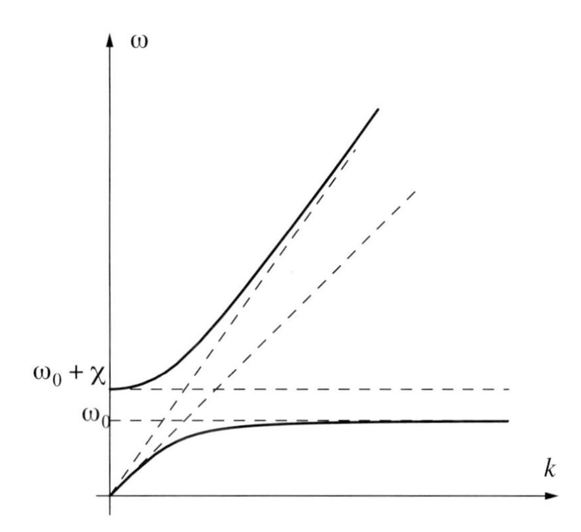

Figure 7.11. Polariton dispersion relation showing upper and lower branch.

polarization also due to the electronic states in the valence band and the conduction band, the so called interband polarization. We may expect that also this polarization couples to the light field in a similar way and indeed the description is very much the same despite marginal differences.

7.6 Inhomogeneities

Inhomogeneities arise where material properties change abruptly. This may be the interface region between two materials or even surfaces. We also count lattice defects for these inhomogeneities. Those are of rather localized character and often randomly distributed. The dimensionality of such inhomogeneities is lower than that of the 3D bulk system. In fact an interface is 2D and an impurity even 0D point-like. Nevertheless these low dimensional objects indeed have influence on the bulk behavior of all phenomena and quasi particles treated up to now. Moreover, some of them might even be described by the quasi-particle formalism in lower dimensions, e.g., surface phonons.

7.6.1 Lattice Defects

Suppose that at specific lattice sites $\boldsymbol{R}_j$ there might be a different atom sitting, an atom missing, or even a built in distortion of the lattice be present. This will lead to the fact that the periodic potential of the crystal lattice that the electrons suffer $V(\boldsymbol{x}-\boldsymbol{R}_j)$ will be different from the unperturbed periodic potential $V_0(\boldsymbol{x}-\boldsymbol{R}_j)$. Let us define the deviation $v(\boldsymbol{x}-\boldsymbol{R}_j) = V(\boldsymbol{x}-\boldsymbol{R}_j) - V_0(\boldsymbol{x}-\boldsymbol{R}_j)$. The additional potential that enters the Schrödinger equation as a perturbation is built by the sum over all impurity sites j. So the transition matrix element between two Bloch waves as defined in (3.92) will yield

$$V_{n'\boldsymbol{k}',n\boldsymbol{k}} = \sum_j \int \exp(i(\boldsymbol{k}-\boldsymbol{k}')\boldsymbol{x})u_{n'}(\boldsymbol{k}',\boldsymbol{x})v(\boldsymbol{x}-\boldsymbol{R}_j)u_n(\boldsymbol{k},\boldsymbol{x})\mathrm{d}^3\mathrm{x} \tag{7.97}$$

where n and n' denote the different bands of the electronic states and $\boldsymbol{k}$ and $\boldsymbol{k}'$ their respective wavevectors.

Let us assume that the impurities are distributed statistically so that we can identify an impurity density, i.e.

$$\overline{\sum_j v(\boldsymbol{x}-\boldsymbol{R}_j)} = \frac{N_I}{N_0}\sum_i v(\boldsymbol{x}-\boldsymbol{R}_i) \tag{7.98}$$

where in (7.98) the averaged sum over j counts all the impurity sites and the sum over i counts all the lattice sites. For a periodic perturbation potential thus the wavevector is a “good” quantum number, i.e., k will be preserved in this scattering, at least in lowest order perturbation theory. Therefore, we have $\overline{V_{n'\boldsymbol{k}',n\boldsymbol{k}}} = N_I v^{(0)}_{n'n}\delta_{\boldsymbol{k}'\boldsymbol{k}}$, where the superscript (0) indicates the lowest order approximation. Inside a band nothing will change due to that. This only affects interband scattering of electron.

We take a quick look at terms of higher order, i.e., averages of the form

$$\overline{\sum_{jj'} v(\boldsymbol{x}-\boldsymbol{R}_j)v(\boldsymbol{x}'-\boldsymbol{R}_{j'})} = \left(\frac{N_I}{N_0}\right)^2 \sum_i v(\boldsymbol{x}-\boldsymbol{R}_i)\sum_{i'} v(\boldsymbol{x}-\boldsymbol{R}_{i'}) + \overline{\sum_j v(\boldsymbol{x}-\boldsymbol{R}_j)v(\boldsymbol{x}'-\boldsymbol{R}_j)} \tag{7.99}$$

where we inserted the deviation from the spatial average $\Delta v = v - \bar{v}$. Suppose that there are no correlations between the single lattice site impurities, i.e., the second term on the r.h.s. of (7.99) vanishes. For a second order perturbation contribution thus we will obtain

$$\overline{|V_{n'\boldsymbol{k}',\, n\boldsymbol{k}}|^2} = N_I^2 |v_{n'n}^{(0)}|^2 \delta_{\boldsymbol{k}'\boldsymbol{k}} + N_I |v_{n'\boldsymbol{k}',\, n\boldsymbol{k}}^{(1)}|^2 \tag{7.100}$$

This means that the transition matrix element that enters the Fermi golden rule will yield changes in the wavevector of the final state after scattering only due to higher order terms and thus ionized impurity scattering is of importance only when the second order term is strong enough. This is the case at low temperatures where the overall importance of the phonon scattering is decreased.

We saw that this effect is due to 0D inhomogeneities and in the end is reduced to a bulk effect. Therefore, it is not as promised above a real lower dimensional effect.

7.6.2 Scattering Near Interfaces (Surface Roughness,

Electrons near interfaces do behave like 2D systems at small enough energies. Imagine an n-doped semiconductor-oxide interface. Similar to the metal semiconductor contact to be described in Section 7.6.5 at the interface the conduction band bends down due to an applied electrostatic potential and therefore gives rise to a potential hole for the electrons on the silicon side, so that localized states form at the interface where the electron density is increased (see Figure 7.12).

Let us assume that the wavefunction at the interface may be described as a product of two functions $\Psi_{nq}(z, \boldsymbol{r}) = \zeta_{nq}(z)\exp(i\boldsymbol{q}\boldsymbol{r})$, a plane wave

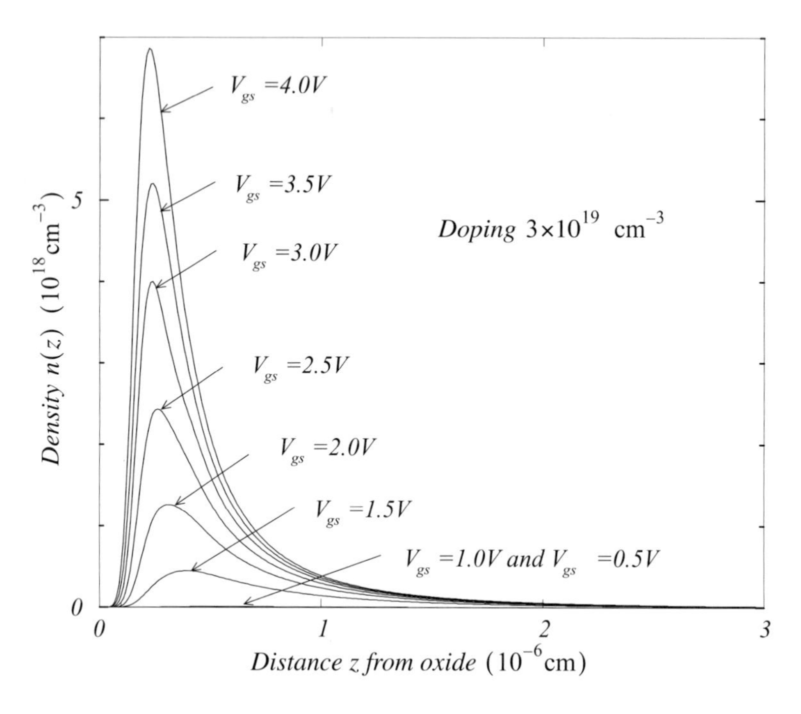

Figure 7.12. Electron densities according to calculated wavefunctions at the oxide-semiconductor interface. Here in the special situation of a Silicon-oxide interface with n-doping and different voltages applied at the outer part of the oxide with respect to the bulk silicon potential, as indicated.

parallel to the interface and an envelope function in the z-direction perpendicular to the interface. Here $\boldsymbol{q}$ denotes the wavevector parallel to the interface and n denotes the quantum number due to the quantizing electronic confinement in the z-direction. The perturbing potential arises from the fact that these interfaces never are perfectly flat but rather vary along the interface. We say that there is a finite roughness present. Following [7.20] a random variation of the interface position gives rise to a random fluctuation of the surface potential. Assume that the roughness may be described by a random displacement $\Delta(\boldsymbol{r})$, where $\boldsymbol{r}$ is the position vector parallel to the interface. The fluctuation of the surface potential may thus be expanded in terms of the displacement, to give

$$\delta V = [V(z+\Delta(\boldsymbol{r}))-V(z)] = -F(z)\Delta(\boldsymbol{r})+\ldots \quad (7.101)$$

where $F(z)$ denotes the electric field perpendicular to the interface. We denote the Fourier transform of $\Delta(\boldsymbol{r})$ by $\Delta(\boldsymbol{q})$. The respective scattering matrix element is then expressed by

$$M(\boldsymbol{k}, \boldsymbol{k}') = e\Delta(\boldsymbol{q})\int_0^\infty F(z)|\zeta_{nq}(z)|^2 dz = e\Delta(\boldsymbol{q})F_{eff} \quad (7.102)$$

In (7.102) the wavefunction is assumed to have its envelope $\zeta_{nq}(z)$ sharply peaked at the interface, which is true the higher the oxide potential is chosen (see Figure 7.12). And thus the integral that was abbreviated by F_{eff} can be taken as the value of the electric field at the interface $F(0)$.

The only parameter unknown to this description is the form of $\Delta(\boldsymbol{q})$. In [7.20] the assumption is made that the roughness is statistically distributed with a Gaussian positional correlation function, i.e., the following average given

$$\langle\Delta(\boldsymbol{r})\Delta(\boldsymbol{r}')\rangle = \Delta^2\exp\left(-\frac{(\boldsymbol{r}-\boldsymbol{r}')^2}{L^2}\right) \quad (7.103)$$

where Δ is the root mean square height of the roughness and L its correlation length. This may be Fourier transformed in order to yield an expression for $\Delta(\boldsymbol{q})$. L and Δ must be measured for the individual technological process and thus are due to fabrication. It is clear that at high carrier densities in the channel of a MOS transistor the Si-SiO2 interface layer quality is crucial for the carrier mobility in this zone of the device, since an increased roughness makes the additional scattering mechanism (7.102) very efficient and thus reduces the carrier mobility.

7.6.3 Phonons at Surfaces

Surface Lattice Dispersion Relation

We consider a 1D lattice with a basis, as we did in Section 2.4., with the important difference that we seek solutions that only exist at the surface of the now semi-infinite chain, see Figure 7.13. A natural way to intro-

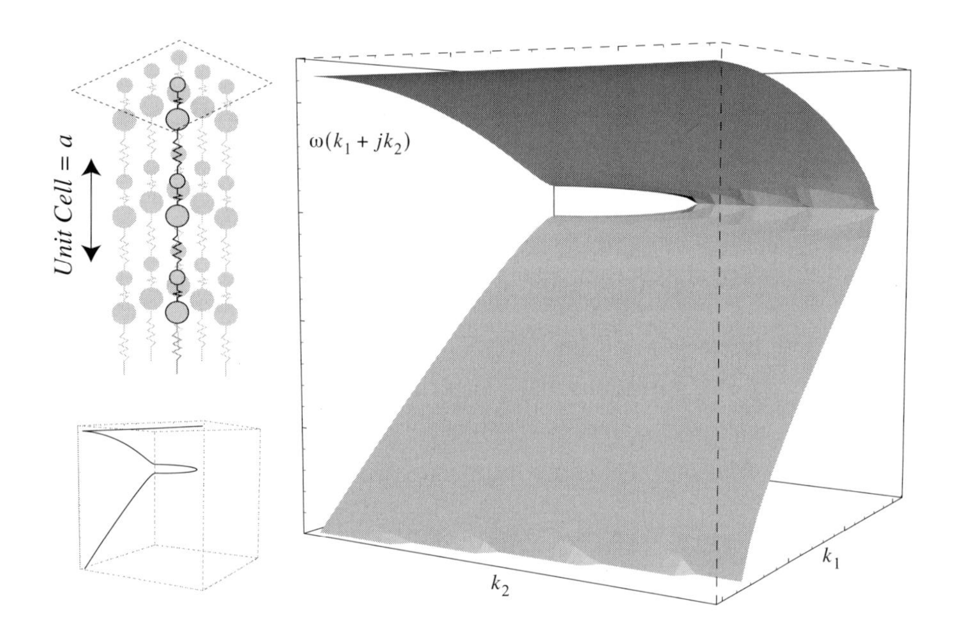

Figure 7.13. Phonon dispersion surface for a 1D surface-terminated uniform lattice with a basis. The upper inset shows the idealized model for the lattice, the lower inset shows the characteristic edges of the dispersion curve. The model is based on a complex wave-vector $k = k_1 + jk_2$.

duce this is to look for real solutions that decay exponentially away from the surface [7.13]. This is achieved by introducing a complex wave vector into the wave ansatz

$$u_m = \frac{1}{\sqrt{m}} c_m \exp\left\{ j\left[(k_1 + jk_2)a\left(i - \frac{1}{4}\right) - \omega t \right] \right\} \quad (7.104a)$$

$$u_M = \frac{1}{\sqrt{M}} c_M \exp\left\{ j\left[(k_1 + jk_2)a\left(i + \frac{1}{4}\right) - \omega t \right] \right\} \quad (7.104b)$$

so that the solutions become

$$\omega^2 = \frac{E}{aMm}[(M+m) \mp \sqrt{(M+m)^2 - 2Mm(1 - \cos[(k_1 + jk_2)a])}] \quad (7.105)$$

With the identity

$$\cos[(k_1 + jk_2)a] = \cos(k_1 a)\cosh(k_2 a) - j\sin(k_1 a)\sinh(k_2 a) \quad (7.106)$$

we find a condition for real ω

$$\sin(k_1 a)\sinh(k_2 a) = 0 \quad (7.107)$$

Clearly, setting $k_2 = 0$ gives the result for a bulk lattice. For $k_2 \neq 0$, we must have that $k_1 = i\pi/a$, $i = 0, \mp 1, \mp 2, \ldots$. Thus the solution becomes

$$\omega^2 = \frac{E}{aMm}[(M+m) \mp \sqrt{(M+m)^2 - 2Mm(1 - \cos(k_1 a)\cosh(k_2 a))}] \quad (7.108)$$

as long as ω is real. This condition is not fulfilled everywhere. In particular, at some value $k_{2\max}$ for $k_1 = \mp\pi/a$ the frequency becomes

$$\omega_{k_{2\max}}^2 = \frac{E}{a}\left[\left(\frac{1}{M} + \frac{1}{m}\right)\right] \quad (7.109)$$

beyond which no frequency is defined. This is clearly seen on the dispersion "surface" of Figure 7.13. In the range $0 \le k_2 \le k_{2\max}$ we observe a completely filled band between the acoustic and optical branch of the dispersion surface. Thus this rather simple idealization of the crystal surface qualitatively reproduces the features of surface phonons, namely

- Exponential amplitude decay into the depth of the crystal, and
- A range of allowed phonon frequencies terminated by the optical and acoustical branches of the dispersion diagram.

Surface Acoustic Waves (SAW)

We now consider what happens when an acoustic wave in a solid travels close to the surface in the long wavelength limit. Recall that the wave equation for a 3D solid is

$$\rho \frac{\partial^2 \boldsymbol{u}}{\partial t^2} = \nabla \bullet \sigma \tag{7.110}$$

which, together with the elastic constitutive equation for a crystal with cubic symmetry and the strain-displacement relation

$$\sigma = C : \varepsilon \tag{7.111a}$$

$$\varepsilon = \frac{1}{2}(\nabla \boldsymbol{u} + (\nabla \boldsymbol{u})^T) \tag{7.111b}$$

results in the wave equations for the displacement field

$$\rho \frac{\partial^2 u_1}{\partial t^2} = C_{11} \frac{\partial^2 u_1}{\partial x_1^2} + (C_{12} + C_{44}) \left(\frac{\partial^2 u_2}{\partial x_1 \partial x_2} + \frac{\partial^2 u_3}{\partial x_1 \partial x_3} \right) + C_{44} \left(\frac{\partial^2 u_1}{\partial x_2^2} + \frac{\partial^2 u_1}{\partial x_3^2} \right) \tag{7.112a}$$

$$\rho \frac{\partial^2 u_2}{\partial t^2} = C_{11} \frac{\partial^2 u_2}{\partial x_2^2} + (C_{12} + C_{44}) \left(\frac{\partial^2 u_1}{\partial x_2 \partial x_1} + \frac{\partial^2 u_3}{\partial x_2 \partial x_3} \right) + C_{44} \left(\frac{\partial^2 u_2}{\partial x_1^2} + \frac{\partial^2 u_2}{\partial x_3^2} \right) \tag{7.112b}$$

$$\rho\frac{\partial^2 u_3}{\partial t^2} = C_{11}\frac{\partial^2 u_3}{\partial x_3^2} + (C_{12}+C_{44})\left(\frac{\partial^2 u_1}{\partial x_3 \partial x_1} + \frac{\partial^2 u_2}{\partial x_3 \partial x_2}\right) + C_{44}\left(\frac{\partial^2 u_3}{\partial x_1^2} + \frac{\partial^2 u_3}{\partial x_2^2}\right) \tag{7.112c}$$

The surface of a material can be used to serve as a mechanical wave guide, and work in much the same way as optical wave guides do. Four geometrical configurations, shown in Figure 7.14, characterize the very

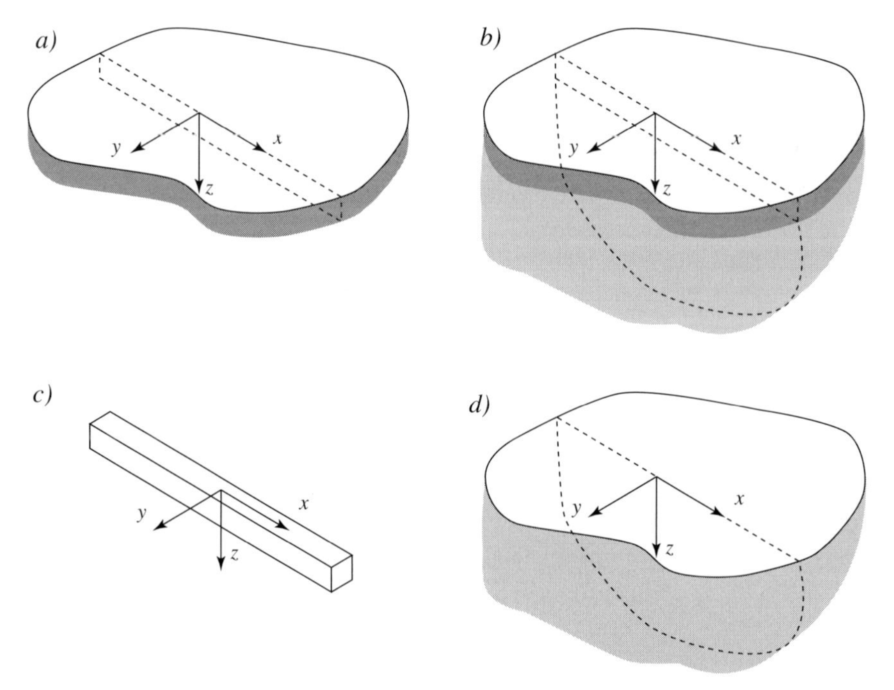

Figure 7.14. Four basic planar surface acoustic waveguide configurations and their wave types. a) Thin mechanical plate: Lamb waves. b) Thin film on top of an elastic half-space: Love waves. c) Mechanical homogeneous beam: Flexural waves. d) Homogeneous elastic half space: Raleigh waves.

different types of waves that we can effectively guide. Some of these surface waves have special names:

- Beams: *Flexural* waves. (Daniel Bernoulli, 08/02/1700–17/03/1782, Leonard Euler, 15/04/1707–1783, and Stephen P. Timoshenko, 22/12/1878–29/05/1972).
- Homogeneous half-space: *Rayleigh* waves (John William Strutt Lord Rayleigh, 12/11/1842–30/06/1919).
- Plates: *Lamb* waves (Horace Lamb, 29/11/1849–4/11/1934).
- Thin layer on an elastic half-space: *Love* waves (Augustus Edward Hough Love, 17/04/1863–5/06/1940).

The basic analysis technique is to first make a unique decomposition of the displacement field into two parts $\boldsymbol{u} = \boldsymbol{u}' + \boldsymbol{u}''$. One component is divergence-free and the other is rotation-free. Because of this separation, we can assume the existence of a scalar potential ϕ and a vector potential $\boldsymbol{\psi}$ (as we did for the electrodynamic potentials ψ and $\boldsymbol{A}$), and hence obtain the unique decomposition

$$\boldsymbol{u} = \nabla\times\boldsymbol{\psi} + \nabla\phi \tag{7.113}$$

Elasticity for Raleigh and Lamb Waves

The analysis is greatly simplified for the Lamb and Raleigh waves if we choose the direction of propagation parallel to a coordinate axis, here the x-axis. In addition, we assume no dependence of $\boldsymbol{u}$ on the y-direction (for the plate this means a plane strain assumption), so that $\boldsymbol{\psi} = (0, \psi, 0)$ and hence

$$\boldsymbol{u} = (u_1, 0, u_3) = \left(-\frac{\partial\psi}{\partial z} + \frac{\partial\phi}{\partial x}, 0, \frac{\partial\psi}{\partial x} + \frac{\partial\phi}{\partial z}\right) \tag{7.114}$$

From elasticity, we recall that the "small" strain is defined componentwise as

$$\varepsilon_{ij} = \frac{1}{2}\left(\frac{\partial u_i}{\partial x_j} + \frac{\partial u_j}{\partial x_i}\right) \tag{7.115}$$

so that we can write that

$$\varepsilon_{11} = -\frac{\partial^2\psi}{\partial x \partial z} + \frac{\partial^2\phi}{\partial x^2} \tag{7.116a}$$

$$\varepsilon_{13} = -\frac{\partial^2\psi}{\partial z^2} + \frac{\partial^2\phi}{\partial z \partial x} \tag{7.116b}$$

$$\varepsilon_{31} = \frac{\partial^2\psi}{\partial x^2} + \frac{\partial^2\phi}{\partial x \partial z} = -\frac{\partial^2\psi}{\partial z^2} + \frac{\partial^2\phi}{\partial x \partial z} \tag{7.116c}$$

$$\varepsilon_{33} = \frac{\partial^2\psi}{\partial z \partial x} + \frac{\partial^2\phi}{\partial z^2} \tag{7.116d}$$

We can write the stress as $\sigma = C:\varepsilon$ in tensor notation or use the more compact Voight engineering notation. For silicon, which possesses cubic crystal symmetry,

$$\begin{bmatrix} \sigma_{11} \\ \sigma_{22} \\ \sigma_{33} \\ \sigma_{23} \\ \sigma_{31} \\ \sigma_{12} \end{bmatrix} = \begin{bmatrix} C_{11} & C_{12} & C_{12} & & & \\ C_{12} & C_{11} & C_{12} & & & \\ C_{12} & C_{12} & C_{11} & & & \\ & & & C_{44} & & \\ & & & & C_{44} & \\ & & & & & C_{44} \end{bmatrix} \cdot \begin{bmatrix} \varepsilon_{11} \\ 0 \\ \varepsilon_{33} \\ 0 \\ 2\varepsilon_{31} \\ 0 \end{bmatrix} \tag{7.117a}$$

$$= \begin{bmatrix} C_{11}\varepsilon_{11} + C_{12}\varepsilon_{33} \\ C_{12}\varepsilon_{11} + C_{12}\varepsilon_{33} \\ C_{12}\varepsilon_{11} + C_{11}\varepsilon_{33} \\ 0 \\ 2C_{44}\varepsilon_{31} \\ 0 \end{bmatrix} \tag{7.117b}$$

Substituting (7.116a)-(7.116d) into (7.117a) gives

$$\begin{bmatrix}\sigma_{11} & \sigma_{22} & \sigma_{33} & \sigma_{23} & \sigma_{31} & \sigma_{12}\end{bmatrix}^T =$$

$$\begin{bmatrix} C_{11}(-\partial^2\psi/\partial x\partial z + \partial^2\phi/\partial x^2) + C_{12}(\partial^2\psi/\partial z\partial x + \partial^2\phi/\partial z^2) \\ C_{12}(-\partial^2\psi/\partial x\partial z + \partial^2\phi/\partial x^2) + C_{12}(\partial^2\psi/\partial z\partial x + \partial^2\phi/\partial z^2) \\ C_{12}(-\partial^2\psi/\partial x\partial z + \partial^2\phi/\partial x^2) + C_{11}(\partial^2\psi/\partial z\partial x + \partial^2\phi/\partial z^2) \\ 0 \\ 2C_{44}(-\partial^2\psi/\partial z^2 + \partial^2\phi/\partial x\partial z) \\ 0 \end{bmatrix} \quad (7.118)$$

Elasticity for Love Waves

We know that Love waves (through hindsight!) are pure shear waves in the plane of the surface of the solid, so that we choose a displacement according to

$$\boldsymbol{u} = (0, u_2(x, z), 0) \quad (7.119)$$

In this case we do not need a scalar and vector potential, since the strain is now

$$\varepsilon = \begin{bmatrix}\varepsilon_{11} & \varepsilon_{22} & \varepsilon_{33} & 2\varepsilon_{23} & 2\varepsilon_{31} & 2\varepsilon_{12}\end{bmatrix}^T = \begin{bmatrix}0 & 0 & 0 & \dfrac{\partial u_2}{\partial z} & 0 & \dfrac{\partial u_2}{\partial x}\end{bmatrix}^T \quad (7.120)$$

which clearly is a pure shear deformation. The stress is

$$\begin{bmatrix}\sigma_{11}\\ \sigma_{22}\\ \sigma_{33}\\ \sigma_{23}\\ \sigma_{31}\\ \sigma_{12}\end{bmatrix} = \begin{bmatrix} C_{11} & C_{12} & C_{12} & & & \\ C_{12} & C_{11} & C_{12} & & & \\ C_{12} & C_{12} & C_{11} & & & \\ & & & C_{44} & & \\ & & & & C_{44} & \\ & & & & & C_{44}\end{bmatrix} \cdot \begin{bmatrix}0\\ 0\\ 0\\ \dfrac{\partial u_2}{\partial z}\\ 0\\ \dfrac{\partial u_2}{\partial x}\end{bmatrix} = \begin{bmatrix}0\\ 0\\ 0\\ C_{44}\dfrac{\partial u_2}{\partial z}\\ 0\\ C_{44}\dfrac{\partial u_2}{\partial x}\end{bmatrix} \quad (7.121)$$

We see that for Raleigh and Lamb waves, two of six stress components in silicon are zero if the displacement occurs in a plane, and in the Love wave case, with displacements in one direction, only two shear stress components in silicon are nonzero.

Timoshenko Beam

To obtain a PDE for the deflection of a beam, we first consider the case of a single plane of forces acting on the beam. Linear superposition can be used to incorporate other directions. The forces may be:

- concentrated shear forces V, acting at the ends of a beam segment, which we denote V_0 and V_L;
- point-wise moments M, and for the derivation we again assume that these forces act at the ends of the beam segment, which we denote M_0 and M_L;
- loads $q(\xi, t)$ distributed over the length L of the beam segment.

We do not add the case of distributed moment loads, although these are straightforward to incorporate. The loads perform work on the beam segment, which we can write as

$$W = \int_0^L q(\xi, t)w(\xi, t)d\xi - M_0\theta(0, t) + M_L\theta(L, t) + V_0 w(0, t) - V_L w(L, t) \tag{7.122}$$

Here $w(\xi, t)$ is the deflection and $\theta(\xi, t)$ angular rotation at position ξ and time t. In general, the gradient of the deflection is generated by two effects, the bending of the beam segment, and the shear of the material at a point, as shown in Figure 7.15. The stored potential energy, assuming that the deflection is elastic, includes the linear addition of terms proportional to the squares of both “displacements”

$$V(\xi, t) = \frac{1}{2}\left\{\int_0^L E(\xi)I(\xi)\left(\frac{\partial\theta(\xi, t)}{\partial\xi}\right)^2 d\xi + \int_0^L k_T(\xi)\left(\frac{\partial w(\xi, t)}{\partial\xi} - \theta(\xi, t)\right)^2 d\xi\right\} \tag{7.123}$$

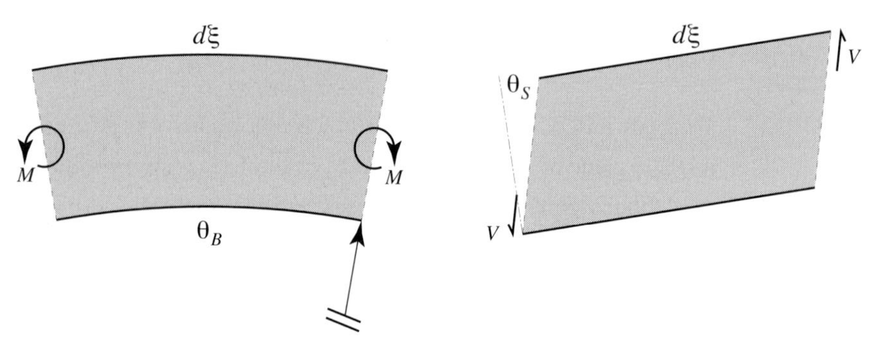

Figure 7.15. The deformation of a differential beam element is generated by a bending deformation and by a shear deformation. Pure bending deformation is generated by a bending moment across the differential element, and a pure shear deformation by a shear force across the element.

The proportionality constants of these two squared displacement terms reflect the resistance of the beam to deflection:

- $E(\xi)I(\xi)$ is the Young modulus of the beam material times the second moment of area (often called the area moment of inertia) of the beam cross section about the axis of bending rotation. Hence this term mixes geometrical shape and material composition, and tells us that prudent choice of shape can more than compensate for a weak material. Recall that we often cannot freely choose materials, due to production constraints, but often can design shapes with considerably more freedom;
- $k_T(\xi)$ is a correction factor that accounts for the non-uniformity of the actual strain over the beam cross section [7.5].

The stored kinetic coenergy is a sum of linear and angular momentum stores

$$T^*(\xi, t) = \frac{1}{2}\left\{\int_0^L \rho_m(\xi)\left(\frac{\partial w(\xi, t)}{\partial t}\right)^2 d\xi + \int_0^L I_m(\xi)\left(\frac{\partial \theta(\xi, t)}{\partial t}\right)^2 d\xi\right\} \quad (7.124)$$

The proportionality constants of these two squared velocity terms reflect the resistance of the beam to movement:

- $\rho_m(\xi) = \rho A$ is the mass per unit length of the beam material and represents the sluggishness or linear inertia of the beam material to being accelerated. The beam has a cross sectional area A and a volumetric mass density ρ;
- $I_m(\xi) = \rho A i_g^2$ is the second moment of mass or mass moment of inertia of the beam material and represents the sluggishness or inertia of the beam cross section to angular acceleration. The radius of gyration i_g represents that radius that would give the same moment of inertia, were all the mass of the cross section concentrated along a circular line. Again, since the geometry enters this term in a squared sense, we can circumvent the constraints of material through prudent geometrical design.

From the Hamilton principle (Box 2.3) for the variation of the action

$$\delta A = \delta \int_{t_1}^{t_2} (T^* - V)dt = W \quad (7.125)$$

we obtain the two partial differential equations of motion [7.5]

$$\frac{\partial}{\partial \xi}\left(E(\xi)I(\xi)\frac{\partial \theta}{\partial \xi}\right) + k_T(\xi)\left(\frac{\partial w(\xi, t)}{\partial \xi} - \theta(\xi, t)\right) - I_m(\xi)\frac{\partial \theta(\xi, t)}{\partial t} = 0 \quad (7.126a)$$

$$\rho_m(\xi)\frac{\partial^2 w(\xi)}{\partial \xi^2} - \frac{\partial}{\partial \xi}\left(k_T(\xi)\left(\frac{\partial w(\xi, t)}{\partial \xi} - \theta(\xi, t)\right)\right) - q(\xi, t) = 0 \quad (7.126b)$$

Equation (7.126a) is subject to either:

- a Dirichlet boundary condition where $\delta\theta = 0$, i.e., the bending angle is fixed at either end of the segment;
- a Neumann boundary condition with an applied moment so that the angular gradient of bending at either end of the segment becomes $\partial\theta/\partial\xi = M/(EI)$ for an applied concentrated moment M.

Similarly, (7.126b) is subject to either of:

- a Dirichlet boundary condition where $\delta w = 0$, i.e., the deflection is fixed at either end of the segment;
- a Neumann boundary condition with an applied moment so that the net deflection gradient at either end of the segment becomes $\partial w/\partial\xi - \theta = Q/k_T$ for a concentrated force Q.

When the beam cross section is uniform, then the geometrical and material quantities E, I, I_m, k_T and ρ_m become independent of the coordinate ξ along the beam. Eliminating the bending angle θ from (7.126a) and (7.126b) in favour of w (by taking the derivative of the first equation w.r.t. ξ, using the second to define θ, and assuming that $k_T \approx 1$), we obtain the classical Timoshenko beam expression

$$\begin{aligned} & EI\frac{\partial^4 w(\xi, t)}{\partial\xi^4} + \rho\frac{\partial^2 w(\xi, t)}{\partial t^2} \\ & -\left(I_m + \frac{EI\rho_m}{k_T}\right)\frac{\partial^4 w(\xi, t)}{\partial\xi^2\partial t^2} + \frac{I_m\rho_m}{k_T}\frac{\partial^4 w(\xi, t)}{\partial t^4} = 0 \end{aligned} \tag{7.127}$$

We take a harmonic displacement assumption $\overline{w}(\xi, t) = w_0\exp[i(k\xi - \omega t)]$ which will be correct for the vibrational modes. The geometrical mode factor w_0 remains undetermined for the following calculation. In general we can then write that

$$\frac{\partial^{(a+b)}\overline{w}(\xi, t)}{\partial\xi^a\partial t^b} = i^{(a+b)}k^a\omega^b\overline{w}(\xi, t) \tag{7.128}$$

and hence (7.127) becomes

$$(EIk^4) - (\rho_m\omega^2) + \left(I_m + \frac{EI\rho_m}{k_T}\right)k^2\omega^2 + \frac{I\rho_m^2}{k_T}\omega^4 = 0 \tag{7.129}$$

We next divide throughout by EIk^4 and make use of $c = \omega/k$ and the bulk wave speed $c_B = \sqrt{E/\rho}$. Also note that $I_m/I = \rho$ and set $s = \sqrt{k_T/\rho}$. This yields

$$1 - \left(\frac{c^2}{c_B^2 i_g^2 k^2}\right) + \left(\frac{1}{c_B^2} + \frac{1}{s^2}\right)c^2 + \left(\frac{1}{c_B^2 s^2}\right)c^4 = 0 \tag{7.130}$$

We are interested in the real roots of this equation, which yield the dispersion curves $c(k)$ of the Timoshenko beam.

Raleigh Waves

Rayleigh waves are harmonic in the plane of the surface of the solid, and decay exponentially into the depth of the material. A convenient ansatz for the displacement (see Figure 7.16) is

$$\boldsymbol{u}' = F(z)e^{i(\boldsymbol{k}_p \cdot \boldsymbol{x}_p - \omega t)} \tag{7.131a}$$

$$\boldsymbol{u}'' = G(z)e^{i(\boldsymbol{k}_p \cdot \boldsymbol{x}_p - \omega t)} \tag{7.131b}$$

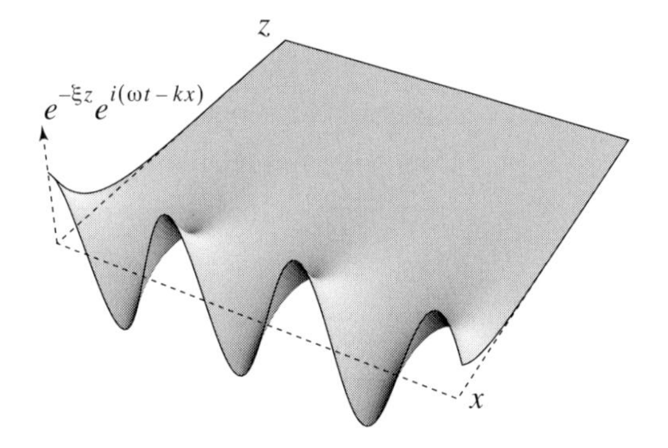

Figure 7.16. A surface acoustic Raleigh wave is represented by a harmonic amplitude function that decays exponentially along the z-direction away from the surface into the solid. The harmonic part of the amplitude strength is plotted as a surface in the vertical direction over the xz-plane.

where $\boldsymbol{x}_p$ and $\boldsymbol{k}_p$ lie in the plane of the surface. The derivatives w.r.t. time and space coordinates are

$$\frac{\partial^2 \boldsymbol{u}'}{\partial t^2} = -\omega^2 \boldsymbol{u}' \qquad \frac{\partial^2 \boldsymbol{u}''}{\partial t^2} = -\omega^2 \boldsymbol{u}'' \tag{7.132a}$$

$$\frac{\partial^2 \boldsymbol{u}'}{\partial x^2} = -k_x^2 \boldsymbol{u}' \qquad \frac{\partial^2 \boldsymbol{u}''}{\partial x^2} = -k_x^2 \boldsymbol{u}'' \tag{7.132b}$$

$$\frac{\partial^2 \boldsymbol{u}'}{\partial y^2} = -k_y^2 \boldsymbol{u}' \qquad \frac{\partial^2 \boldsymbol{u}''}{\partial y^2} = -k_y^2 \boldsymbol{u}'' \tag{7.132c}$$

$$\frac{\partial^2 \boldsymbol{u}'}{\partial z^2} = \frac{\partial F(z)}{\partial z} e^{i(\boldsymbol{k}_p \cdot \boldsymbol{x}_p - \omega t)} \qquad \frac{\partial^2 \boldsymbol{u}''}{\partial z^2} = \frac{\partial G(z)}{\partial z} e^{i(\boldsymbol{k}_p \cdot \boldsymbol{x}_p - \omega t)} \tag{7.132d}$$

$$\frac{\partial^2 \boldsymbol{u}'}{\partial x \partial y} = -k_x k_y \boldsymbol{u}' \qquad \frac{\partial^2 \boldsymbol{u}''}{\partial x \partial y} = -k_x k_y \boldsymbol{u}'' \tag{7.132e}$$

$$\frac{\partial^2 \boldsymbol{u}'}{\partial x \partial z} = i k_x \frac{\partial F(z)}{\partial z} \boldsymbol{u}' \qquad \frac{\partial^2 \boldsymbol{u}''}{\partial x \partial z} = k_x \frac{\partial G(z)}{\partial z} \boldsymbol{u}'' \tag{7.132f}$$

$$\frac{\partial^2 \boldsymbol{u}'}{\partial z \partial y} = i \frac{\partial F(z)}{\partial z} k_y \boldsymbol{u}' \qquad \frac{\partial^2 \boldsymbol{u}''}{\partial z \partial y} = i \frac{\partial G(z)}{\partial z} k_y \boldsymbol{u}'' \tag{7.132g}$$

since we assume no dependence of the motion on the y-direction.

Lamb Waves in Mindlin Plates

Plates are idealizations used when a geometrical object is flat and very much thinner than its other dimensions. In this case we can conveniently describe the geometry using two coordinates only, say x and y. For the following formulation we closely follow [7.5]. We define the averaged plate stresses (or bending moments and shear forces) by

$$\{M_{xx}, M_{yy}, M_{xy}\} = \int_{-\frac{t}{2}}^{\frac{t}{2}} \sigma z dz \tag{7.133a}$$

$$\{Q_{xz}, Q_{yz}\} = \int_{-\frac{t}{2}}^{\frac{t}{2}} \sigma dz \tag{7.133b}$$

if we assume that the plate lies symmetrically about its neutral axis and placed at $z = 0$. In order to preserve a simple algebraic structure later on, we define the averaged plate strains by

$$\{\overline{\varepsilon_{xx}}, \overline{\varepsilon_{yy}}, \overline{\varepsilon_{xy}}\} = \left(\frac{12}{h^3}\right)\int_{-\frac{t}{2}}^{\frac{t}{2}}\{\varepsilon_{xx}, \varepsilon_{yy}, \varepsilon_{xy}\}zdz \tag{7.134a}$$

$$\{\overline{\varepsilon_{xz}}, \overline{\varepsilon_{yz}}\} = \left(\frac{2}{h}\right)\int_{-\frac{t}{2}}^{\frac{t}{2}}\{\varepsilon_{xz}, \varepsilon_{yz}\}dz \tag{7.134b}$$

We now join the above relations to obtain the plate-specific constitutive relation for an isotropic material, namely

$$M_{xx} = D(\overline{\varepsilon_{xx}} + \nu\overline{\varepsilon_{yy}}) \tag{7.135a}$$

$$M_{yy} = D(\overline{\varepsilon_{yy}} + \nu\overline{\varepsilon_{xx}}) \tag{7.135b}$$

$$M_{xy} = (1-\nu)D\overline{\varepsilon_{xy}} \tag{7.135c}$$

and

$$Q_{xz} = \mu' h R_{xz} \tag{7.136a}$$

$$Q_{yz} = \mu' h R_{yz} \tag{7.136b}$$

where the plate stiffness is $D = Eh^3/(12(1-\nu^2))$ and the modified shear stiffness is $\mu' = k_M\mu$.

The first z moment of the first two mechanical equations of motion, represented by (7.110), with the third equation of motion combine to give the plate equations of motion

$$\frac{\partial M_{xx}}{\partial x} + \frac{\partial M_{xy}}{\partial y} - Q_{xz} = \rho\int_{-\frac{t}{2}}^{\frac{t}{2}}\ddot{u}_x zdz \tag{7.137a}$$

$$\frac{\partial M_{xy}}{\partial x} + \frac{\partial M_{yy}}{\partial y} - Q_{yz} = \rho\int_{-\frac{t}{2}}^{\frac{t}{2}}\ddot{u}_y zdz \tag{7.137b}$$

$$\frac{\partial Q_{xz}}{\partial x} + \frac{\partial Q_{yz}}{\partial y} + \int_{-\frac{t}{2}}^{\frac{t}{2}} \frac{\partial \sigma_{zz}}{\partial z} dz = \rho \int_{-\frac{t}{2}}^{\frac{t}{2}} \ddot{u}_z dz \qquad (7.137c)$$

Next we make the Mindlin-specific assumption about the deformation through the thickness of the plate, namely that

$$\begin{bmatrix} u_x & u_y & u_z \end{bmatrix}(x, y, z) = \begin{bmatrix} zU_x(x, y) & zU_y(x, y) & W(x, y) \end{bmatrix}, \qquad (7.138)$$

i.e., we assume that the entire 3D deformation field can be described by the three averaged deformation values U_x, U_y and W defined only in the plane of the plate, and that variation of the lateral deflection (in the x- and y-directions) depends linearly on the distance from this "neutral" plane of the plate. Inserting this assumption and the plate constitutive relations (7.135a)–(7.136b) into (7.137a)–(7.137b) give the isotropic plate equations of motion

$$D\left((1+\nu)\left(\frac{\partial^2 U_x}{\partial x^2} + \frac{\partial^2 U_y}{\partial x \partial y}\right) + (1-\nu)\nabla^2 U_x\right) - \mu' h\left(U_x + \frac{\partial W}{\partial x}\right) = \frac{\rho}{12} h^3 \ddot{U}_x \qquad (7.139a)$$

$$D(1-\nu)\frac{\partial \overline{\varepsilon_{xy}}}{\partial x} + D\frac{\partial(\overline{\varepsilon_{yy}} + \nu \overline{\varepsilon_{xx}})}{\partial y} - \mu' h R_{yz} = \frac{\rho}{12} h^3 \ddot{U}_y \qquad (7.139b)$$

$$\mu' h\left(\frac{\partial R_{xz}}{\partial x} + \frac{\partial R_{yz}}{\partial y}\right) + \int_{-\frac{t}{2}}^{\frac{t}{2}} \frac{\partial \sigma_{zz}}{\partial z} dz = \rho \int_{-\frac{t}{2}}^{\frac{t}{2}} \ddot{u}_z dz \qquad (7.139c)$$

Love Waves

By placing a thin material layer such as silicon dioxide or silicon nitride on a semi-infinite half-space made of silicon (or thick enough so that the theory applies) we form a waveguide for the surface waves. In fact, many of the mathematical techniques used for optical waveguides are applicable here as well. To keep this section as short as possible, we only consider one layer on top of the half space, even though practical situations may involve more layers [7.5], [7.6]. We number the substrate as $i = 1$

and the thin upper layer of thickness d as $i = 2$. In each material the wave equation holds

$$\nabla^2 u_{yi} - \frac{1}{c_{\beta i}} \frac{\partial^2 u_{yi}}{\partial t^2} = 0 \qquad i = 1, 2 \tag{7.140}$$

We assume here that the wave travels in the x-direction, that z points into the depth of the half-space, and (through hindsight!) that we only have transverse vibrations. Showing that this is so requires more space than available. On the free surface of the layer we have zero traction, hence

$$\left.\frac{\partial u_{y2}}{\partial z}\right|_{z = -d} = 0 \tag{7.141}$$

Between the thin layer and the half-space we require displacement continuity, hence

$$u_{y2}\big|_{z = 0} = u_{y1}\big|_{z = 0} \tag{7.142}$$

In addition, we also require that the stress field is continuous at this point, so that

$$\rho_2 c_{\beta 2}^2 \left.\frac{\partial u_{y2}}{\partial z}\right|_{z = 0} = \rho_1 c_{\beta 1}^2 \left.\frac{\partial u_{y1}}{\partial z}\right|_{z = 0} \tag{7.143}$$

The wave is assumed to have a shape $f(z)$ in the z-direction that travels in the x-direction, so that we make the following ansatz for the displacement

$$u_{yi} = f_i(z)\exp(\omega t - k_l x) \qquad i = 1, 2 \tag{7.144}$$

which we insert into (7.140) to obtain

$$\frac{\partial^2 f_1(z)}{\partial z^2} - (k_l \beta)^2 f_1(z) = 0 \qquad \beta = \left(\frac{c_l}{c_{\beta 2}}\right)^2 - 1 \tag{7.145a}$$

$$\frac{\partial^2 f_2(z)}{\partial z^2} - (k_l\alpha)^2 f_2(z) = 0 \qquad \alpha = 1 - \left(\frac{c_l}{c_{\beta 1}}\right)^2 \tag{7.145b}$$

with $c_{\beta 2} < c_l < c_{\beta 1}$. Candidate functions that satisfy these conditions are

$$f_1(z) = C\exp(-\beta k_l z) + D\exp(\beta k_l z) \tag{7.146a}$$

$$f_2(z) = A\sin(\alpha k_l z) + B\cos(\alpha k_l z) \tag{7.146b}$$

We now use the boundary conditions to determine the constants A, B, C and D. In the limit as $z \to \infty$ we must have that $u_{y1}(z) = 0$, so that $D = 0$. From the displacement continuity at $z = 0$ we obtain that

$$[f_1(0) = f_2(0)] \Rightarrow [B = C] \tag{7.147}$$

and from the stress continuity at $z = 0$ we obtain that

$$A = -C\frac{\rho_1 c_{\beta 1}^2 \beta}{\rho_2 c_{\beta 2}^2 \alpha} \tag{7.148}$$

Combining the terms we obtain the following solution structure

$$u_{y1} = C\exp(-\beta k_l z)\exp(\omega t - k_l x) \tag{7.149a}$$

$$u_{yi} = \left[\cos(\alpha k_l z) - \frac{\rho_1 c_{\beta 1}^2 \beta}{\rho_2 c_{\beta 2}^2 \alpha}\sin(\alpha k_l z)\right]\exp(\omega t - k_l x) \tag{7.149b}$$

The zero traction boundary condition at the top surface finally gives us the speed conditions for a general solution, i.e.,

$$\tan(\alpha k_l d) = \frac{\rho_1 c_{\beta 1}^2 \beta}{\rho_2 c_{\beta 2}^2 \alpha} \tag{7.150}$$

7.6.4 The PN Junction

The technologically most important PN junctions are formed through doping a semiconductor with two types of impurity atoms, called the donor and acceptor atoms. For example, a uniformly doped silicon substrate, e.g., with the donor arsenic (As), is subsequently doped with the acceptor boron (B), in select regions, to form PN junctions. We will not concern ourselves with the fabrication methods of PN junctions, but it is important to note that they have a strong influence on the spatial distribution of foreign ions in the silicon lattice. More important for the analysis here is the level of doping. It is technologically possible to vary the concentration level of foreign dopants in the range $10^{20} - 10^{26}$ atoms per cubic meter; typical values, however, are $10^{21} - 10^{24}$ dopant atoms per cubic meter. The 1D doping profile for a typical PN junction is shown in Figure 7.17.

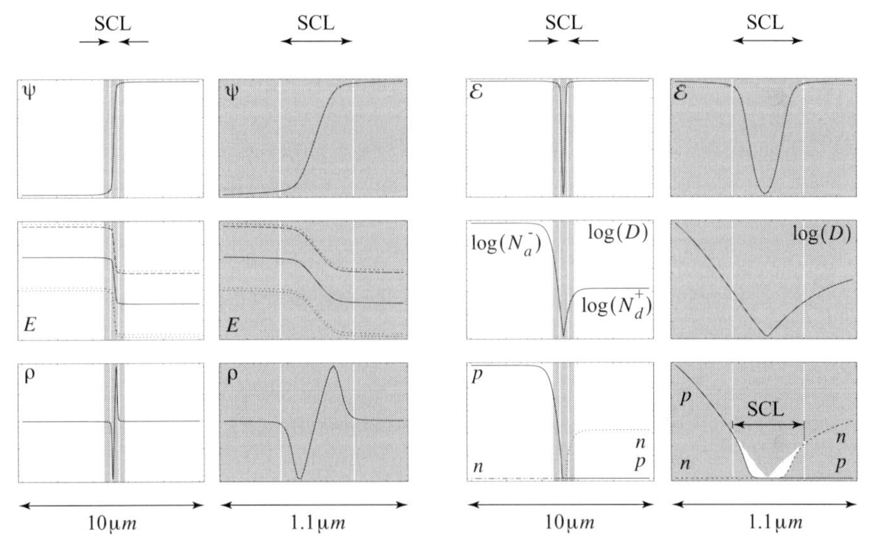

Figure 7.17. Seven related quantities for a one-dimensional PN junction: the electrostatic potential ψ, the energy band diagram E, the charge density ρ, the electric field $\mathcal{E}$, the net impurity doping $D = N_d^+ - N_a^-$ and the carrier concentrations shown for the electrons n and holes p. The quantities are displayed in pairs of plots. The horizontal axis of each left-hand plot is over the diode's full length of $10\mu m$. The gray stripe corresponds to the plot on the immediate right, which provides more detail in the region of the space-charge layer (SCL). The curves were computed using the author's 1D device simulation program.

Unbiased Junction

For an unbiased junction, we expect charge neutrality at distances far from the junction where the potential is constant, i.e., where $\nabla^2\psi = 0$. This can be expressed as

$$p + N_d^+ = n + N_a^- \tag{7.151}$$

i.e., the sum of the number of holes and ionized donors (the positive charges in the crystal) must equal the sum of the number of electrons and ionized acceptors. The ionization is governed by fermion statistics, which leads us to write (7.151) as

$$N_c \mathrm{Exp}\left(\frac{-E_F}{k_B T}\right) + \frac{N_d}{1 + \mathrm{Exp}\left(-\frac{E_d - E_F}{k_B T}\right)} = \left(N_v \mathrm{Exp}\left(\frac{E_F - E_g}{k_B T}\right) + \frac{N_a}{1 + \mathrm{Exp}\left(\frac{E_a - E_F}{k_B T}\right)}\right) \tag{7.152a}$$

$$\text{where } N_c = 2\left(\frac{2\pi m_e k_B T}{h^2}\right)^{\frac{3}{2}} \text{ and } N_v = 2\left(\frac{2\pi m_h k_B T}{h^2}\right)^{\frac{3}{2}} \tag{7.153a}$$

This equation is nonlinear in E_F and must be solved graphically (see Figure 7.18) or numerically. Also note that E_F is measured here from the

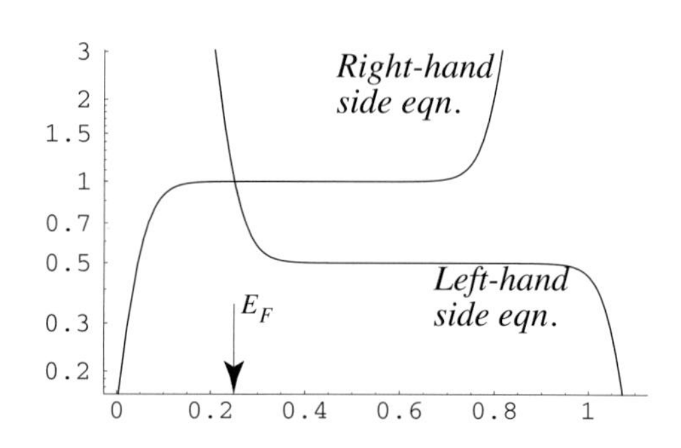

Figure 7.18. Illustration of the graphical solution of (7.152a). Each curve represents (the logarithm of) one side of the equation. The vertical projection of the intersection is the required Fermi energy level.

edge of the valence band. Once we have obtained either the Fermi level or the potential in the device, we can easily compute the other quantities required for the band diagram, as was done for Figure 7.17.

At the Junction

We now zoom in on the junction, see Figure 7.19, which is where all the

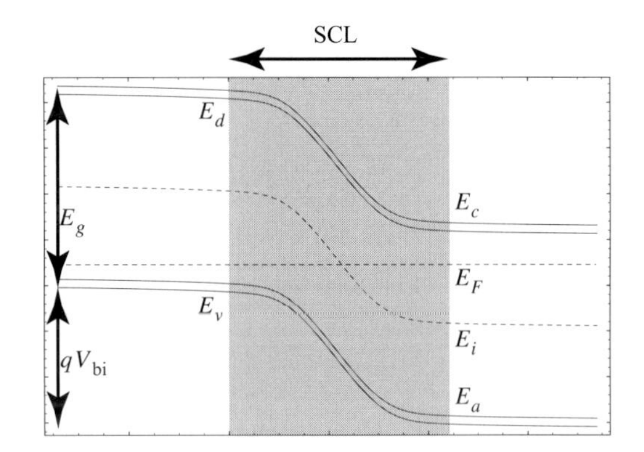

Figure 7.19. The labeled band diagram at the junction of the diode. The vertical axis is the energy, the horizontal axis is distance along the diode. The shaded region marks the space-charge layer (SCL) that forms around the junction. The junction results in a built-in potential V_{bi} at thermodynamic equilibrium.

"action" is. Because the Fermi level is constant in a material, a basic consequence of thermodynamics, but removed from the band-gap center in doped materials, the PN junction causes a transition in the energy level of the band edge as we move from one side to the other. This energy "jump" is associated with an electrostatic potential "jump", which we call the built-in potential.

Carriers on the two sides of the junction experience large concentration gradients and hence a diffusion driving force. Hence, in Figure 7.19, electrons move from the right to the left, and holes from left to right, leaving the vicinity devoid of free charges, hence the space-charge region. The resulting positive region on the right and negative region on the left of Figure 7.19 form a double-layer of charges and hence a "hump" in electric field pointing from right to left. It is this electric field that eventually causes the diffusion process to stop, for it acts on the charges as a drift-like driving force opposing the diffusion current. An electrically disconnected diode is in a delicate dynamic equilibrium. Here we can use the value of the Fermi energy predicted by (7.152a) only as a reasonable starting value with which to iterate

$$\nabla^2\psi = \frac{q}{\varepsilon}(n - p + N_d^+ - N_a^-) \tag{7.154}$$

For reasonably doped (non-degenerate) diodes, i.e., where the Fermi level is not closer than $3k_BT$ to either band edge, we can use Boltzmann statistics, giving

$$\nabla^2 E_v = \frac{q^2}{\varepsilon}\left(N_v \mathrm{Exp}(\frac{E_v}{kT}) - N_c \mathrm{Exp}(\frac{-E_v + E_g}{kT}) + N_d^+ - N_a^-\right) \tag{7.155a}$$

Otherwise the doping is degenerate, and we have to use Fermi statistics, so that (7.154) becomes

$$\nabla^2 E_v = \frac{q^2}{\varepsilon}\left(N_v \frac{2}{\pi} \mathrm{F}_{\frac{1}{2}}(\frac{E_v}{kT}) - N_c \mathrm{F}_{\frac{1}{2}}(\frac{-E_v + E_g}{kT}) + N_d^+ - N_a^-\right) \tag{7.155b}$$

where $\mathrm{F}_{1/2}(x) = \int_0^\infty((\sqrt{y})/(1 + \mathrm{Exp}(y - x)))dy$ is the Fermi integral of order $1/2$. We again have obtained an equation nonlinear in the energy, but now it also is spatially coupled.

For 1D junctions (7.155a) can be solved numerically in a straightforward manner, as shown in Figure 7.17. Many of the features of these plots, however, can be obtained through assumptions that greatly simplify the calculations. Two models are in use:

- If we assume that the doping profile is abrupt, and that the space-charge has a piece-wise constant profile (often called a top-hat function). The value of the space charge density is N_a^- on the P side and N_d^+ on the N side of the junction. In this case the electric field in the SCL is a piece-wise linear “hat” function, and the potential is piece-wise quadratic.

- If we assume that the doping profile at the junction within the SCL is linear, which is in fact quite reasonable, then the space-charge is also linear, making the electric field quadratic and the electric potential a piece-wise cubic function within the profile.

Abrupt Junction Built-in Potential

From the abrupt junction model we can quickly estimate the width of the SCL and the value of the built-in potential. From the relation for the majority carrier concentrations on each side of the junction, i.e.,

$$\text{P-type material } p = n_i \mathrm{Exp}\left(\frac{E_i - E_F}{k_B T}\right) \text{ or } E_i - E_F = k_B T \ln\left(\frac{p}{n_i}\right) \quad (7.156a)$$

$$\text{N-type material } n = n_i \mathrm{Exp}\left(\frac{E_F - E_i}{k_B T}\right) \text{ or } E_F - E_i = k_B T \ln\left(\frac{n}{n_i}\right) \quad (7.156b)$$

we obtain, by adding, the total energy drop over the junction and hence the built-in potential

$$\begin{aligned}(E_F - E_i)_{\text{N-type}} + (E_i - E_F)_{\text{P-type}} &= k_B T\left[\ln\left(\frac{n}{n_i}\right) + \ln\left(\frac{p}{n_i}\right)\right] \\ &= qV_{\text{bi}}\end{aligned} \quad (7.157)$$

and hence

$$V_{\text{bi}} = \frac{k_B T}{q} \ln\left[\frac{np}{n_i^2}\right] \quad (7.158)$$

The relation can be simplified by making use of the fact that, at distances removed from the junction, where charge neutrality is simply enforced as shown in Figure 7.17, the number of carriers exactly equal the numbers of ionized impurities, i.e., in the N region $n = N_d^+$ and in the P region $p = N_a^-$, so that we may write that

$$V_{\text{bi}} = \frac{k_B T}{q} \ln\left[\frac{N_d^+ N_a^-}{n_i^2}\right] \quad (7.159)$$

Abrupt Junction Space Charge Layer Width

We expect the junction as a whole also to be charge neutral. From Figure 7.17 we see that the space charge is negative on the P side and positive on the N side. For an abrupt profile model, the space charge has a constant negative value over the entire P depletion region and a constant positive value over the entire N depletion region, so that for overall charge neutrality we require that

$$x_P N_a^- = x_N N_d^+ \tag{7.160}$$

that is, the density of negative space charge on the left of the junction in the P region times the left width equals the density of positive space charge to the right of the junction in the N region times the right width. In general, the depletion width is

$$W = x_N + x_P \tag{7.161}$$

Hence both widths are to be determined, which means that we need an additional equation. This is provided by the Poisson equation (7.154) applied to the depletion region only. We already know the voltage drop over the depletion region, the built-in voltage V_{bi}. Therefore, it is prudent to integrate the 1D Poisson equation over the depletion region, to obtain the voltage drop

$$V(x) = \int_0^x\left(\int_0^x \frac{\partial^2\psi}{\partial x^2}dx\right)dx = \int_0^x\left(\int_0^x \rho dx\right)dx = V_{bi} \tag{7.162}$$

Due to the simple form of the SCL, we then obtain, after integration, that

$$\left(\frac{qN_a^-}{2\varepsilon}\right)x_P^2 = \left(\frac{-qN_a^-}{2\varepsilon}\right)x_N^2 + (V_{bi})_{Abrupt} \tag{7.163}$$

Using (7.160) to eliminate one of the variables from (7.163), solving the resultant quadratic equation and then back substituting this value into (7.160) for the other variable, we obtain that

$$(x_N^2)_{\text{abrupt}} = \frac{2\varepsilon(V_{\text{bi}})_{\text{Abrupt}}}{q} \frac{N_a^-}{N_d^+(N_d^+ + N_a^-)} \tag{7.164a}$$

$$(x_P^2)_{\text{abrupt}} = \frac{2\varepsilon(V_{\text{bi}})_{\text{Abrupt}}}{q} \frac{N_d^+}{N_a^-(N_a^- + N_d^+)} \tag{7.164b}$$

As an example we calculate parameters for the simple diode of Figure 7.17, where $N_a^- = 2.5\times10^{22}$ and $N_d^+ = 5.8\times10^{22}$. For this case, the relations (7.159) and predict a built-in potential of $V_{\text{bi}} = 0.77$ Volt and a depletion width of $W = 0.14\mu\text{m}$. The numerical prediction, based on a self-consistent solution of (7.154), gives $V_{\text{bi}} = 0.8$ and $W = 0.4\mu\text{m}$. The differing accuracy is reasonable, since the built-in voltage depends on the amount of charge in the depletion region, whereas the width of the depletion region is more sensitive to the exact distribution of charge, which we see from Figure 7.17 is more spread out than the abrupt model assumes. More accuracy can be achieved by assuming a linear variation of doping and hence of space charge across the junction, which Figure 7.17 confirms to be closer to reality. In this case we do not repeat the calculations, but just state the results

$$(V_{\text{bi}})_{\text{Linear}} = \frac{k_B T}{q} \ln\left[\frac{a(W)_{\text{Linear}}}{2n_i}\right] \tag{7.165a}$$

$$(W)^3_{\text{Linear}} = \frac{12\varepsilon}{qa}(V_{\text{bi}})_{\text{Linear}} \tag{7.165b}$$

The two equations are nonlinearly interdependent, and can be solved graphically, numerically, but most often the values are presented on convenient plots for a range of reasonable values of the doping.

Currents

In thermal equilibrium the current through the diode is a delicate balance of motion motivated by the electrostatic force on carriers $q\boldsymbol{E}$, which we term the drift current, and motion from a higher to a lower concentration gradient and thus proportional to $-\nabla n$ and $-\nabla p$, which we call the diffusion current, to give a net current of zero. Thus the following current relations hold in thermal equilibrium

$$j = j_n + j_p = (j_{n,\,\mathrm{Drift}} + j_{n,\,\mathrm{Diff}}) + (j_{p,\,\mathrm{Drift}} + j_{p,\,\mathrm{Diff}}) = 0 \quad (7.166a)$$

$$j_n = j_{n,\,\mathrm{Drift}} + j_{n,\,\mathrm{Diff}} = 0 \qquad j_p = j_{p,\,\mathrm{Drift}} + j_{p,\,\mathrm{Diff}} = 0 \quad (7.166b)$$

$$j_n = q\mu_n n\boldsymbol{E} + qD_n\nabla n = 0 \qquad j_p = q\mu_p p\boldsymbol{E} - qD_p\nabla p = 0 \quad (7.166c)$$

Biased Junction

When the thermally equilibrated junction is biased, i.e., a voltage is applied across the junction, it either adds to or subtracts from the built-in voltage. This has the effect of either raising or lowering the band edges on one side of the junction w.r.t. the other:

- When the band edges on the N side of the junction are lowered w.r.t the P side, we say that the junction is *reverse* biased; calculations suggest that a small limiting leakage current, due mainly to drift, flows through the diode for all reverse bias voltages. In this mode the diode appears to have a very high series resistance.
- When the band edges on the N side of the junction are raised w.r.t the P side, we say that the junction is *forward* biased; calculations suggest that the current through the diode, now mainly due to diffusion, grows as an exponential function of the forward bias voltage. In this mode the diode appears to have a very low series resistance.

It is the above asymmetry makes the PN-junction diode so very useful technologically, because, used alone and even better in pairs, these diodes can be used to rectify the current flow through a particular circuit branch, as illustrated in Figure 7.20. To simplify the bookkeeping, we now introduce further notation. Since the diode has a P and an N doped side, and in the vicinity of the junction we expect to find both holes and electrons on both sides, we will term the P side holes p_p and the N side electrons n_n the *majority* carriers, and the P side electrons n_p and the N side holes p_n the *minority* carriers. The derivation is straightforward [7.16] if we remember the following:

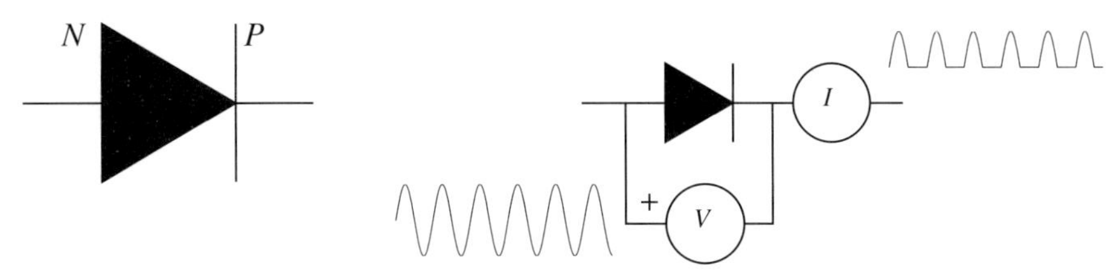

Figure 7.20. The diode is a pure PN-junction device with the symbol convention shown on the left. In the simplest of circuit arrangements, an alternating voltage (an AC signal) applied to the diode results in a rectified current, i.e., the full current can only flow from right to left when the left N terminal is at a more positive voltage than the right P terminal. By placing two diodes in parallel, but aligned in opposite directions, we can achieve full rectification of the current. Such pairs are used to generate a direct current (DC) from an alternating current. Many more sophisticated circuit arrangements exist.

- In the bulk regions, the electric field is essentially constant. Clearly this is true for the step and linearly-graded junction, but it is also largely true for other doping profiles, as long as the bulk region is uniformly doped, for example for the profile of Figure 7.17. For 1D steady state in the bulk semiconductor regions, with vanishing drift field and no generation of carriers, current continuity requires that

$$\frac{\partial n}{\partial t} = 0 = \nabla \cdot \boldsymbol{j}_n = D_n \frac{\partial^2 n}{\partial x^2} + R_n \tag{7.167a}$$

$$R_n = -\frac{n - n_p}{\tau_n} = -\frac{\Delta n_p}{\tau_n} \tag{7.167b}$$

$$\frac{\partial p}{\partial t} = 0 = \nabla \cdot \boldsymbol{j}_p = D_p \frac{\partial^2 p}{\partial x^2} + R_p \tag{7.167c}$$

$$R_p = -\frac{p - p_n}{\tau_p} = -\frac{\Delta p_n}{\tau_p} \tag{7.167d}$$

The recombination terms R are simple lifetime terms.

- The current at all points in the device is constant. From this we can derive separate expressions for the bulk and depletion regions. In the bulk the sum of majority and minority currents of one type add up to the total current so that in general

$$J = \text{constant} = J_N(x) + J_P(x) \tag{7.168a}$$

In the depletion region, since we will assume no carrier recombination here, the minority carrier densities add up to give the total current so that

$$J = \text{constant} = J_N(x_n) + J_P(-x_p) \tag{7.168b}$$

- We obtain closure of the relations by obtaining the carrier concentrations and their slopes at the edges of the depletion region. These four values become boundary conditions for the above expressions. We obtain an expression for the electric field by considering the drift-diffusion equation in the depletion region together with the Einstein relation

$$J_N \approx 0 = q\mu_n n\boldsymbol{E} + qD_N\frac{dn}{dx} \tag{7.169a}$$

$$\boldsymbol{E} = -\frac{D_N}{\mu_n}\left(\frac{1}{n}\right)\left(\frac{dn}{dx}\right) = -\frac{k_BT}{q}\left(\frac{1}{n}\right)\left(\frac{dn}{dx}\right) \tag{7.169b}$$

Since the voltage drop over the depletion region is simply minus the integral of the electric field, we obtain

$$V_{\text{depletion}} = V_{bi} - V_{\text{Applied}} = -\int_{-x_p}^{x_n}\boldsymbol{E}dx$$
$$= \frac{k_BT}{q}\int_{-x_p}^{x_n}\left(\frac{dn}{n}\right) = \frac{k_BT}{q}\ln(n)\Big|_{-x_p}^{x_n} = \frac{k_BT}{q}\ln\left[\frac{n_n(x_n)}{n_p(-x_p)}\right] \tag{7.170}$$

In thermal equilibrium, where $V_{\text{Applied}} = 0$, we have from (7.158) that

$$\exp\left[-\frac{qV_{bi}}{k_BT}\right] = \frac{n_i^2}{n_{n0}p_{p0}} \tag{7.171}$$

which can be combined with (7.170) to give

$$n_p(-x_p) = n_n(x_n)\frac{n_i^2}{n_{n0}p_{p0}}\exp\left[\frac{qV_{\text{Applied}}}{k_BT}\right] = n_{p0}\exp\left[\frac{qV_{\text{Applied}}}{k_BT}\right] \quad (7.172)$$

so that we may write that

$$\Delta n(-x_p) = n_{p0}\left(\exp\left[\frac{qV_{\text{Applied}}}{k_BT}\right] - 1\right) \quad (7.173)$$

By a completely analogous route we obtain a similar expression for the holes

$$\Delta p(x_n) = p_{n0}\left(\exp\left[\frac{qV_{\text{Applied}}}{k_BT}\right] - 1\right) \quad (7.174)$$

We now solve (7.167a)-(7.167d) in both bulk regions, taking the origin at the junction

$$p_n(x - x_n) = p_{n0} + p_{n0}\left(\exp\left[\frac{qV_{\text{Applied}}}{k_BT}\right] - 1\right)\exp\left[-\frac{x - x_n}{\sqrt{D_P\tau_p}}\right] \quad (7.175a)$$

$$n_p(x_p - x) = n_{p0} + n_{p0}\left(\exp\left[\frac{qV_{\text{Applied}}}{k_BT}\right] - 1\right)\exp\left[-\frac{x_p - x}{\sqrt{D_N\tau_n}}\right] \quad (7.175b)$$

It is usual to write $\sqrt{D_P\tau_p} = L_P$ and $\sqrt{D_N\tau_n} = L_N$, the minority carrier diffusion lengths for holes and electrons. We insert (7.175a) and (7.175b) into the expressions for the minority currents at the depletion edges, since these hold throughout the depletion region

$$\begin{aligned} J_P(x_n) &= -qD_P\left.\frac{d\Delta p_n}{d(x - x_n)}\right|_{x = x_n} \\ &= q\frac{D_P}{L_P}p_{n0}\left(\exp\left[\frac{qV_{\text{Applied}}}{k_BT}\right] - 1\right) \end{aligned} \quad (7.176a)$$

$$\begin{aligned} J_N(-x_p) &= -qD_N\left.\frac{d\Delta n_p}{d(-x + x_p)}\right|_{x = x_p} \\ &= q\frac{D_N}{L_N}n_{p0}\left(\exp\left[\frac{qV_{\text{Applied}}}{k_BT}\right] - 1\right) \end{aligned} \quad (7.176b)$$

Adding these two expressions together yields the Shockley ideal diode equation

$$\begin{aligned} J &= J_P(x_n) + J_N(-x_p) \\ &= q\left\{\frac{D_P}{L_P}p_{n0} + \frac{D_N}{L_N}n_{p0}\right\}\left(\exp\left[\frac{qV_{\text{Applied}}}{k_BT}\right] - 1\right) \\ &= J_0\left(\exp\left[\frac{qV_{\text{Applied}}}{k_BT}\right] - 1\right) \end{aligned} \tag{7.177}$$

where the reverse saturation current density is $J_0 = q\{p_{n0}D_P/L_P + n_{p0}D_N/L_N\}$. Equation (7.177) is plotted in Figure 7.21.

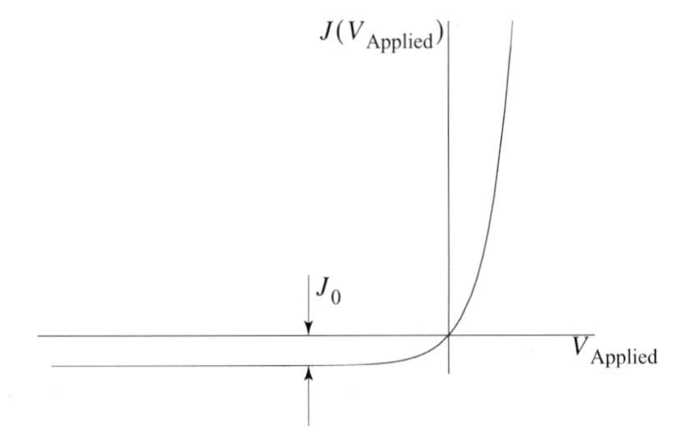

Figure 7.21. The shape of the Shockley ideal diode equation, (7.177), showing the reverse saturation current density J_0.

Junction Capacitance

The space-charge region, with its separated positive and negative charges, represents a nonlinear capacitor, i.e., a capacitance dependent on the applied voltage. In general, the nonlinear capacitance is defined as $C = dQ/dV$, i.e., the differential change in charge Q that would result for a differential change in voltage V. The total negative charge in the space-charge region, per unit abrupt junction length, is:

$$Q = qx_PN_a^- = \sqrt{(2\varepsilon)\frac{N_d^+N_a^-q}{(N_a^- + N_d^+)}}\sqrt{(V_{\text{bi}})_{\text{Abrupt}} \mp V_{\text{Applied}} - \frac{2k_BT}{q}} \tag{7.178}$$

The derivative of the charge w.r.t. the applied voltage gives

$$C = \frac{dQ}{dV} = \frac{1}{\sqrt{(V_{\mathrm{bi}})_{\mathrm{Abrupt}} \mp V_{\mathrm{Applied}} - \frac{2k_B T}{q}}} \sqrt{\left(\frac{\varepsilon}{2}\right) \frac{N_d^+ N_a^- q}{(N_a^- + N_d^+)}} \tag{7.179}$$

for the abrupt junction.

7.6.5 Metal-Semiconductor Contacts

In order to electrically interface the semiconductor devices to circuitry in the outside world, we have to connect metallic wires to its surface. Typically, metal pads are deposited onto the semiconductor surface at positions where the doping has been increased to ensure that good electrical contact is made. Bonding wires are then attached to the metal pads to connect them to the metal pins of the chip package. The whole geometrical arrangement is illustrated in Figure 7.22. As we will soon see, the

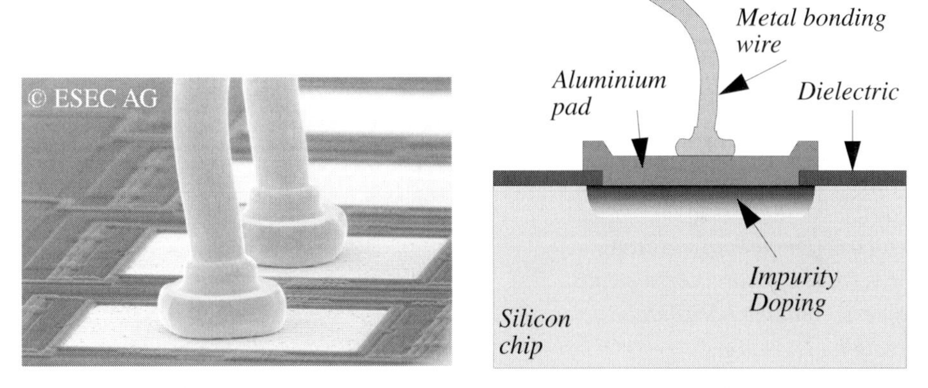

Figure 7.22. Schematic and scanning electron micrograph (SEM) of the geometry at a semiconductor-metal contact pad. SEM © ESEC AG, Cham [7.21].

metal pads can be operated both as good ohmic contacts to the semiconductor surface, as well as to create a high-speed diode. In a third application, metal contacts can be used to characterize the underlying semiconductor surface and extract parameters such as the doping concentration.

Work Function

An electron taken from an infinitely removed position in vacuum towards a clean metallic surface that is at the Fermi potential of the ensemble of metallic valence electrons, needs to pass through a potential, called the work function Φ_M, that is a fundamental property of the metal. It is simply equal to the work required to move an electron from the Fermi level to an energy level where it is free from the crystal. The Fermi surface of the metal represents a "sea" of non-localized valence electrons that are available for conduction, unhampered by a forbidden bandgap. Typical values of metallic work functions are $\Phi_{\mathrm{Al}} = 4.25\mathrm{eV}$, $\Phi_{\mathrm{W}} = 4.5\mathrm{eV}$ and $\Phi_{\mathrm{Pt}} = 5.3\mathrm{eV}$.

The band diagram of a metal is shown in Figure 7.23. One consequence

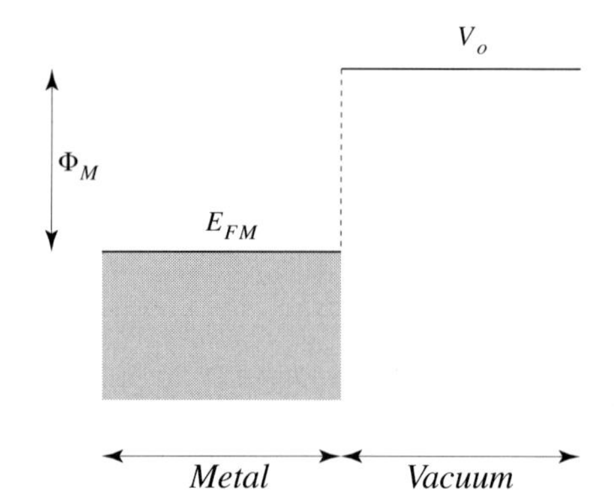

Figure 7.23. Band structure of a metal as a function of position. The vertical direction corresponds to energy, and the horizontal direction to position.

of the metal's band structure is that it represents an effectively unlimited source of conduction electrons, so that its Fermi surface does not deform noticeably in response to applied potentials. Hence we often treat metal contacts as being at a single electrostatic potential. Clearly this assumption only holds as long as we consider phenomena that are slow w.r.t.

metallic relaxation times. If we bring the electron close to the metallic surface, we notice an interesting effect.

Image Charge and Image Potential

The electron induces an equal but opposite charge on the metal's surface, and the interaction of the electron with its image charge becomes appreciable close to the metal's surface. At infinity, the electron experiences the vacuum potential, so that $V(x)|_{x \to \infty} = V_0$. Closer to the surface, the electron experiences a potential that varies as [7.15]

$$V(x) = V_0 - \frac{1}{4\pi\varepsilon_0(x - x_0)}, \; x > 0 \tag{7.180}$$

where x_0 represents the position of the image charge, and is often called the image plane. The potential of (7.180) approaches a value of minus infinity close to the metal's surface, and conflicts with experimental results. In fact, there is no unique information of the potential close to the surface. The following model fits experimental data better [7.15]

$$V(x) = lx, \; 0 > x > x_{im} \tag{7.181a}$$

$$V_{\mathrm{im}}(x) = V_0 - \frac{1}{4\pi\varepsilon_0(x - x_0)}, \; x > x_{im} \tag{7.181b}$$

$$\left.\frac{\partial V_{\mathrm{im}}(x)}{\partial x}\right|_{x_{im}} = l \tag{7.181c}$$

and is plotted in Figure 7.24. Here the value x_{im} represents a cut-off distance that represents the lower limit of the image force that the electron experiences. The image potential is very important in characterizing barrier heights at real metal-semiconductor junctions (see right-hand side plot in Figure 7.24).

Electron Affinity

Imagine now a clean metallic surface and a clean semiconductor surface, infinitely removed, in vacuum. The Fermi level of the metal lies at Φ_M below the vacuum level V_0. The semiconductor's conduction band edge lies at a level χ below the vacuum level V_0. The electron affinity χ is a fundamental property of the semiconductor, for example $\chi_{\mathrm{Si}} = 4.05\mathrm{eV}$.

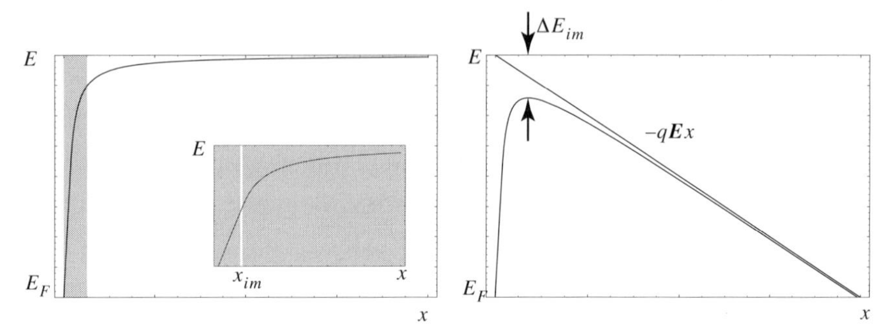

Figure 7.24. The left-hand curve shows the image potential $E(x)$ *plotted vs. distance as measured from a metallic surface. The potential models an electron approaching the metallic surface. The insert expands the left gray area horizontally and shows the cutoff distance with a white line. To the left of the cutoff we assume a linear potential rise. To the right of the cutoff* x_{im} *we assume a Coulomb interaction of the electron with its positive image charge. The right-hand plot shows the effect of a linearly varying vacuum potential due to a constant imposed electric field, and* ΔE_{im} *is indicative of the Schottky effect, a reduction of an energy barrier due to an electron's image charge.*

The position of the semiconductor's Fermi level w.r.t. the conduction band edge is defined by the impurity doming level. We can write the following equation for the work function Φ_S of the semiconductor [7.16]

$$\Phi_S(D) = \chi - [E_c - E_{FS}(D)] \tag{7.182}$$

We now slowly bring the two materials into contact, and as they approach, their Fermi potentials will adjust to keep the combined system at thermodynamic equilibrium, see Figure 7.25. Since in general $\Phi_M \neq \Phi_S$, the band of the semiconductor will bend in the vicinity of the junction so as to keep its Fermi level aligned with the metal's Fermi level. In other words, the metal-semiconductor junction forms a one-sided junction that looks like a $p^{++}n$ junction. Depending on the choice of metal and the doping in the semiconductor, we can achieve either an *ohmic contact* with the semiconductor, or a *diode* that is usually referred to as a Schottky contact or Schottky barrier in honour of its inventor. If a

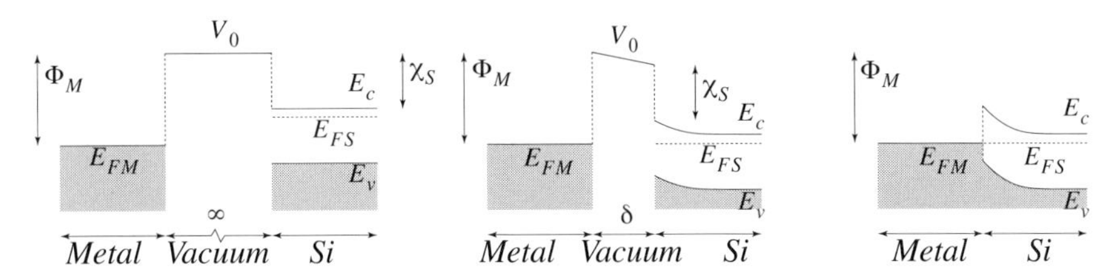

Figure 7.25. Three steps in bringing a metal and a semiconductor together. Notice how the band of the semiconductor is curved in order that the Fermi level of the joined metal and semiconductor can remain constant throughout.

slab of doped semiconductor is metallized on both free ends, we can expect an equilibrium electrostatic state for Shottky contacts as depicted in Figure 7.26.

Ohmic Contact

Ohmic contacts, i.e., contacts that have a very small potential drop and hence a low contact resistance, are very important for the formation of general semiconductor circuits. To form closed circuits, we use the metal of the fabrication process to define interconnect wiring. Clearly, since we expect to have many contacts in series, it is of utmost importance that the contact resistance be minimal. Because of the small difference between the electron affinity of silicon and the work function of aluminium, i.e., $\Phi_M - \Phi_S = 0.2\text{eV}$, we can control the nature of the contact between these two materials using only the doping level in the silicon. Hence this material system is very popular for CMOS and Bipolar processes. We now consider two cases [7.16].

Tunneling Ohmic Contact

- The tunneling ohmic contact is formed due to the fact that, with a highly-doped n-type semiconductor substrate, the deformed band edge permits permanent tunneling of electrons from the metal into and out of the semiconductor, see Figure 7.27a. For tunneling to happen, we must have that $\Phi_M > \Phi_S$. This is achieved by raising the Fermi level of the semiconductor towards the conduction band edge through degenerate n^+ doping. For silicon this means donor doping

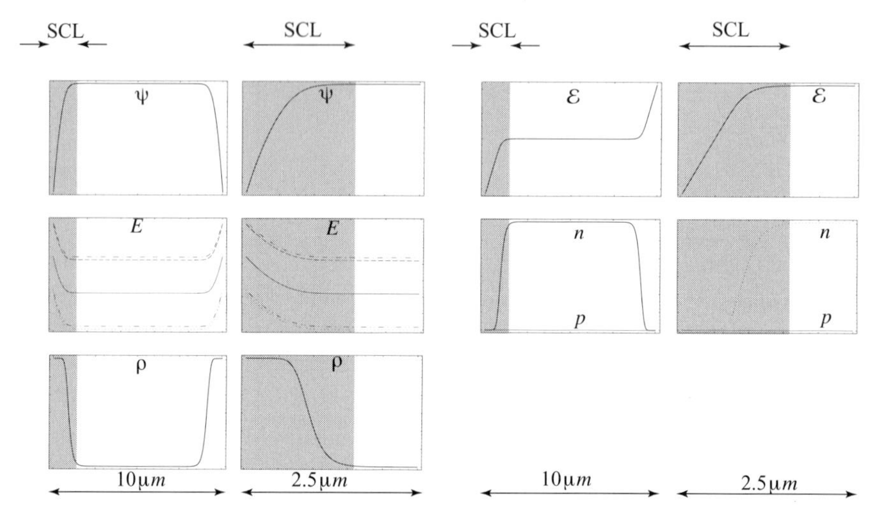

Figure 7.26. Six related quantities for a one-dimensional slab of homogeneously-doped semiconductor with metallic Schottky contacts: the electrostatic potential ψ *, the energy band diagram* E *, the charge density* ρ *, the electric field* $\mathcal{E}$ *, the carrier concentrations of electrons* n *and holes* p *. The quantities are displayed in pairs of plots. The horizontal axis of each left-hand plot is over the semiconductor slab's full length of* $10\mu m$ *. The gray stripe corresponds to the space-charge layer (SCL) in the semiconductor and directly adjacent to the metal. The curves were computed using a custom 1D device simulation program.*

levels of around $10^{25}\ m^{-3}$. In this way we achieve a large positive built-in potential just as for a semiconductor PN-junction. The built-in potential is computed as [7.16]

$$V_{\text{bi}} = \frac{(\chi + (E_c - E_F)_{\text{Bulk Si}} - \Phi_{\text{M}})}{q} = \frac{\Phi_{\text{B}} - (E_c - E_F)_{\text{Bulk Si}}}{q} \quad (7.183)$$

The barrier potential Φ_{B} is simply the difference between the two material work functions [7.16]

$$\Phi_{\text{B}} = \chi - \Phi_{\text{M}} \quad (7.184)$$

and hence is independent of the doping levels. The shape of the band at the junction forms a sharp point, see Figure 7.28, and in reality is

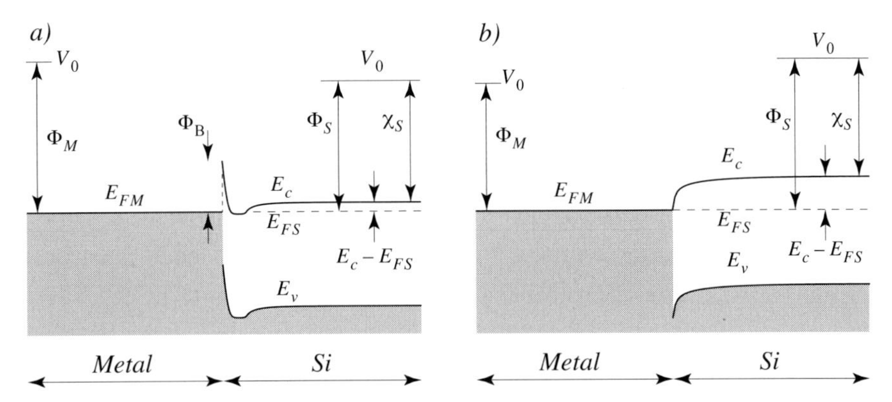

Figure 7.27. Computed curves for ohmic metal-semiconductor contacts. a) Tunneling junction. An n+ layer directly adjacent to the metal ensures a sharp gradient in the band, creating the characteristic narrow barrier through which carriers can tunnel. b) Low work-function metal contact. Carriers experience no barrier and can easily cross the metal-semiconductor junction.

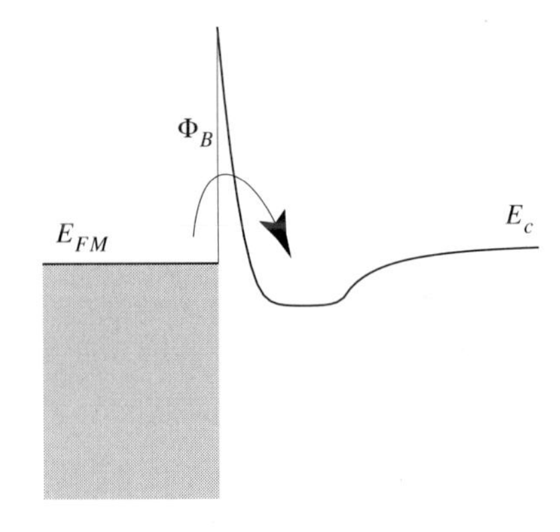

Figure 7.28. Detail of the barrier of the tunnel junction of Figure 7.27a. The choice of metal work function and impurity doping profile forces the conduction band of the semiconductor to lie below the metal's Fermi level, but behind a narrow barrier. Electrons can tunnel through this barrier, after which the field accelerates them away.

somewhat lower than our calculations indicate. The most important reason for its lowering is the image charge potential, (7.181a)-(7.181c), which we have to subtract from the energy diagram on the semiconductor side [7.17]. This correction was first considered by Schottky, and hence is called Schottky barrier lowering. Since the tunneling current through the barrier is exponential in the barrier height, this correction is important. Another important effect is the presence of surface defects [7.17]. In fact, we find many additional "disturbances" at the band edges, mainly due to impurity states and surface reconstruction states, all of which contribute to modify the curvature of the band and height of the barrier at the metal-semiconductor interface. We will not consider these modifications here, but refer the reader to [7.17].

Metal-like Ohmic Contact

- We can also form an ohmic contact with a metal-like semiconductor, where $\Phi_S > \Phi_M$ for an n-type semiconductor, see Figure 7.27b. This is achieved by adjusting the doping level to moderate levels so that the built-in potential is negative. In this case the electrons in the semiconductor experience no barrier towards the metal and can flow freely. In turn the correct choice of metal can significantly reduce the barrier height and hence permit metal electrons to easily enter the semiconductor states. In this case the built-in voltage is negative

$$qV_{bi} = \chi + (E_c - E_{FM}) - \Phi_M < 0 \tag{7.185}$$

Schottky Barrier

The Schottky contact can be considered to be a diode, and is often used as a rectifier. Its switching speed is much higher than that of a semiconductor PN-junction. This is because:

- the Schottky barrier has very little minority carrier storage, i.e.,
- the main conduction mechanism is via majority carriers, the minority carriers do not contribute appreciably to the current;
- the metal has a very large supply of majority carriers ready to enter unoccupied semiconductor states;
- the metal Fermi band edge does not appreciably change when electrons are drained into the semiconductor, due to the extremely short dielectric relaxation time of the metal;
- the Schottky barrier has almost no diffusion capacitance, so that the depletion capacitance is almost independent of frequency at normal

operating frequencies. Here the limitation is the dielectric relaxation time of the semiconductor.

Biased Junction

For the Schottky contact diode, under forward biasing conditions, three transport mechanisms are identified:

- Conduction band electrons can drop over the potential barrier into the metal;
- Conduction band electrons can tunnel through the potential barrier into the metal;
- Recombination can occur between conduction band electrons and holes that are injected from the metal into the semiconductor. Recombination can take place in the space-charge region, or immediately adjacent to the space-charge region, also known as the neutral region.

Thermionic Emission

The first mechanism above dominates the transport process, and is explained as thermionic emission of electrons into the metal. We will not derive the expression, but shortly explain the idea following [7.17]. Conduction band electrons that have energies higher than the barrier can be injected. If we assume that the available electrons have all their energy in the form of kinetic energy, and an effective mass given by the energy band shape, then we can convert the electron density expression for the distribution of electrons from a dependency of energy, i.e., $dn = N(E)F(E)dE$ to one of velocity $dn = N(v)F(v)dv$. Now, using $J_{S\to M} = \int_{(E_F+\Phi_B)}^{\infty} qvdn$ and a an estimate for the electron distribution, we obtain the current

$$J_{S\to M} = A^*T^2\exp\left[-\frac{\Phi_B}{k_BT}\right]\exp\left[\frac{qV}{kT}\right] \tag{7.186}$$

where $A^* = 4\pi qm^*k^2/h^3$ is called the effective Richardson constant. For n-type $\langle 111\rangle$ silicon, $A^*_{\langle 111\rangle} = 264$, and for n-type $\langle 100\rangle$ silicon, $A^*_{\langle 100\rangle} = 252$.

In the other direction, from the metal to the semiconductor, there is no dependence on applied voltage, because all carriers only see the barrier energy, so that

$$J_{M \to S} = -A^* T^2 \exp\left[-\frac{\Phi_B}{k_B T}\right] \tag{7.187}$$

Summing up (7.186) and (7.187), we obtain the total current flowing through the junction

$$J = A^* T^2 \exp\left[-\frac{\Phi_B}{k_B T}\right]\left(\exp\left[\frac{qV}{kT}\right] - 1\right) = J_o\left(\exp\left[\frac{qV}{kT}\right] - 1\right) \tag{7.188}$$

where $J_o = A^* T^2 \exp[-\Phi_B/(k_B T)]$.

Barrier Tunneling

To proceed with a theoretical model of the tunnel current, the quantum transmission coefficient $T(\eta)$ of electrons, where η measures energy downward from the tip of the barrier, should be computed using a quantum-mechanical approach, and we find that $T(\eta) \propto \exp[-\eta/E_0]$, for some datum energy E_0. The current is then proportional to an integral over the energy height of the barrier, of the products of transmission coefficient, the occupation probability F_M of electrons that are available for tunneling (a Fermi distribution), and the unoccupation probability of states F_S that the electrons will tunnel into (another Fermi distribution). These concepts are clarified by Figure 7.29. As a result of all these terms, we find that the tunneling current has the following form [7.17]

$$J_t \propto \exp\left[-2\Phi_B \Big/ \left(q\hbar\sqrt{\frac{N_D}{\varepsilon_S m^*}}\right)\right] \tag{7.189}$$

i.e., it depends exponentially on the root of the impurity concentration.

Parameter Extraction

The capacitance of a metal-semiconductor contact depends on both the built-in potential and the doping level, and can be measured to extract these values. The depletion capacitance can be modelled by a parallel-

Figure 7.29. a) The energy band diagram of a tunneling metal-semiconductor junction and its movement under bias. b) Illustration of the product of an electron quantum barrier transmission coefficient with the leaving electron availability distribution and the arrival state availability distribution. The shaded region on the right represents the electron density that can tunnel through.

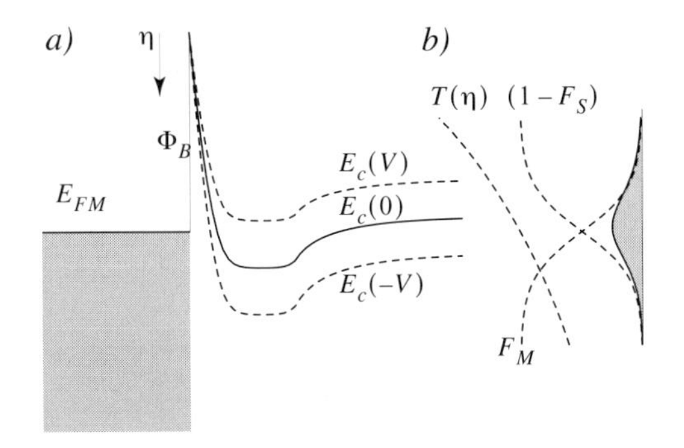

plate ideal capacitor, with the plate separation equal to the depletion layer width x_n

$$C = \frac{\varepsilon_r \varepsilon_0 A}{x_n} \tag{7.190}$$

The depletion layer width is the same as for a one-sided PN-junction

$$x_n = \sqrt{\frac{2\varepsilon_r \varepsilon_0}{qN_D}(V_{bi} - V_{\text{applied}})} \tag{7.191}$$

We can combine (7.190) and (7.191) to obtain an expression for the inverse squared capacitance

$$\frac{1}{C^2} = \frac{2(V_{bi} - V_{\text{applied}})}{qN_D \varepsilon_r \varepsilon_0 A^2} \tag{7.192}$$

The trick is to measure the capacitance as a function of the applied voltage and to plot the inverse squared capacitance as the applied voltage, which should be a straight line. On this curve, the zero applied voltage point will give us the built-in voltage. From the slope of the curve, $a = 2/(qN_D \varepsilon_r \varepsilon_0 A^2)$, we can recover the value for the average doping in the space-charge region

$$N_D = \frac{aq\varepsilon_r\varepsilon_0 A^2}{2} \tag{7.193}$$

In other words, we can use Schottky diodes as a technique with which to characterize doping profiles, but can only expect good results for uniform doping at levels where a reasonable space-charge capacitor forms.

7.7 Summary for Chapter 7

In this chapter we have taken a closer look at the interactions that arise when the three particle systems, phonons, electrons, and photons interact with each other. The phenomena, or constitutive equations, described include heat conductivity, dielectric permittivity, piezoresistivity, the thermoelectric effects, and many more. It was shown that the in-homogeneity of a crystal can also lead to useful effects, for example the PN junction and surface acoustic waveguides. Most effects described in this chapter have lead to useful microelectronic devices, including solid state sensors, actuators, lasers and transistors. Most of these devices can be built using silicon technology.

7.8 References for Chapter 7

7.1 D. F. Nelson, *Electric, Optic and Acoustic Interactions in Dielectrics,* John Wiley and Sons, New York (1979)

7.2 Neil W. Ashcroft, N. David Mermin, *Solid State Physics*, Saunders College Publishing, Philadelphia (1988)

7.3 Charles Kittel, *Introduction to Solid State Physics*, Second Edition, John Wiley and Sons, New York (1953)

7.4 Baltes, H., Paul, O., Korvink, J. G., Schneider, M., Bühler, J., Schneeberger, N., Jaeggi, D., Malcovati, P., Hornung, M., Häberli, A., von Arx, M. and Funk, J., *IC MEMS Microtransducers*, (Invited Review), IEDM Technical Digest (1996) 521–524

7.5 Abraham I. Beltzer, *Acoustics of Solids*, Springer Verlag, Berlin (1988)

7.6 Igor Aleksandrovich Viktorov, *Raleigh and Lamb Waves: Physical Theory and Applications*, Plenum Press, New York (1967)

7.7 Peter Y. Yu, Manuel Cordona, *Fundamentals of Semiconductors*, Springer Verlag, Berlin (1996)

7.8 T. Manku, A. Nathan, *Valence energy-band structure for strained group-IV semiconductors*, J. of Applied Physics 73 (1993) 1205

7.9 S. Middelhoek, S. A. Audet, *Silicon Sensors*, Academic Press, London (1989)

7.10 Otfried Madelung, *Introduction to Solid-State Theory*, Springer-Verlag, Heidelberg (1981)

7.11 Herbert B. Callen, *Thermodynamics and an Introduction to Thermostatistics*, Second Edition, John Wiley and Sons, New York (1985)

7.12 J. F. Nye, *Physical Properties of Crystals*, Oxford University Press, Oxford, 1985

7.13 Marie-Catharin Desjonquères, D. Spanjaard, *Concepts in Surface Physics*, 2nd Ed., Springer Verlag, Berlin (1996)

7.14 Martin Bächtold, Jan G. Korvink, Jörg Funk, Henry Baltes, *Simulation of p-n Junctions in Highly Doped Semiconductors at Equilibrium Conditions*, PEL Report (public) No. 97/6, ETH Zurich (1997)

7.15 Sydney Davidson, Maria Streslicky, *Basic Theory of Surface States*, Oxford Science Publications, Oxford University Press, Oxford (1992)

7.16 Gerold W. Neudeck, *The PN Junction Diode*, Volume II of Modular Series on Solid State Devices, G. W. Neudeck and R. F. Pierret, Editors, 2nd Ed., Addison-Wesley Publishing Company, Reading, Massachusetts (1989)

7.17 Simon M. Sze, *Physics of Semiconductor Devices*, 2nd Edition, John Wiley & Sons, New York (1981)

7.18 J.A. Kash and J. C. Tsang, *Hot Luminescence from CMOS Circuits: A Picosecond Probe of Internal Timing*, Phys. Stat. Sol. (b) **204,** 507 (1997)

7.19 R. W. Dixon, *Photoelastic Properties of Selected Materials and Their Relevance for Applications to Acoustic Light Modulators and Scanners*, Journal of Applied Physics **38,** 5149 (1967)

7.20 David K. Ferry, *Semiconductors,* Macmillan Publishing Company, Singapore (1991)

7.21 Michael Mayer, Private communication

Index

D

E

Q